John David Crawford
1987

Instabilities and Nonequilibrium Structures

Mathematics and Its Applications

Instabilities and Nonequilibrium Structures

edited by

Enrique Tirapegui

Departamento de Física,
Facultad de Ciencias Físicas y Matemáticas,
Universidad de Chile, Santiago, Chile

and

Danilo Villarroel

Departamento de Física, Facultad de Ciencias,
Universidad Técnica Federico Santa María,
Valparaíso, Chile

D. Reidel Publishing Company

A MEMBER OF THE KLUWER ACADEMIC PUBLISHERS GROUP

Dordrecht / Boston / Lancaster / Tokyo

Library of Congress Cataloging in Publication Data

CIP

Instabilities and nonequilibrium structures.

(Mathematics and its applications)
"Contains the lectures given at the First International Workshop on Instabilities and Nonequilibrium Structures which took place in the Universidad Técnica Federico Santa Maria, Valparaíso, Chile, December 16–21, 1985"—Frwd.
Includes index.
1. Fluid dynamics—Congresses. 2. Stochastic processes—Congresses. 3. Stability—Congresses. I. Tirapegui, Enrique. II. Villarroel, D. III. International Workshop on Instabilities and Nonequilibrium Structures (1st : 1985 : Universidad Técnica Federico Santa María) IV. Series: Mathematics and its applications (D. Reidel Publishing Company).
QA911.I52 1987 510 86-31392
ISBN 90-277-2420-2

Published by D. Reidel Publishing Company,
P.O. Box 17, 3300 AA Dordrecht, Holland.

Sold and distributed in the U.S.A. and Canada
by Kluwer Academic Publishers,
101 Philip Drive, Assinippi Park, Norwell, MA 02061, U.S.A.

In all other countries, sold and distributed
by Kluwer Academic Publishers Group,
P.O. Box 322, 3300 AH Dordrecht, Holland.

Printed in The Netherlands

Series Editor's Preface

Approach your problems from the right end and begin with the answers. Then one day, perhaps you will find the final question.

'The Hermit Clad in Crane Feathers' in R. van Gulik's *The Chinese Maze Murders*.

It isn't that they can't see the solution. It is that they can't see the problem.

G.K. Chesterton. *The Scandal of Father Brown* 'The point of a Pin'.

Growing specialization and diversification have brought a host of monographs and textbooks on increasingly specialized topics. However, the "tree" of knowledge of mathematics and related fields does not grow only by putting forth new branches. It also happens, quite often in fact, that branches which were thought to be completely disparate are suddenly seen to be related.

Further, the kind and level of sophistication of mathematics applied in various sciences has changed drastically in recent years: measure theory is used (non-trivially) in regional and theoretical economics; algebraic geometry interacts with physics; the Minkowsky lemma, coding theory and the structure of water meet one another in packing and covering theory; quantum fields, crystal defects and mathematical programming profit from homotopy theory; Lie algebras are relevant to filtering; and prediction and electrical engineering can use Stein spaces. And in addition to this there are such new emerging subdisciplines as "experimental mathematics", "CFD", "completely integrable systems", "chaos, synergetics and large-scale order", which are almost impossible to fit into the existing classification schemes. They draw upon widely different sections of mathematics. This programme, Mathematics and Its Applications, is devoted to new emerging (sub)disciplines and to such (new) interrelations as exempla gratia:

- a central concept which plays an important role in several different mathematical and/or scientific specialized areas;
- new applications of the results and ideas from one area of scientific endeavour into another;
- influences which the results, problems and concepts of one field of enquiry have and have had on the development of another.

The Mathematics and Its Applications programme tries to make available a careful selection of books which fit the philosophy outlined above. With such books, which are stimulating rather than definitive, intriguing rather than encyclopaedic, we hope to contribute something towards better communication among the practitioners in diversified fields.

For some time now mathematics and physics and ... have been increasingly aware that nonlinear models and phenomena offer opportunities and a richness undreamt of in the linear world. Of course nonlinear in itself means little and is too understructured to be of use. What we, as scientists, are currently engaged in, may perhaps be described as isolating a substantial number of kinds of nonlinear models which can be managed in some way; at the same time the necessary (mathematical) tools and concepts are developed. Certainly the opportunities are fantastic. As I have had occasion to remark several times before, a purely linear world would be a sad place to live in. And certainly if our surrounding universe were linear you would not be reading these lines.

Instabilities bifurcations, nonequilibrium structures, and pattern formation, and such related concepts as chaotic systems and metastability, form a strongly intertwined set of aspects and concepts within the general nonlinear field with a well-developed set of mathematical tools, rich in applications, and also vigorous in generating new concepts and new phenomena demanding understanding and analysis.

These are what this book is about. And, if the editors and organisors of this workshop and summerschool maintain their current high standards in selecting lectures and authors of survey papers, I have no doubt that further equally interesting volumes on these topics will also appear in this series.

The unreasonable effectiveness of mathematics in science ...

Eugene Wigner

Well, if you know of a better 'ole, go to it.

Bruce Bairnsfather

What is now proved was once only imagined.

William Blake

As long as algebra and geometry proceeded along separate paths, their advance was slow and their applications limited.

But when these sciences joined company they drew from each other fresh vitality and thenceforward marched on at a rapid pace towards perfection.

Joseph Louis Lagrange.

Bussum, December 1986

Michiel Hazewinkel

TABLE OF CONTENTS

FOREWORD

This book contains the lectures given at the First International Workshop on Instabilities and Nonequilibrium Structures which took place in the Universidad Técnica Federico Santa María, Valparaíso, Chile, from December 16 to 21, 1985.

The idea of this meeting was to give the scientific community of Chile, and more generally of Latin America, the opportunity of receiving first hand information on the latest developments in a field which has known an impressive progress in the last decade. We think that this is one possible and important way to help developing countries in their efforts to reduce their scientific gap and we are very much grateful to all our lecturers who accepted with great enthusiasm to participate in this Workshop. It is our intention to organize this meeting every two years and consequently the second Workshop will take place in December 1987.

We want to thank here our sponsors and supporters whose interest and help was essential for the success of the meeting. But the physical realization of the Workshop would have not been possible without the decisive collaboration of the Universidad Técnica Federico Santa María. We acknowledge here this support. We are also especially grateful to Ms. Michele Goldstein, Scientific Advisor in the French Embassy in Santiago, for her valuable and constant cooperation.

We are very much indebted to the members of the Departments of Physics and Mathematics of the Universidad Santa María, and more generally to all the people who helped in the preparation of the Conference. Ms. Antonina Salazar deserves a special mention for her assistance in the organization of the meeting and also for her work in the material preparation of this book. Finally we express our genuine thanks to Professor Michiel Hazewinkel for accepting this book in his series and to Dr. David Larner and the Reidel Publishing Company for their efficient collaboration.

E. Tirapegui
D. Villarroel

LIST OF SPONSORS OF THE WORKSHOP

- Academia Chilena de Ciencias
- Universidad Técnica Federico Santa María
- Ministère Français des Affaires Etrangères
- Proyecto Fortalecimiento del Desarrollo de la Física en Chile (PNUD-UNESCO)
- Departamentos de Investigación y Bibliotecas y de Relaciones Internacionales de la Universidad de Chile.
- Comisión Nacional de Investigación Científica y Tecnológica (CONICYT)
- Departamento de Física de la Universidad de Santiago.
- Sociedad Chilena de Física
- Sociedad Chilena de Matemáticas
- Centro de Estudios Científicos de Santiago.

LIST OF SUPPORTERS OF THE WORKSHOP

- Petroquímica Dow S.A.
- Shell Chile S.A.C.E.I.

BRIEF INTRODUCTION TO THE CONTENTS OF THE BOOK

The purpose of this Introduction is essentially to help the reader giving him an account of the matters treated here and a justification of the order we have chosen to present the different contributions. Our feeling is that the papers included here give a fair and complete account of several subjects which are to-day of current interest in the field, insisting in review aspects to make them reasonably self-contained.

We have ordered the works presenting first papers which treat mainly deterministic aspects of instability problems and then works which are oriented towards the stochastic behavior near instabilities. This division is of course artificial, however we think that it can be useful to the reader since deterministic treatments are in general more familiar and intuitive.

We start with two papers of very general nature and mathematical character. The first by G. Iooss refers to the elimination of fast variables near a bifurcation point. The center manifold theorem is discussed and method are given to compute this manifold as well as normal forms. A very interesting section on the consequences of the symmetries of the system on its normal form is included here. The second work by J.M. Gambaudo, P. Glendinning and C. Tresser treats in a rigorous way a general bifurcation of dissipative dynamical systems producing signals which can be interpreted as rotation compatible sequences.

We proceed then to the paper of S. Fauve which reviews the important problems of phase instabilities arising in periodic patterns. Symmetry considerations play here a primordial role and it is shown that they determine the nonlinear terms of the phase equations. The next work of H.M. Nussenzveig refers to the physical interpretation of bifurcations. Indeed the very general techniques used to attack bifurcation problems, such as the center manifold theorem and normal forms described in detail in the paper of G. Iooss, have the disadvantage of making often the physical interpretation difficult or indirect and this is a drawback since experience tells us that a clear physical understanding suggests analogies, new developments and new problems. The standard model for optical bistability in a laser is discussed in great detail and it is shown that the bifurcations which occur here can be understood by the mechanism of the generation and amplification of sidebands in parametric processes.

The next two papers refer to fluid dynamics, a subject which has been so rich in the production of instability problems. The first work by F. Lund treats the interaction of sound and vorticity and presents some important new results for the motion of a vortex in a prescribed external flow. The second paper by P. Huerre gives a very clear and nice review of the evolution of spatio-temporal instabilities in fluid media.

The very important problems of patterns formation is the subject of the two works which follow. First P. Coullet and D. Repaux study patterns transitions with an interesting method:

E. Tirapegui and D. Villarroel (eds.), Instabilities and Nonequilibrium Structures, 1–2.

the equation giving the static patterns, which is an ordinary differential equation in one space dimension, is treated as a (spatial) dynamical system thus allowing the transposition to this situation of many concepts and methods of usual bifurcation theory. It is in fact shown in some simple models that these spatial dynamical systems can exhibit complicated patterns such as quasiperiodic or chaotic. In the second work D. Walgraef gives a very complete and exciting review of patterns and defects in far from equilibrium systems together with examples in physics, chemistry and biology. The sections treating patterns selection in anisotropic media and dislocation patterns in solids are of special interest. The role of fluctuations on these structures is also considered here mainly through the phenomenological introduction of Langevin forces and prepares the transition to the second set of papers of the book.

In the first of these C.P. Enz treats correlation effects in nonequilibrium systems under the influence of small applied gradients, i.e. sufficiently far from instabilities. The correlation functions are obtained here by suitable averages over the fluctuations which are included in the equations for the macrovariables by adding Gaussian white noises. In the next paper the fundamental problem of competing locally stable states and the transitions between them due to fluctuations is given a complete and fascinating review by N.G. van Kampen in the framework of the Fokker-Planck equation.

The following contribution by R. Graham is concerned with the very important question of the existence of nonequilibrium potentials and their statistical interpretation. In far from equilibrium systems one has intrinsic fluctuations which are often minor perturbations as in equilibrium thermodynamics and one can then naturally think that here also one can find extremum principles characterizing through potentials their behavior. Recent progress in these problems is reviewed in the weak noise limit and several models are studied illustrating the complexity of the situation.

We present next a paper by R. Dandoloff and Y. Pomeau in which the examination of a clever "gedanken" experiment shows that in metastable systems fluctuations may lack time reversal symmetry contrarily to what one finds for equilibrium thermal fluctuations. The following work by C. Elphick and E. Tirapegui treats the adiabatic elimination of fast variables in stochastic differential equations and a very general situation is studied using the method of normal forms.

Finally three shorter contributions are included. In the first E. Meyer argues that spontaneous nucleation can be assumed to occur in a large adiabatic system at constant external pressure with increasing entropy. Comparison with experimental data is given supporting this claim. Then A.S. Fernandes, M.V.A.P. Heller and I.L. Caldas study disruptive instabilities in tokamaks and in the third paper M.O. Cáceres considers a model of coupled random walk to describe correlated Brownian motion with anisotropic scattering.

REDUCTION OF THE DYNAMICS OF A BIFURCATION PROBLEM USING NORMAL FORMS AND SYMMETRIES

G. Iooss
Université de Nice
Faculté des Sciences de Nice, Parc Valrose 06034
Nice CEDEX, France

1. REDUCTION OF THE DIMENSION

1.1 The Hyperbolic Case.

Let us consider a differential system

$$\frac{dX}{dt} = F(X) \qquad (1.1)$$

where $X(t)$ lies in $\mathbb{R}^n$ or in some functional Hilbert space E, suitable for our physical problem. In this latter case we have to assume in what follows that the initial data problem has a unique solution in some real time interval $[0,T]$, with all the usual regularity properties with respect to initial data and parameters.

We assume that

$$F(o) = 0 \qquad (1.2)$$

This means that we know a priori a steady solution of our problem. Now we make an assumption of the stability of this solution. Let us denote by

$$L = D_X F(o) \qquad (1.3)$$

the nonlinear operator obtained by linearizing around the steady solution 0. Then we assume that the spectrum of L is well-separated by the imaginary axis. This means that, for the type of evolution problems we consider (reaction-diffusion, fluid mechanics,...) a finite number of eigenvalues lies on the right part of the compex plane while the others have a negative real part, staying in a sector centered on the real axis (see fig. 1). The distance of the spectrum to the imaginary axis is not zero: this is by definition an hyperbolic situation.

E. Tirapegui and D. Villarroel (eds.), Instabilities and Nonequilibrium Structures, 3–40.

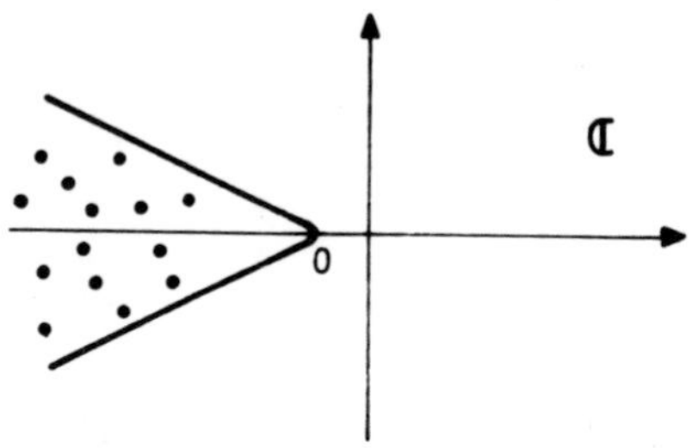

Figure 1. Spectrum of L.

This situation is the simplest in a dynamical sense. In fact if we consider the linear system

$$\frac{dX}{dt} = LX \quad , \quad X\big|_{t=0} = X_0 \tag{1.4}$$

the dynamics of the solution

$$X(t) = e^{Lt} X_0 \tag{1.5}$$

is shown in fig. 2. Let us decompose X in two parts:

$$X = X_1 + X_2 \tag{1.6}$$

where X_1 is in the subspace E_1 belonging to all eigenvalues with positive real part, and X_2 is in the subspace E_2 belonging to the remaining eigenvalues. Both subspaces are invariant under L and if $X_0 \varepsilon E_2$ than $X(t)$ tends exponentially towards 0 when $t \to \infty$, while if $X_0 \varepsilon E_1$ then $\|X(t)\|$ tends exponentially to infinity.

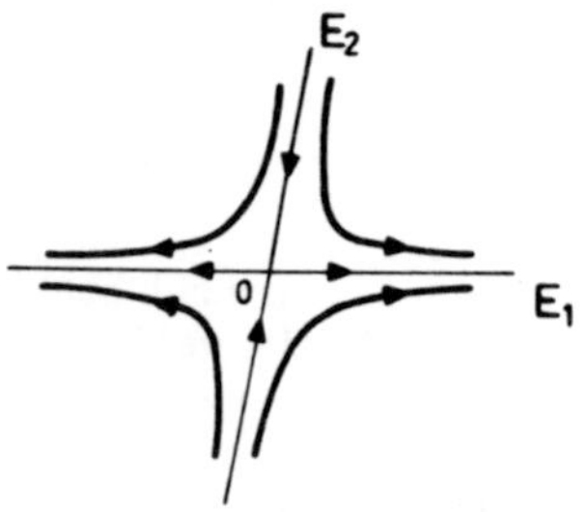

Figure 2. Linear hyperbolic case.

We have in general

$$X(t) = e^{L_1 t} X_{01} + e^{L_2 t} X_{02} \tag{1.7}$$

where

$$\| e^{L_1 t} X_{01} \| \underset{t\to\infty}{\longrightarrow} \infty \quad \text{and} \quad e^{L_2 t} X_{02} \underset{t\to\infty}{\longrightarrow} 0 \quad ,$$

and where $L_1 = L\,|_{E_1}$, $L_2 = L\,|_{E_2}$ are the restrictions of the linear operator L to the invariant subspaces.

The main fact is that, for the nonlinear problem (1.1), we have just to consider that the previous situation is maintained, but slightly deformed, in such a way that there are two invariant manifolds: M_2, tangent to E_2 at 0, is the stable manifold, while M_1, tangent to E_1 at 0 is the unstable manifold. These manifolds replace the spaces E_1 and E_2 occuring in the linear case (see fig. 3).

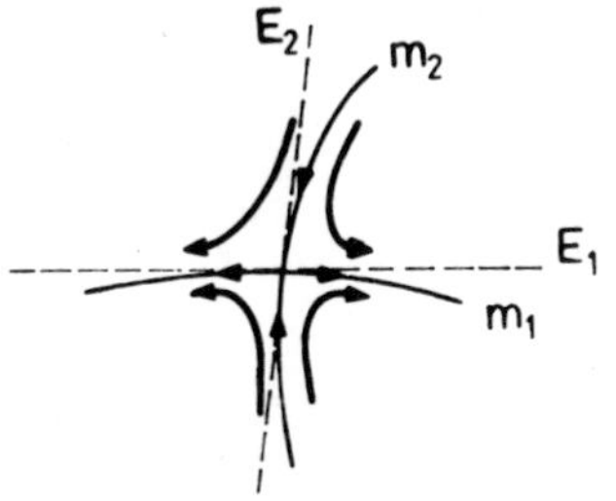

Figure 3. Hyperbolic case.

This property is only local, valid in a neighborhood of 0, which depends on the distance of the spectrum of L to the imaginaring axis. If same eigenvalue of L is very close to the imaginary axis, then the situation of fig. 3 remains valid in a very small neighborhood of 0 (its size is of the order of the distance to the imaginary axis).

The mathematical proofs of these results can be find in [1-4]. In what follows we shall give an idea of why it is possible to prove this, and what type of difficulties occur when some eigenvalue is close to the imaginary axis.

As a corollary we can say that if all eigenvalues of L have a negative real part, then the 0 solution is exponentially attracting, at least in a neighborhood of 0.

1.2 The Central Case.

1.2.1 Position of the Problem. We start again with the system (1.1), satisfying (1.2), but the assumption on the spectrum of L is now the following: a finite number of eigenvalues lie on the imaginary axis: $\sigma_1, \dots \sigma_N$, each being of finite multiplicity, and the remaining part of the spectrum of L has a strictly negative real part, lying in a sector centered on the real axis.

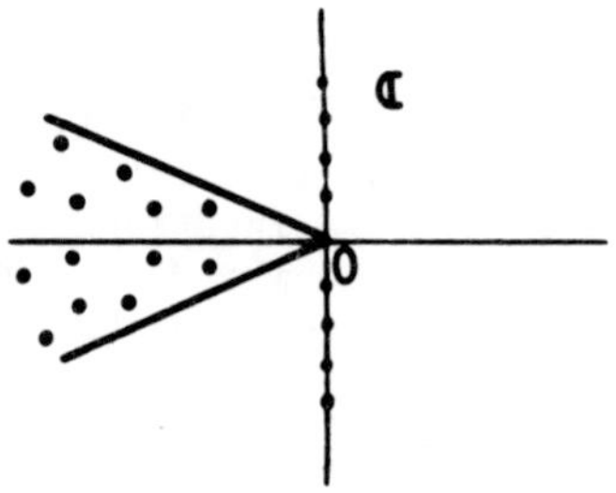

Figure 4. Spectrum of L in the central case.

The dynamics of the linear situation is of course very different from the one of the hyperbolic case (see fig. 5). If we decompose the space as in 1.1

$$E = E_1 \oplus E_2 \;,\; X = X_1 + X_2 \;,\; X_j \in E_j$$

we have again

$$X(t) = e^{L_1 t} X_{01} + e^{L_2 t} X_{02} \tag{1.8}$$

with $e^{L_2 t} X_{02} \underset{t \to \infty}{\longrightarrow} 0$ exponentially. But now the part $X_1(t) = e^{L_1 t} X_{01}$ is no longer exploding exponentially. If all eigenvlaues of L_1 are semi-simple (i.e. L_1 is diagonalizable), $X_1(t)$ is oscillating when t varies, while if some eigenvalue of L_1 is not semi-simple then $\|X_1(t)\|$ may grow as a polynomial in t (depending on X_{01}).

The fundamental idea for treating the nonlinear situation (1.1) is to notice that in the linear case all the dynamics reduces to the one on E_1 when $t \to \infty$. Consequently we want to eliminate the transient behavior of trajectories attracted towards some manifold replacing E_1 in the nonlinear situation. Since this manifold is expected to be tangent to the subspace E_1, we want to obtain it in the form

$$X_2 = \Phi(X_1) = 0(\|X_1\|^2) \quad . \tag{1.9}$$

Let us give an idea of what is going on for a proof of this result.

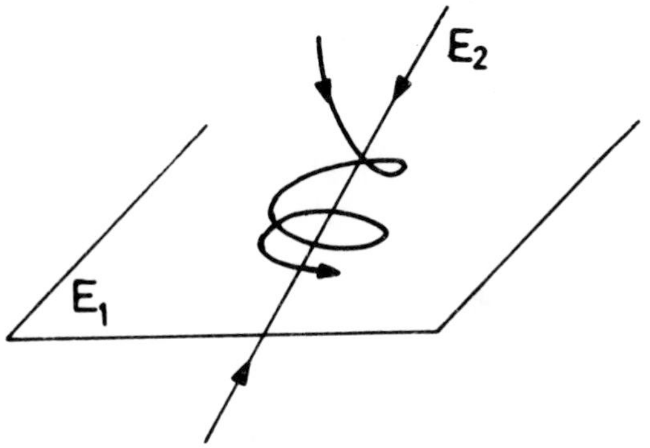

Figure 5. Linear central case.

1.2.2 Idea of the problem and of the result. Let us decompose (1.1) on E_1 and E_2 as follows:

$$\frac{dX_1}{dt} = L_1X_1 + P_1(X_1,X_2) \tag{1.10a}$$

$$\frac{dX_2}{dt} = L_2X_2 + P_2(X_1,X_2) \tag{1.10b}$$

where the linear operators L_1 and L_2 have the properties described in 1.2.1, and where $P_j, j = 1,2$, are nonlinear functions of X_1, X_2, taking values in E_j, such that

$$\| P_j(X_1,X_2) \| = 0(\| X_1 \| + \| X_2 \|)^2 \quad . \tag{1.11}$$

Now the idea is to make change of variables on X_2 in such a way as to eliminate step by step the dependence of (1.10b) in X_1! Let us write P_2 as follows

$$P_2(X_1,X_2) = B_2(X_1,X_1) + B_3(X_1,X_1,X_1) + \dots$$
$$+ B_N(X_1,\dots X_1) + 0(\| X_1 \|^{N+1} + \| X_1 \| \| X_2 \| + \| X_2 \|^2) \tag{1.12}$$

where $B_k(X_1,\dots X_1)$ is k-linear symmetric in its arguments, and comes from the Taylor expansion of $P_2(X_1,0)$.

Let us change variables nonlinearly:

$$X_2' = X_2 + \Gamma_k(X_1,X_1,\dots X_1) \tag{1.13}$$

where Γ_k is k-linear symmetric in its arguments in E_1, taking values in E_2. We obtain for the differential equation in X_2':

$$\frac{dX_2'}{dt} = L_2X_2' + \sum_{j=2}^{k-1} B_j(X_1,\dots X_1) + B_k'(X_1,\dots X_1)$$
$$+ O(\| X_1 \|^{k+1} + \| X_1 \| \| X_2' \| + \| X_2' \|^2) \tag{1.14}$$

with

$$B_k'(X_1,\dots X_1) = B_k(X_1,\dots X_1) + k\Gamma_k(X_1,\dots,L_1X_1) - L_2\Gamma_k(X_1,\dots X_1) \quad . \tag{1.15}$$

We observe that we only modify the terms of order k in X_1, and higher order. The main fact here is that we can choose Γ_k such that $B_k' = 0$. So we solve the equation

$$k\Gamma_k(X_1,\dots,L_1X_1) - L_2\Gamma_k(X_1,\dots X_1) + B_k(X_1,\dots X_1) = 0 \tag{1.16}$$

for any X_1 in E_1. If we denote $\{e_j\}$ the set of eigenvectors of L_1 belonging to the eigenvalues $i\omega_j$, we obtain

$$\left(\sum_{\ell=1}^{k} i\omega_{j_\ell} - L_2\right)\Gamma_k(e_{j_1},\dots e_{j_k}) + B_k(e_{j_1},\dots e_{j_k}) = 0 \tag{1.17}$$

which determines $\Gamma_k(e_{j_1},\dots e_{j_k})$ for any family $\{e_{j_1},\dots e_{j_k}\}$ of eigenvectors of L_1, since L_2 has no pure imaginary eigenvalue. For generalized eigenvectors $e_j^{(1)}$ which satisfy

$$L_1e_j^{(1)} = i\omega_je_j^{(1)} + e_j \tag{1.18}$$

we may use in the same way

$$\left(\sum_{\ell=1}^{k} i\omega_{j_\ell} - L_2\right)\Gamma_k(e_{j_1},\dots e_{j_{k-1}},e_j^{(1)}) + k\Gamma_k(e_{j_1},\dots e_{j_{k-1}},e_j)$$
$$+ B_k(e_{j_1},\dots e_{j_{k-1}},e_j^{(1)}) = 0 \quad , \tag{1.19}$$

since the only unknown here is $\Gamma_k(e_{j_1},\dots e_{j_{k-1}},e_j^{(1)})$, and so on...
Hence, starting by eliminating B_2, we modify B_k, $k \geq 3$, but we then eliminate B_3 (the new one),... and so on. So the result is, after all, a change of variables

$$X_2' = X_2 - \Phi_N(X_1) \tag{1.20}$$

where ϕ_N is a polynomial in the vector variable X_1, of degree N, and in the new variables (1.10) is now written as follows:

$$\begin{aligned} \frac{dX_1}{dt} &= L_1X_1 + P_1(X_1,\Phi_N(X_1)) + O(\| X_1 \| \| X_2' \| + \| X_2' \|^2) \; , \\ \frac{dX_2'}{dt} &= L_2X_2' + O(\| X_1 \|^{N+1} + \| X_1 \| \| X_2' \| + \| X_2' \|^2) \; . \end{aligned} \tag{1.21}$$

Let us pose

$$X_2' = \| X_1 \|^N Y$$

then

$$\frac{dY}{dt} = L_2Y + O(\| X_1 \|) \quad , \tag{1.22}$$

hence if $X_1(t)$ is bounded by some δ (close to 0), since L_2 has a spectrum on the negative side, this shows that Y(t) will be bounded by $O(\delta)$ when $t \to \infty$. Coming back to (1.20), (1.21), this shows that

$$\frac{dX_1}{dt} = L_1X_1 + P_1(X_1,\Phi_N(X_1)) + O(\delta^{N+1}) \tag{1.23}$$

and that the dynamics will be attracted by the manifold

$$X_2 = \Phi_N(X_1) \tag{1.24}$$

at the order δ^{N+1}.

Here we do not prove the existence of a "center manifold" $X_2=\Phi(X_1)$ locally attracting and locally invariant (see [1], [2], [3] for such a result), but we see that there is a region, very thin if N is large, where the dynamics is locally attracted. The center manifold whose existence is proved by the theory lies inside this region. It is then clear that $\Phi_N(X_1)$ is the Taylor expansion at the order N of Φ, but we have to realize that there is non uniqueness of the center manifold, even though its Taylor expansion is unique.

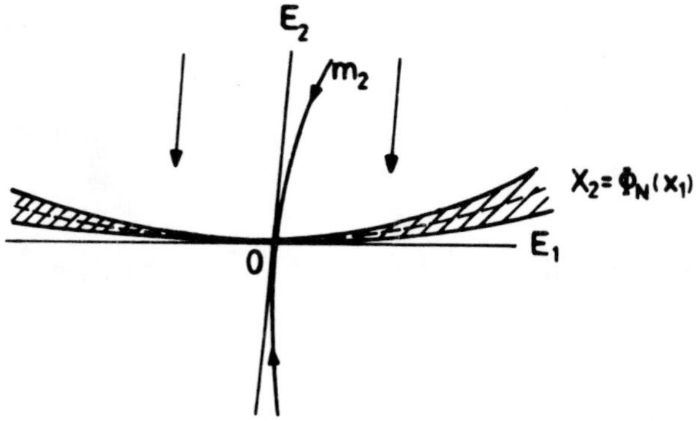

Figure 6. The region $X_2 - \Phi(X_1) = O(\| X_1 \|^{N+1})$.

To illustrate this fact, let us give an example (see [2]). We consider in $\mathbb{R}^3$ the following differential system:

$$\frac{dx_1}{dt} = -y_1$$

$$\frac{dy_1}{dt} = 0 \tag{1.25}$$

$$\frac{dx_2}{dt} = -x_2 + h(x_1) \quad .$$

In this example $X_1 = (x_1, y_1)$, $X_2 = x_2$, 0 is a double non semi-simple eigenvalue of L_1 ; while -1 is the only eigenvalue of L_2.

$$L_1 = \begin{pmatrix} 0 & -1 \\ 0 & 0 \end{pmatrix}$$

It is easy to check that the surface $x_2 = \varphi(x_1, y_1)$ defined by

$$\varphi(x_1,y_1) = \int_{-\infty}^{0} e^s \tilde{h}(x_1 - sy_1)ds \tag{1.26}$$

is a center manifold, with $\tilde{h}$ defined by

$$\tilde{h}(x) = h(x) \quad \text{for} \quad |x| \leq \delta$$

$$\tilde{h}(x) \text{ bounded for } \quad |x| > \delta .$$

Exercise: (i) prove that $\frac{d}{dt}\varphi[x_1(t),y_1(t)] = \frac{dx_2(t)}{dt}$, if $|x_1(t)| \leq \delta$.

(ii) integrate (1.25) and show that

$$x_2(t) - \varphi[x_1(t),y_1(t)] = e^{-t}[x_{20} - \varphi(x_{10},y_{10})], \text{ if } |x_1(t)| \leq \delta,$$

where $x_1(0) = x_{10}, x_2(0)] = x_{20}, y_1(0) = y_{10}$.

The nonuniqueness of (1.26) is obvious. Moreover the Taylor expansion of φ is such that

$$\varphi(x_1,y_1) = \sum_{p+q\geq 2} \varphi_{pq} x_1^p y_1^q \quad , \quad \varphi_{pq} = \frac{(p+q)!}{p!} h_{p+q}$$

where $h(x) = \sum_{p\geq 2} h_p x^p$, here it is also clear that, even though h is a analytic, φ is not necessarily analytic.

1.3 Bifurcation problems.

1.3.1 Position of the problem. Let us assume that the system depends on parameters, that we denote by μ in $\mathbb{R}^p$. So we have now:

$$\frac{dX}{dt} = F(\mu,X), \text{ with } F(0,0) = 0, \tag{1.27}$$

and $D_X F(0,0) = L$ has a spectrum as described in section 1.2.1. The idea is to consider a new system

$$\frac{d}{dt}(X,\mu) = (F(\mu,X),0) \tag{1.28}$$

which enters into the frame of section 1.2. The corresponding linear operator is now

$$\mathcal{A} = \begin{pmatrix} L & \xi \\ 0 & 0 \end{pmatrix} \tag{1.29}$$

where $\xi = D_\mu F(0,0)$. The study of the spectrum of $\mathcal{A}$ is easy: its eigenvalues are those of L plus $\{0\}$. In the space E_1 we add the variable μ, since its corresponds to an additional eigenvalue 0.

Remark. If 0 is already an eigenvalue of L, then 0 is an eigenvalue of $\mathcal{A}$ with a greater multiplicity (excercise left to the reader). The index of the eigenvalue 0 is the same if ξ belongs to the image of L (case when $\mu \in \mathbb{R}$).

Hence, a center manifold for the system (1.28) has the form

$$X_2 = \Phi(X_1,\mu) = O(\|X_1\|^2 + |\mu|)^* \tag{1.30}$$

and we are only interested in the sections of (1.30) where μ = constant! (since $\frac{d\mu}{dt} = 0$). On the center manifold (1.30) the dynamics of the system (1.29) becomes

$$\frac{dX_1}{dt} = F(\mu,X_1) = L_1 X_1 + O(\|X_1\|^2 + |\mu|) , \tag{1.31}$$

where $F(\mu,X_1) = F_1[\mu,X_1 + \Phi(X_1,\mu)]$, and $F = F_1 + F_2$ is the decomposition of F on $E_1 \oplus E_2$.

Remark 1. If 0 is a steady solution of (1.27) for any μ in a neighborhood of $\mu = 0$, then it can be proved that

$$\Phi(0,\mu) = 0 \quad ,$$

hence
$$\Phi(X_1,\mu) = O[\|X_1\|(|\mu| + \|X_1\|)] \quad . \tag{1.32}$$

Remark 2. Notice that the dynamics is attracted by the center manifold in a neighborhood of 0, which is independent of μ close enough to 0. This is in fact the main point here, since when μ is not 0, in general we get a weakly hyperbolic situation, with eigenvalues close to the imaginary axis (at the order μ in general). This shows that the unstable (hyperbolic) manifold might only exist in a neighborhood of 0 of size $|\mu|$ (see section 1.1). This is not sufficient for studying the dynamics of (1.27) since we shall see that most of the bifurcated solutions have size $\sqrt{|\mu|}$.

---

* If we have no term of order μ in the component of F on E_2, then $\Phi = O(\|X_1\| + |\mu|)^2$.

Remark 3. Assume that we have a "minimal" invariant set γ_μ such as a point or a periodic orbit for instance isolated on each possible center manifold M_μ. Then, due to the local attractivity, γ_μ belongs to all center manifolds, hence it is also isolated in the full space E. This shows that one should not worry with the non uniqueness of such a manifold.

1.3.2 Computation of the center manifold. We saw in section 1.2.2, that the Taylor expansion of all center manifolds is uniquely determined. Let us write an a priori series:

$$X_2 = \Phi(X_1,\mu) = \sum_{p+q\geq 1} \mu^p \Phi_{pq}(X_1,\ldots X_1) \quad , \quad \Phi_{01} = 0 \quad , \quad (1.33)$$

where Φ_{pq} is q-linear symmetric in X_1. We wish to compute the operators Φ_{pq}. To do so, we may just identify powers of μ and X_1 in

$$D_{X_1}\Phi(X_1,\mu)\cdot F_1[\mu,X_1 + \Phi(X_1,\mu)] = F_2[\mu,X_1 + \Phi(X_1,\mu)] \ , \quad (1.34)$$

where we decomposed (1.27) into

$$\frac{dX_j}{dt} = F_j(\mu,X_1 + X_2) \ , \quad F_j \ ,X_j \in E_j \quad . \quad (1.35)$$

Let us precise the notations:

$$F_j\ (\mu,X) = \sum_{p+q\geq 1} \mu^p\ F^{(j)}_{pq}(X,\ldots X) \quad , \quad F^{(j)}_{01} = L_j \quad . \quad (1.36)$$

Successive identifications in (1.34) lead to:

$$\begin{aligned}
& - L_2\Phi_{10} = F^{(2)}_{10} \\
& 2\Phi_{02}(X_1,\ L_1X_1) - L_2\ \Phi_{02}(X_1,X_1) = F^{(2)}_{02}(X_1,X_1) \qquad (1.37) \\
& \Phi_{11}L_1\ X_1 - L_2\Phi_{11}X_1 = -2\Phi_{02}[X_1,\ F^{(1)}_{10}] + F^{(2)}_{11}X_1 \\
& \qquad\qquad + 2\ F^{(2)}_{02}\ [X_1,\Phi_{10}] \\
& - L_2\Phi_{20} = -\Phi_{11}F^{(1)}_{10} + F^{(2)}_{20} + F^{(2)}_{11}\ \Phi_{10} + F^{(2)}_{02}\ (\Phi_{10},\Phi_{10}) \\
& \ldots\ldots\ldots\ldots\ldots\ldots\ldots \\
& q\Phi_{pq}(X_1,\ldots,\ L_1\ X_1) - L_2\Phi_{pq}(X_1,\ldots X_1) \\
& = -(q+1)\Phi_{p-1,q+1}[X_1,\ldots,\ F^{(1)}_{10}\ ] + \text{known terms depending on}
\end{aligned}$$

$F^{(j)}_{nm}$ and $\Phi_{n'm'}$ with $n + m \leq p + q$, $n' + m' \leq p + q - 1$.

Since we studied how to solve (1.16), it is now clear how to proceed to obtain any Φ_{pq} in (1.33)!

1.3.3 Presence of a symmetry in the system. Let us assume that we have some symmetry in our system (1.27). We translate this property by saying that a linear operator S commutes with $\mathcal{F}$:

$$\mathcal{F}(\mu,SX) = S\,\mathcal{F}(\mu,X) \quad . \tag{1.38}$$

As a consequence, we can show that the Taylor expansion (1.33) of a center manifold commutes with S too.

In fact, since S commutes with $\mathcal{F}$, it commutes with all its derivatives at $X = 0$ (differentiate (1,38)). For instance we have

$$\mathcal{L}S = S\mathcal{L} ,$$

$$\mathcal{F}_{pq}(SX,\ldots SX) = S\mathcal{F}_{pq}(X,\ldots,X) \quad , \quad q \geq 0 \quad . \tag{1.39}$$

This shows that the subspaces E_j are invariant under S and

$$\mathcal{L}_j S = S\mathcal{L}_j \quad \text{on each } E_j \quad .$$

Looking now at (1.37), we see that $\Phi_{pq}\circ S$ and $S\circ\Phi_{pq}$ satisfy the same equations (make a recursion on the right hand side of (1.37)). Since the result is unique, we obtain necessarily:

$$\Phi_{pq}(SX_1,\ldots SX_1) = S\Phi_{pq}(X_1,\ldots X_1) \quad . \tag{1.40}$$

Now replacing X_2 by $\Phi(X_1,\mu)$ into the X_1 equation, we obtain the dynamics on the center manifold and

$$\frac{dX_1}{dt} = F(\mu,X_1) = \mathcal{F}_1[\mu,X_1 + \Phi(X_1,\mu)] \quad . \tag{1.41}$$

It is then easy to see that F commutes with S too:

$$SF(\mu,X_1) = \mathcal{F}_1[\mu,SX_1 + S\Phi(X_1,\mu)] = \mathcal{F}_1[\mu,SX_1 + \Phi(SX_1,\mu)]$$
$$= F(\mu,SX_1) \quad . \tag{1.42}$$

In fact we have only proved this equivariance property for the Taylor expansion of F! It can be proved that if E_1 is finite dimensional and if S is an isometry in E_1, then this property is true for the full function F. We shall see in a special chapter all the consequences of such a result. For the moment we just notice that S can be a member of a discrete group of symmetries as well as a member of a continuous group.

2. AMPLITUDE EQUATIONS - GENERIC CASES

2.1 Saddle-node bifurcation.

Since we know that there is no surprise when the situation is hyperbolic (see section 1.1), let us now consider the simplest case of a non hyperbolic situation, depending on one real parameter μ.

We assume that we have an evolution problem in a space E:

$$\frac{dX}{dt} = \mathcal{F}(\mu,X) \quad , \tag{2.1}$$

where

$$\mathcal{F}(0,0) = 0 \text{ , and } D_X \mathcal{F}(0,0) = L \tag{2.2}$$

is such that 0 is a simple eigenvalue of L, and other eigenvalues have strictly negative real part.

Coming back to section 1.3, we see that the dynamics of (2.1) reduces to a one dimensional differential equation:

$$\frac{dx}{dt} = F(\mu,x) \text{ ,} \tag{2.3}$$

where $F(0,0) = 0$ and $D_X F(0,0) = 0$.

To compute explicitly F, it is necessary to define the eigenvector ξ_0 belonging to the 0 eigenvalue :

$$L\xi_0 = 0 \text{ ,} \tag{2.4}$$

and to define the projection P_0 on the one dimensional subspace of E spanned by ξ_0 such that $Q_0 = Id - P_0$ has the range(*) of L:

$$X = x\xi_0 + y \text{ , } x\xi_0 = P_0X, \; y = Q_0X = (Id - P_0)X$$

$$y = \sum_{p+q\geq 1} \mu^p x^q \Phi_{pq} \quad \text{center manifold } (\Phi_{01} = 0) \text{ ,}$$

we note $P_0 \mathcal{F}_{pq} = f_{pq}\xi_0$. Then we obtain easily:

$$\frac{dx}{dt} = \mu f_{10} + \mu x[f_{11}(\xi_0) + 2f_{02}(\xi_0,\Phi_{10})] + x^2 f_{02}(\xi_0,\xi_0) +$$

(*) Since the eigenvalue 0 of L is simple, the range of L has a codimension 1 for the usual cases [5] and we can construct the projections P_0 and Q_0. If there is an adjoint for L , then ξ_0 being the normalized eigenvector $L^*\xi_0^* = 0$, $x = \langle X,\xi_0^*\rangle$.

$$+ \mu^2[f_{20} + f_{11}(\Phi_{12}) + f_{02}(\Phi_{10},\Phi_{10})] + \ldots \tag{2.5}$$

where Φ_{10} is defined by

$$L_2\ \Phi_{10} = -Q_0 F_{10} \quad .$$

The steady solutions of (2.5) in the neighborhood of 0 are given by the resolution of $F(\mu,x) = 0$. If we assume

$$f_{10} \neq 0 \ , \tag{2.6}$$

then we can solve $F(\mu,x) = 0$ with respect to μ and we find

$$\mu = -\frac{f_{02}(\xi_0,\xi_0)}{f_{10}} x^2 + 0(x^3) \qquad \text{(see figure 7).} \tag{2.7}$$

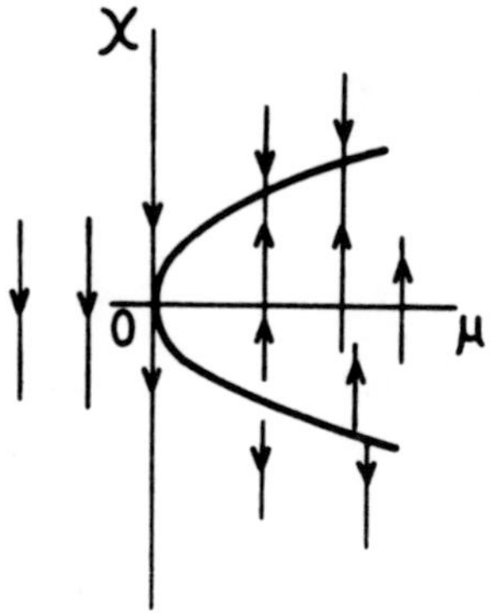

Figure 7. Dynamics of (2.5) for $f_{02}(\xi_0,\xi_0) < 0$, $f_{10} > 0$.

The stability analysis of the corresponding branch of steady solutions of (2.1)

$$\mu = g(x_0) \ , \ y = \Phi[x_0,g(x_0)] \qquad (x_0 \text{ is a parameter on the branch}) \tag{2.8}$$

is easy since its reduces to the stability analysis on (2.5). We have

$$D_X F[\mu(x_0),x_0] = 2x_0 f_{02}(\xi_0,\xi_0) + 0(x_0^2) \tag{2.9}$$

which shows that for a fixed μ, the two steady solutions have opposite stabilities since x_0 takes two values with opposite sings. The dynamics for (2.1) is shown in figure 8 (neighborhood of 0).

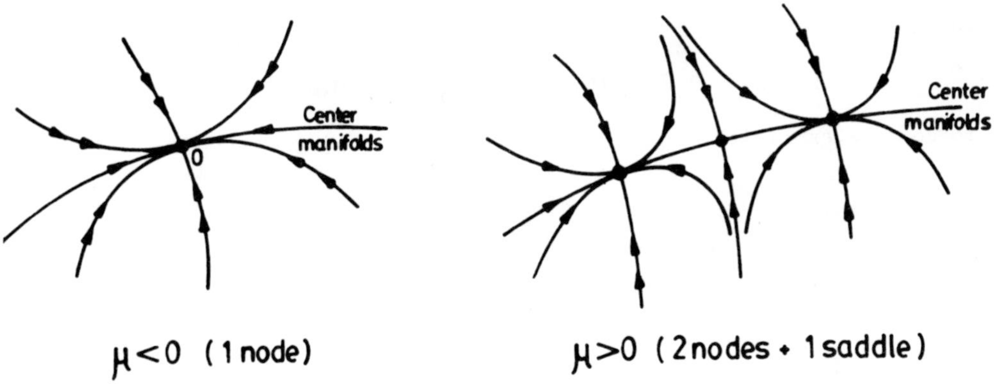

Figure 8. Dynamics in E, same assumptions as in figure 7.

2.2 Hopf bifurcation - normal form.

Here we consider the other generic one parameter case which consists in taking the system (2.1) - (2.2) and assuming that L has two simple eigenvalues $\pm i\omega_0$, other eigenvalues being of strictly negative real part.

We first remark that, since 0 is not an eigenvalue of L, we can compute a family of steady solutions $X(\mu)$ such that $X(0) = 0$, using identification of Taylor expansions (implicit function theorem). We now take the origin at this point, hence we have from now on:

$$F(\mu,0) = 0 \quad . \tag{2.10}$$

The study of section 1.3 shows that the dynamics of (2.1) reduces to a two dimensional differential system. Let us pose

$$X = z\zeta_0 + \bar{z}\bar{\zeta}_0 + Y \quad , \tag{2.11}$$

and let us construct a projection P_0 as in 2.1 such that $Q_0 = Id - P_0$ has a range which is the intersection of the ranges of $L - i\omega_0$ and of $L + i\omega_0$. If L has an adjoint, we can obtain ζ_0^* such that

$$(L^* + i\omega_0)\zeta_0^* = 0 \ , \ \langle\zeta_0\ \zeta_0^*\rangle = 1$$

and $z = \langle X,\zeta_0^*\rangle$ while $\langle Y,\zeta_0^*\rangle = \langle Y,\bar{\zeta}_0^*\rangle = 0$.

Then we may decompose F in the following way:

$$F = f\zeta_0 + \bar{f}\bar{\zeta}_0 + G \tag{2.12}$$

and we look for a differential equation on the center manifold $Y = \Phi(z,\bar{z},\mu)$, in the following form:

$$\frac{dz}{dt} = F(\mu,z,\bar{z}) \ , \text{ in } \mathbb{C}, \tag{2.13}$$

where

$$F(\mu,0,0) = 0 \ , \ D_z F(0,0,0) = i\omega_0 \ , \ D_{\bar{z}} F(0,0,0) = 0 \ . \qquad (2.14)$$

To compute explicitly F, let us define the following expansions

$$\Phi(z,\bar{z},\mu) = \sum_{\substack{p+q+r\geq 2 \\ q+r\geq 1}} \mu^p z^q \bar{z}^r \Phi_{pqr} \ , \ F(\mu,X) = LX + \mu L_1 X$$

$$+ F_{02}(X,X) + \ldots$$

$$F(\mu,z,\bar{z}) = \sum_{q+r\geq 1} \mu^p z^q \bar{z}^r F_{pqr} \quad .$$

Then we obtain by identifications of powers of $\mu,z,\bar{z}$ in (2.1):

$$F_{110}\zeta_0 + \bar{F}_{101}\bar{\zeta}_0 - L_1\zeta_0 = (L - i\omega_0)\Phi_{110} \ , \ \Phi_{101} = \bar{\Phi}_{110} \ ,$$

$$F_{020}\zeta_0 + \bar{F}_{002}\bar{\zeta}_0 - F_{02}(\zeta_0,\zeta_0) = (L - 2i\omega_0)\Phi_{020} \ , \ \Phi_{002} = \bar{\Phi}_{020}$$

$$F_{011}\zeta_0 + \bar{F}_{011}\bar{\zeta}_0 - 2F_{02}(\zeta_0,\bar{\zeta}_0) = L\Phi_{011} \ , \ (\Phi_{011} \text{ is real}) \ . \qquad (2.15)$$

..................,

$$F_{021}\zeta_0 + F_{012}\bar{\zeta}_0 + [2F_{011} + \bar{F}_{011} + F_{020} + 2\bar{F}_{002}]\zeta_0$$

$$+ [2\bar{F}_{011} + F_{011} + \bar{F}_{002} + 2\bar{F}_{020}]\bar{\zeta}_0 - 2F_{02}(\zeta_0,\Phi_{011})$$

$$- 2F_{02}(\bar{\zeta}_0,\Phi_{020}) - 3F_{03}(\zeta_0,\zeta_0,\bar{\zeta}_0) = (L - i\omega_0)\Phi_{021} \quad .$$

Since Φ_{pqr} lies in a codimension 2 subspace, this leads to two compatibility conditions from which we can obtain F_{pqr} and F_{prq}. It is clear that if we really have all possible coefficients in F, then it is hard to see something for the dynamics!

So, the next idea is to simplify this equation by making suitable change of variables (nonlinear ones).

Let us show that we can treat step by step all increasing orders in $z,\bar{z}$. We rewrite (2.13) in the following form

$$\frac{dz}{dt} = i\omega_0 z + \mu A_1(\mu,z,\bar{z}) + A_2(\mu,z,\bar{z}) + A_3(\mu,z,\bar{z}) + \ldots \qquad (2.16)$$

where $A_k(\mu,z,\bar{z})$ is homogeneous of degree k in $(z,\bar{z})$ and depends regularly on μ.

Let us change variables in the following way:

$$z' = z + \mu\Gamma_1(\mu,z,\bar{z}) \quad , \tag{2.17}$$

where Γ_1 is linear in $(z,\bar{z})$. Let us be more precise posing

$$A_1(\mu,z,\bar{z}) = \alpha_{10}z + \alpha_{01}\bar{z}$$

$$\Gamma_1(\mu,z,\bar{z}) = \gamma_{10}z + \gamma_{01}\bar{z} \ ;$$

the inversion of (2.17) leads to

$$z = \frac{(1+\mu\bar{\gamma}_{10})z' - \mu\gamma_{01}\bar{z}'}{(1+\mu\gamma_{10})^2 - \mu^2(\gamma_{01})^2} = [1 - \mu\gamma_{10} + O(\mu^2)]z' - [\mu\gamma_{01} + O(\mu^2)]\bar{z}' \ , \tag{2.18}$$

and replacing this expression in $\frac{dz'}{dt}$ we obtain a new $A_1'(\mu,z',\bar{z}')$ where

$$\alpha'_{10} = \alpha_{10} + O(\mu)$$
$$\alpha'_{01} = \alpha_{01} - 2i\omega_0\gamma_{01} + O(\mu) \quad . \tag{2.19}$$

This shows that we can choose γ_{01} such that $\alpha'_{01} = O(\mu)$, but we cannot cancel α'_{10} since the coefficients of γ_{10} and γ_{01} are $O(\mu)$ in the first equation. In fact we can do more since we can choose $\gamma_{10} = 0$ and find γ_{01}, using the implicit function theorem in such a way that

$$A_1'(\mu,z',\bar{z}') = \alpha'_{10}(\mu)z' \ , \tag{2.20}$$

$(i\omega_0 + \alpha'_{10}(\mu) = \sigma(\mu)$ eigenvalue of $L_\mu = L + \mu L_1 + \ldots)$, i.e. we suppress directly <u>all</u> terms of the form $\mu^p\bar{z}'$, $p \geq 1$.
Excercise: show that we have to solve the equation

$$\alpha_{01} + \gamma_{01}[-2i\omega_0 + \mu(\bar{\alpha}_{10} - \alpha_{10})] - \mu^2\bar{\alpha}_{01}\gamma_{01}^2 = 0 \ , \tag{2.21}$$

to eliminate all terms in $\bar{z}'$ in A_1'.

Let us write z instead of z' in the new differential equation. We can now make nonlinear change of variables:

$$z' = z + \Gamma_n(\mu,z,\bar{z}) \tag{2.22}$$

where Γ_n is homogeneous of degree n in $(z,\bar{z})$, $n \geq 2$.

The inversion of (2.22) leads to

$$z = z' - \Gamma_n(\mu,z',\bar{z}') + O(|z'|^{n+1}) \tag{2.23}$$

as it can be shown by a simple iterative process. In replacing z by z' in (2.16) we only modify terms of order n and of higher orders. It is easy to check that the new n-order terms have the form

$$A'_n(\mu,z',\bar{z}') = A_n(\mu,z',\bar{z}') - \sigma(\mu)\Gamma_n(\mu,z',\bar{z}') + \sigma(\mu)z'\frac{\partial\Gamma_n}{\partial z}(\mu,z',\bar{z}') + \bar{\sigma}(\mu)\bar{z}'\frac{\partial\Gamma_n}{\partial\bar{z}}(\mu,z',\bar{z}') \,. \tag{2.24}$$

Let us precise Γ_n:

$$\begin{aligned} \Gamma_n(\mu,z,\bar{z}) &= \sum_{p+q=n} \gamma_{pq} z^p \bar{z}^q \\ A_n(\mu,z,\bar{z}) &= \sum_{p+q=n} \alpha_{pq} z^p \bar{z}^q \end{aligned} \tag{2.25}$$

then (2.24) leads to:

$$\alpha'_{pq} = \alpha_{pq} - [(p-1)\sigma + q\bar{\sigma}]\gamma_{pq} \quad , \tag{2.26}$$

and the coefficient of γ_{pq}, for $\mu = 0$, is just $(p-1-q)i\omega_0$. This shows that if $p \neq q+1$, we can smoothly cancel α'_{pq} in choosing suitable γ_{pq}. The terms such that $p = q+1$ are the "resonant terms" which cannot be removed by this change of variables.

Hence after successive change of variables, making this in increasing orders $p+q$ (since at each step we modify higher orders), it finally remains a system of the form

$$\frac{dz}{dt} = \sigma(\mu)z + \sum_{p\geq 1}^{N-1} \alpha_p(\mu)z|z|^{2p} + O(|z|^{2N+1}) \tag{2.27}$$

where we wrote z instead of the new variable. The form (2.27) of the system is called the "normal form". It is in general impossible to put (2.27) fully into the normal form (with no $O(|z|^{2N+1})$), since the change of variables reduces at each step the neighborhood of 0 where (2.27) holds.

Nevertheless, the truncated normal form of (2.27) (without $O(|z|^{2N+1})$), gives us sufficiently many results to make us happy... and prove that these results also hold for the full system (2.27).

Let us study truncated normal form, and pose

$$z = re^{i\theta} \,. \tag{2.28}$$

Then this leads to

$$\frac{dr}{dt} = \xi r + \sum_{p=1}^{N-1} a_p r^{2p+1} \tag{2.29}$$

$$\frac{d\theta}{dt} = \eta + \sum_{p=1}^{N-1} b_p r^{2p} \tag{2.30}$$

where we noted $\sigma(\mu) = \xi(\mu) + i\eta(\mu)$, $\eta(0) = \omega_0$, $\xi(0) = 0$, and

$$\alpha_p = a_p + ib_p \quad .$$

We see now why it is simple, since the radial part of eq.(2.29) is uncoupled from the angular part.

The steady solution $r = 0$ of (2.29) corresponds to the 0 solution of (2.1). Now, if $\xi'(0) \neq 0$, i.e. if the eigenvalue $\sigma(\mu)$ crosses the imaginary axis at the first order in μ, we can find a bifurcating steady solution of (2.29) such that

$$\xi(\mu) + \sum_{p=1}^{N-1} a_p r_0^{2p} = 0 \ , \tag{2.31}$$

hence $r_0^2 = -\dfrac{\xi(\mu)}{a_1(\mu)} + \ldots = -\mu\dfrac{\xi'(0)}{a_1(0)} + O(\mu^2)$, if $a_1(0) \neq 0$. The stability of this family of steady solutions of (2.29) is easy to determine:

$$\frac{d}{dr}(\xi_r + \Sigma a_p r^{2p+1})\Big|_{r=r_0} = \xi + 3a_1 r_0^2 + O(r_0^4) = -2\xi + O(r_0^4), \tag{2.32}$$

where $\xi = O(\mu) = O(r_0^2)$. Assuming now that

$$\xi(\mu) \text{ has the sign of } \mu \text{ for } \mu \text{ close to } 0, \tag{2.33}$$

we have the result that if $a_1(0) < 0$, the bifurcated solution is supercritical ($\mu > 0$) and r_0 is attractive (with a strength -2ξ), while if $a_1(0) > 0$, the bifurcation is subcritical ($\mu < 0$) and the solution is repulsive (with a strength $-2\xi > 0$). This solution of (2.29)-(2.30) corresponds to a periodic solution of frequency $\Omega = \eta(\mu) + \sum_{p=1}^{N-1} b_p(\mu) r_0^{2p}$:

$$z(t) = r_0 e^{i(\Omega t + \varphi_0)} \quad . \tag{2.34}$$

Now, for the full system (2.27), it can be shown that a periodic solution with principal part (2.34) remains, with the same stability property as for (2.29) - (2.30) (see [3] for a complete proof). This is due to the smallness of the terms $O(|z|^{2N+1})$ with respect to the strength of attraction (or repulsion) in the neighborhood of the periodic orbit (2.34).

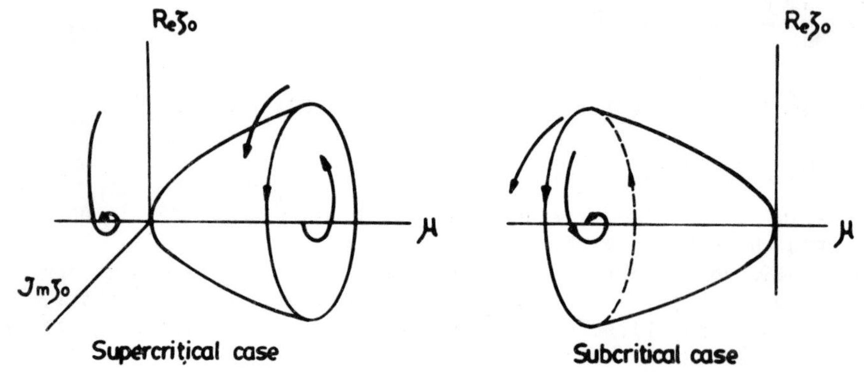

Figure 9.

Remark. The successive changes of variables (2.22) have the effect to make this periodic orbit more and more circular. The dynamics is shown in figure 9 for the bifurcation diagram, and in figure 10 for the phase portrait.

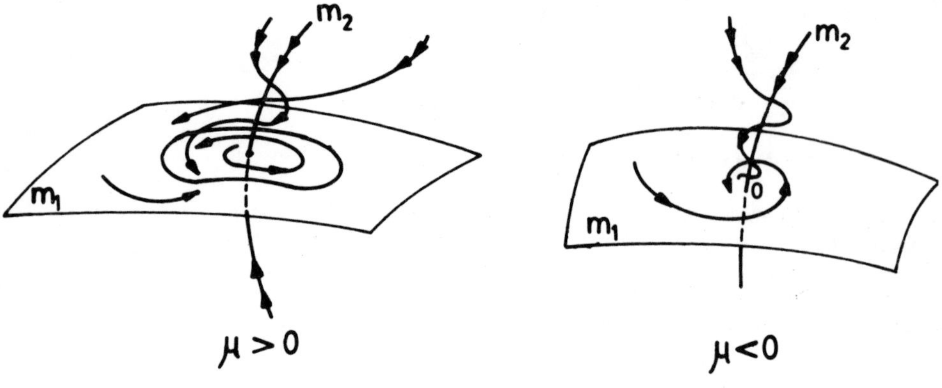

Figure 10. Case of a supercritical Hopf bifurcation.

The problem is now to compute efficiently the dynamics on the center manifold, by going faster to the relevant coefficients of the system (2.24). We can realize that when we make the change of variables (2.17) and (2.22) we obtain a new decomposition of X:

$$X = z'\zeta_0 + \bar{z}'\bar{\zeta}_0 + Y'$$

where $Y' = \Gamma'(\mu,z',\bar{z}')\zeta_0 + \bar{\Gamma}'(\mu,z',\bar{z}')\bar{\zeta}_0 + \Phi(z'+\Gamma',\bar{z}'+\bar{\Gamma}',\mu)$, and Γ' is defined by the inversion of Γ:

$$z' = z + \Gamma(\mu,z,\bar{z})$$

$$z = z' + \Gamma'(\mu,z',\bar{z}') \ .$$

This shows that we no longer require that Y' belongs to the range of the projection Q_0; the strong requirement here is to make disappear as far as possible any term in (2.27). Doing so, we obtain immediately in the same way as in (2.15):

$$\begin{aligned}
&F_{101} = 0 \ , \ F_{020} = F_{002} = F_{011} = 0 \ , \ F_{030} = F_{012} = F_{003} = 0, \\
&F_{110}\zeta_0 - L_1\zeta_0 = (L - i\omega_0)\Phi_{110} \ , \ \Phi_{101} = \bar{\Phi}_{110} \ , \\
&- F_{02}(\zeta_0,\zeta_0) = (L - 2i\omega_0)\Phi_{020} \ , \ \Phi_{002} = \bar{\Phi}_{020} \ , \\
&- 2F_{02}(\zeta_0,\bar{\zeta}_0) = L\Phi_{011} \\
&F_{021}\zeta_0 - 2F_{02}(\zeta_0\ \Phi_{011}) - 2F_{02}(\zeta_0,\Phi_{020}) - 3F_{03}(\zeta_0,\zeta_0,\bar{\zeta}_0) \\
&= (L - i\omega_0)\Phi_{021} \qquad (2.35) \\
&\ldots\ldots\ldots\ldots\ldots\ldots
\end{aligned}$$

and the coefficients F_{110}, F_{021},... are obtained by a unique compatibility condition since the range of $L - i\omega_0$ has codimension one. We choose Φ_{110}, Φ_{021}, in the range of $L - i\omega_0$ to fix the determination of the "amplitude" z, while Φ_{020}, Φ_{011} are well defined since $L - 2i\omega_0$ and L are invertible.

2.3 Theory of normal forms.

The idea is now to avoid the pedestrian way for computing a normal form such as (2.27) and to give a method to predict a priori the structure of the simplest form of the differential system on the center manifold.

We notice that the truncated normal form (2.27) is invariant under the plane rotations: it commutes with the operator

$$z \to ze^{i\varphi} \ .$$

It is clear that if this property could be predicted, we could avoid two pages of computations. It is perhaps not very economic here, but if we consider problems with more than one parameter μ, and more than one pair of pure imaginary eigenvalues at criticality then it is much more interesting!

We first consider a system

$$\frac{dX}{dt} = F(X) \ , \ F(0) = 0 \ , \ D_X F(0) = L_0 \qquad (2.36)$$

in $\mathbb{R}^n$, and such that L_0 has all its eigenvalues with a zero real part. We want to make nonlinear change of variables to simplify F. For this we define the term $A_k(X,\ldots,X)$ of degree k in F, A_k is a k-linear symmetric operator in its arguments. The change of variables is written

$$X' = X + \Gamma_k(X,X,\ldots X) \quad , \tag{2.37}$$

where Γ_k is k-linear symmetric too. The new system may be written

$$\frac{dX'}{dt} = F'(X')$$

and the term of degree k in F' is now such that

$$A'_k(X,\ldots X) = A_k(X,\ldots X) + k\Gamma_k(X,\ldots X,L_0X) - L_0\Gamma_k(X,\ldots X) \quad . \tag{2,38}$$

Of course this change of variables modify higher order terms too. The unknown in (2.38) is Γ_k, and we wish A'_k as simple as possible. Let us rewrite (2.38) into the following form:

$${}^k\mathrm{ad}L_0(\Gamma_k) + A_k = A'_k \tag{2.39}$$

where we define a linear operator ${}^k\mathrm{ad}L_0$ acting in the vector space H_k of k-linear symmetric operators. By definition

$$[{}^k\mathrm{ad}L_0(\Gamma_k)](X,\ldots X) = k\Gamma_k(X,\ldots X,L_0X) - L_0\Gamma_k(X,\ldots X) \quad . \tag{2.40}$$

The idea is to have in A'_k "the part of A_k" which is not in the image in H_k of ${}^k\mathrm{ad}L_0$.

For this purpose, let us introduce the group generated by L_0:

$$R_t = e^{L_0 t} \tag{2.41}$$

and define in H_k, the operator R_{t*} by

$$[R_{t*}\Gamma_k](X,\ldots X) = R_{-t}\Gamma_k(R_tX,\ldots R_tX) \tag{2.42}$$

then, by construction

$$\frac{d}{dt}(R_{t*}\Gamma_k) = {}^k\mathrm{ad}L_0(R_{t*}\Gamma_k) \quad ,$$

hence $R_{t*}\Gamma_k$ is constant if and only if Γ_k is in the kernel of ${}^k\mathrm{ad}L_0$. Let us note this kernel G_k:

$$G_k = \{\Gamma_k \in H_k \ ; \ R_{t*}\Gamma_k = \Gamma_k\} \text{ (elements invariant under } R_{t*}). \tag{2.43}$$

We can show that we have the decomposition:

$$H_k = G_k \oplus \text{Image } ({}^k\text{adL}_0) \ , \tag{2.44}$$

provided that we can define the limit

$$P(\Gamma_k) = \lim_{T \to \infty} \frac{1}{T} \int_0^T R_{t*}\Gamma_k dt, \text{ for any } \Gamma_k \in H_k. \tag{2.45}$$

In fact, assume that (2.45) holds, then P is a <u>projection</u> operator on G_k since it is clear that G_k is in the image of P and

$$R_{\tau*}P(\Gamma_k) = \lim_{T \to \infty} \frac{1}{T} \int_0^T R_{t+\tau*}\Gamma_k dt = \lim_{T \to \infty} \frac{1}{T} \int_0^T R_{t*}\Gamma_k dt = P(\Gamma_k).$$

Now it is a classical result of finite dimensional linear algebra that

$$H_k = \ker P \oplus G_k$$

and to prove (2.44), it is sufficient to prove that Image $({}^k\text{adL}_0) \subset \ker P$, since $\dim G_k + \dim \text{Im}({}^k\text{adL}_0) = \dim H_k$. But it is not hard to ckeck that for any $\Gamma_k \in H_k$ we have

$$\int_0^T R_{t*}\, {}^k\text{adL}_0(\Gamma_k)dt = \int_0^T {}^k\text{adL}_0[R_{t*}(\Gamma_k)]dt = {}^k\text{adL}_0(\int_0^T R_{t*}\Gamma_k dt) \ ,$$

hence $P[{}^k\text{adL}_0(\Gamma_k)] = {}^k\text{adL}_0[P(\Gamma_k)] = 0$.

Remark 1. The property (2.45) is certainly not true for the cases when some eigenvalue of L_0 is not semi-simple. In those cases there are polynomials of t in the integral (2.45), hence there is no limit when $T \to \infty$.

Remark 2. In the case studied in section 2.2 (Hopf bifurcation), for $\mu = 0$, we have one pair $\pm i\omega_0$ of simple eigenvalues for L_0. The group generated is the group of rotations of the plane and G_k will be the set of Γ_k invariant under rotations. This is exactly what we obtained. Check in exercise that (2.45) holds.

Conclusion: If the property (2.45) holds, we just keep in the normal form, the terms in G_k, $k = 2,3,\ldots,$ invariant under R_{t*}.

Let us now consider a bifurcation problem with parameters $\mu \in \mathbb{R}^p$:

$$\frac{dX}{dt} = F(\mu,X) \text{ in } \mathbb{R}^n \tag{2.46}$$

$$F(0,0) = 0, \ D_X F(0,0) = L_0 \ , \tag{2.47}$$

where L_0 has only eigenvalues of zero real part. In incorporating μ into the changes of variables, we observe that we could consider here changes of degree 0 and 1 in X. The problem is the same as in (2.40):

$$^k\mathrm{ad}L_\mu(\Gamma_k) + A_k = A'_k \ , \ k \geqq o \tag{2.48}$$

and for k = 0 and 1 A_k is $O(\mu)$.

Here again we have to study the image of $^k\mathrm{ad}L_0$ since we can write (2.48) in the form:

$$^k\mathrm{ad}L_0(\Gamma_k) + \mu Q_0 {}^k\mathrm{ad}L_1(\Gamma_k) + Q_0 A_k = 0$$

$$P_0(A_k - A'_k) + \mu P_0 {}^k\mathrm{ad}L_1(\Gamma_k) = 0$$

where Q_0 and P_0 are the projections on the image and on a supplementary space. This shows that Γ_k is uniquely determined with Q_0A_k, and that $A'_k = P_0(A_k + O(\mu))$ lies in the supplementary space. If the property (2.45) holds, we have here again the normal form which is invariant under R_{t*}.

Remark 3. For k = 0, ker ($^0\mathrm{ad}L_0$) is just the kernel of L_0, so we only keep in $F(\mu,0)$ terms belonging to the kernel of L_0. For k = 1, ker($^1\mathrm{ad}L_0$) is the space of linear operators commuting with L_0, this restricts very simply the allowed linear terms in the perturbated normal form.

Remark 4. In the case when L_0 has no semi-simple eigenvalues (non diagonalizable), then it is shown in [6] that

$$H_k = \mathrm{Image}(^k\mathrm{ad}L_0) \oplus \ker(^k\mathrm{ad}L_0^*)$$

where L_0^* is the adjoint of L_0 in $\mathbb{R}^n$.

The proof is based on the existence of a scalar product such that

$$(^k\mathrm{ad}L_0)^* = {}^k\mathrm{ad}L_0^* \quad .$$

Hence, in all cases, if we substract the part L_0X from the normal form the remaining part F satisfies

$$\tilde{F}(\mu, e^{L_0^* t}X) = e^{L_0^* t}\tilde{F}(\mu,X) \ ,$$

which is of fundamental interest for explicit computations.

Exercise. Examine the case of two pairs of simple eigenvalues $\pm i\omega_0$, $\pm i\omega_1$. Show that if $\omega_0/\omega_1 \notin Q$ then the normal form is invariant under the group of the torus T^2. What happens when $\omega_0/\omega_1 = p/q$? Show that this adds terms with respect to the previous case, and that the terms of lowest degree which are added have degree $p + q - 1$.

General references for section 2 are [7,8].

3. PRESENCE OF SYMMETRIES

3.1 Simple symmetry - Pitchfork bifurcation.

Let us start with the following evolution problem in a space E:

$$\frac{dX}{dt} = F(\mu,X) \quad , \quad \mu \in \mathbb{R} \ , \quad X(t) \in E \tag{3.1}$$

with

$$F(0,0) = 0 \quad \text{and} \quad D_X F(0,0) = L \quad . \tag{3.2}$$

We assume moreover that there is a linear operator S such that

$$S^2 = Id \quad , \quad S \neq Id \quad , \tag{3.3}$$

$$F(\mu,SX) = S F(\mu,X) \quad . \tag{3.4}$$

Now we start again with the assumption made in section 2.1: 0 is a simple eigenvalue of L , other eigenvalues having strictly negative real part.

Let us note ξ_0 the eigenvector belonging to 0: $L\xi_0 = 0$, then all the discussion relies upon the property

$$S\xi_0 = \pm \xi_0 \quad . \tag{3.5}$$

This is just a consequence of the simplicity of the eigenvalue, and of (3.3).

1. Let us first assume that

$$S\xi_0 = \xi_0 \tag{3.6}$$

then due to the results of section 1.3.3 me know that the center manifold (same notation as in 2.1)

$$Y = \Phi(x,\mu) \tag{3.7}$$

is invariant under S, since $S(x\xi_0) = x\xi_0$. Hence the family of steady solutions found in section 2.1 is fully invariant under S.

2. Let us asuume now that

$$S\xi_0 = -\xi_0 \tag{3.8}$$

which means that the eigenvector breaks the symmetry of the problem. As a consequence of the results of 1.3.3 we have on the center manifold:

$$S\Phi(x,\mu) = \Phi(-x,\mu) \quad . \tag{3,9}$$

Let us go now to the differential equation (one dimensional) on the center manifold

$$\frac{dx}{dt} = F(\mu,x) \quad , \tag{3.10}$$

the propagation of the invariance under S gives us a new property in this case. Since $S(x\xi_0) = -x\xi_0$, we should have (see eq. (1.42):

$$F(\mu,-x) = -F(\mu,x) \quad . \tag{3.11}$$

Hence $x = 0$ is a steady solution of (3.10), and as a consequence we obtain the steady solution of (3.1)

$$x = \Phi(0,\mu) \tag{3.12}$$

which is invariante under S.

Now computing the Taylor expansion of $F(\mu,x)$ close to 0, we obtain

$$F(\mu,x) = a\mu x + bx^3 + \dots \tag{3.13}$$

So, if $a \neq 0$ (i.e. if $f_{11}(\xi_0) + 2f_{02}(\xi_0,\Phi_{10}) \neq 0$), we have another branch of steady solutions of (3.10), on only one side of $\mu = 0$ (see figure 11)

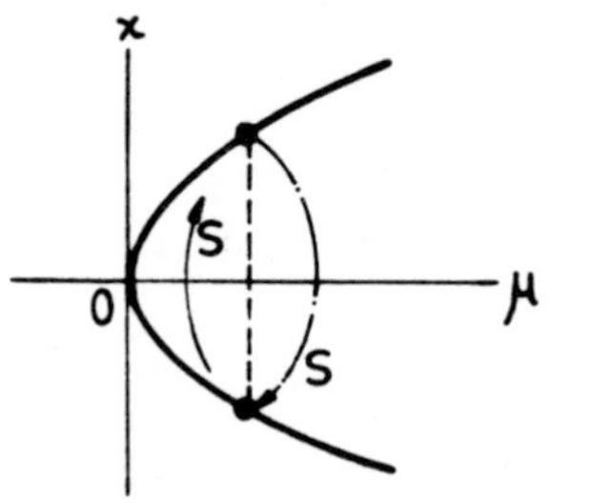

Figure 11. Pitchfork bifurcation case $ab < o$.

We observe now that μ is an even function of x, hence $\pm x$ correspond to the same value of μ.

Exercise. Show that the coefficient b in (3.13) is determined by the formula (note that a lot of coefficients are zero in (2.5))

$$b = 2f_{02}(\xi_0,\Phi_{02}) + f_{03}(\xi_0,\xi_0,\xi_0) \tag{3.14}$$

where Φ_{02} is defined by

$$L\,\Phi_{02} = -Q_{02}\,F_{02}(\xi_0,\xi_0) \quad .$$

The branch of steady solutions we have just defined gives us a family of solutions of (3.1) of the form

$$X = x\xi_0 + \Phi(x,\mu)$$

and

$$SX = -x\xi_0 + \Phi(-x,\mu)$$

shows that changing x in $-x$ with the same μ exchanges the two steady solutions which can be deduced one from the other by the symmetry S.

The stability analysis of these steady solutions can be reduced to a one dimensional study on the center manifold. Assume that $a > 0$, which means that the symmetric solutions looses its stability when μ increases across 0. Then, if we note that $a\mu + bx_0^2 + O(|\mu| + x_0^2)^2 = 0$, we have

$$D_x F(\mu(x_0),x_0) = -2a\mu + O(\mu^2) \tag{3.15}$$

hence if the bifurcation is supercritical (for $\mu > 0$), the bifurcated branch is stable, while if it is subcritical (for $\mu < 0$) the branch is unstable (see figures 12 and 13).

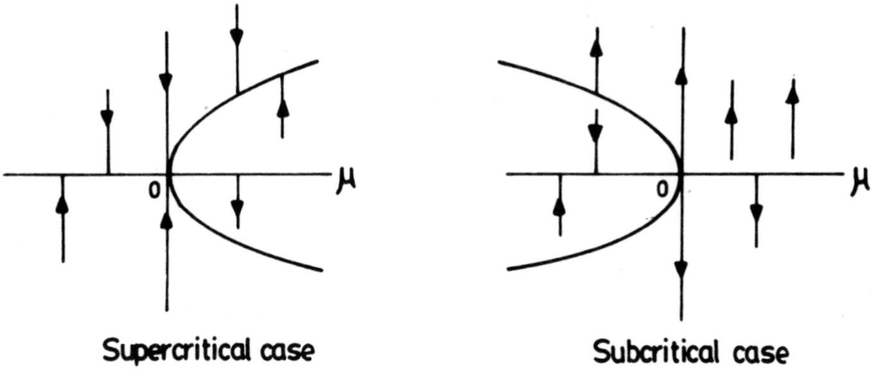

Figure 12. Bifurcation diagrams.

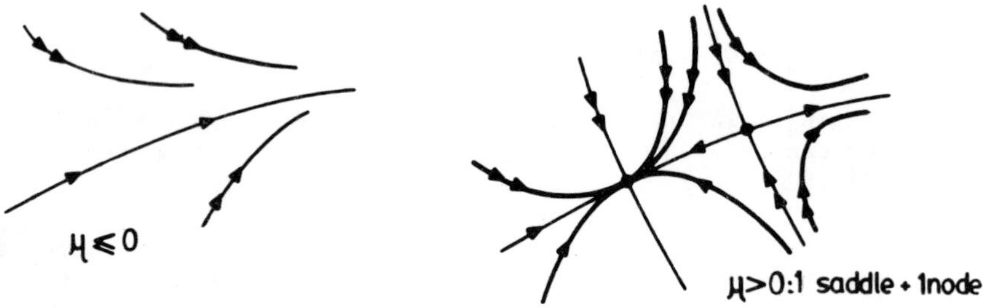

Figure 13. Dynamics in E, in the supercritical bifurcation case.

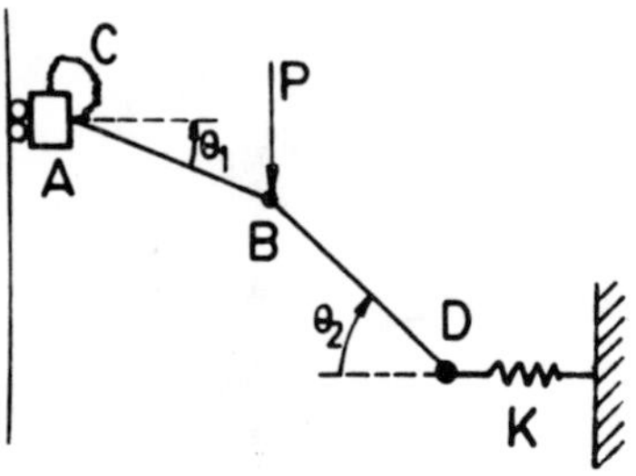

Figure 14.

We treat now a simple mechanical example. Potential energy (see figure 14, $AB = BD = \ell$):

$$V = \frac{1}{2}c\dot{\theta}_1 + \frac{1}{2}K\ell^2(\cos\theta_1 + \cos\theta_2 + \cos\alpha - 1)^2 - P\ell\sin\theta_2 \quad ,$$

natural equilibrium: $\theta_1 = 0$, $\theta_2 = \alpha$. In posing $\mu = P\ell/c$, $k = K\ell^2/c$, $x = (\theta_1,\theta_2)$ the system $F(\mu,X) = 0$ which gives the equilibrium positions is given by the two equations: $\partial_{\theta_1}V = \partial_{\theta_2}V = 0$. Show that

$$F(\mu,SX) = SF(\mu,X)$$

where

$$S(\theta_1,\theta_2) = (-\theta_1\ \theta_2) \quad .$$

The symmetric solutions are such that $\theta_1 = 0$. Show that they can be represented as on figure 15.

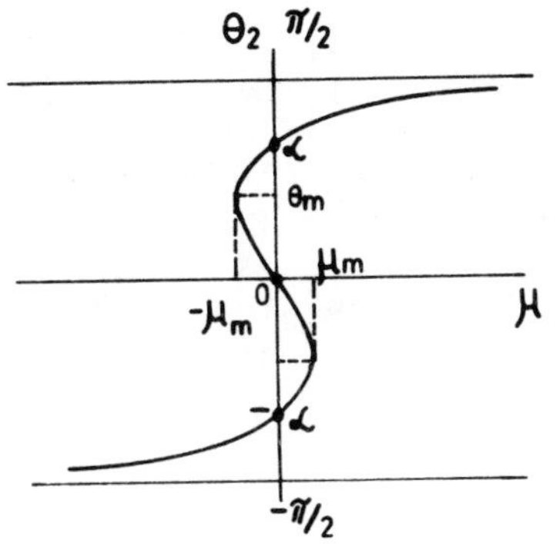

Figure 15. Symmetric solutions.

The stability of any equilibrium of this system is determined by the eigenvalues of $D_X F(\mu,X)$ for μ, X satisfying $F(\mu,X) = 0$. It is stable if both eigenvalues are positive since then the potential energy will be a minimum.

Show that if $\cos\alpha + \frac{1}{k} < 1$, then from the symmetric solution just defined bifurcates a branch of non - symmetric solutions. Discuss the stability depending on the sign of $\cos\alpha - (\cos\alpha + 1/k)^2$!

Remark. In this example there are two saddle node bifurcations and two pitchfork bifurcations.

3.2 SO(2) invariance.

A system (3.1) has SO(2) invariance if it commutes with the representation R of the group of plane rotations. In fact this means that there is a family R of linear operators in E such that:

$$R_{\varphi_1} R_{\varphi_2} = R_{\varphi_1 + \varphi_2}, \quad \varphi_1, \varphi_2 \in \mathbb{R} \tag{3.16}$$

$$R_0 = R_{2\pi} = \mathrm{Id} \tag{3.17}$$

$$F(\mu, R_\varphi X) = R_\varphi F(\mu,X) \quad . \tag{3.18}$$

A physical system may easily have this invariance: consider for instance a system invariant under one dimensional translations in some direction, where we assume, to simplify the analysis, that it is spatially periodic. It happens very often that some critical spatial period occurs after some stability analysis, for instance in hydrodynamic stability problems. Another possibility to have this invariance is to consider a system with a cylindrical geometry but where we have not the symmetry $\theta \to -\theta$, for instance if we break this invariance by external conditions (rotating the cylinder for instance).

Assuming again that we have (3.2) and (3.5), we necessarily have

$$R_\varphi \xi_0 = \xi_0 \tag{3.19}$$

since $R_\varphi \xi_0$ is colinear to ξ_0 and the only real function $k(\varphi)$ such that $k(\varphi_1)\, k(\varphi_2) = k(\varphi_1 + \varphi_2)$, $k(0) = 1$, $k(2\pi) = 1$, is $k(\varphi) \equiv 1$. The same argument as the one developed in 3.2 shows that we now have a branch of steady solutions which are invariant under R_φ ("axisymmetric solutions"). A much more interesting case is when we assume

$\pm i\omega_0$ are simple eigenvalues of L and other eigenvalues are of negative real part. (3.20)

Let us note ζ_0 the eigenvector belonging to $i\omega_0$:

$$L\zeta_0 = i\omega_0\zeta_0 \tag{3.21}$$

then we necessarily have

$$R_\varphi \zeta_0 = e^{i\ell\varphi}\zeta_0 \text{ , with some } \ell \in \mathbb{Z}. \tag{3.22}$$

In fact, $R_\varphi \zeta_0$ is colinear to ζ_0 with a factor exponential in φ, which is 2π-periodic. If $\ell = 0$, we have eigenvectors ζ_0 and $\bar{\zeta}_0$ invariant under R_φ and the study of section 2.2 shows that the bifurcated periodic solution is pointwise invariant under R_φ, as well as the center manifold. In this case the selfoscillating solution is "axisymmetric".

Let us consider now the most interesting case, which is when $\ell \neq 0$ in (3.22). To understand the structure of the differential equation on the center manifold (see (2.13)):

$$\frac{dz}{dz} = F(\mu,z,\bar{z}) \tag{3.23}$$

we use the commutation property proved in section 1.3.3.We have

$$R_\varphi(z\zeta_0 + \bar{z}\bar{\zeta}_0) = ze^{i\ell\varphi}\zeta_0 + \overline{(ze^{i\ell\varphi})}\bar{\zeta}_0 \quad ,$$

hence we necessarily have

$$F(\mu,e^{i\ell\varphi}z,e^{-i\ell\varphi}\bar{z}) = e^{i\ell\varphi}F(\mu,z,\bar{z}) \ . \tag{3.24}$$

This very strong property shows that $F(\mu,z,\bar{z}) = zG(\mu,r)$, where $z = re^{i\theta}$ and with G <u>even</u> in r. Let us write (3.23) more precisely in polar coordinates:

$$\frac{dr}{dt} = r\mathcal{R}e\,G(\mu,r) = \xi(\mu)r + a(\mu)r^3 + O(r^5) \tag{3.25}$$

$$\frac{d\theta}{dt} = \mathcal{I}m\,G(\mu,r) = \eta(\mu) + b(\mu)r^2 + O(r^4) \tag{3.26}$$

We see that (3.23) is directly in normal form. The consequence of this fact is that we have an exact periodic solution of (3.23) of the form

$$z(t) = r_0 e^{i(\Omega t + \varphi_0)} \quad . \tag{3.27}$$

Hence, coming back to $X(t) = z(t)\zeta_0 + \bar{z}(t)\bar{\zeta}_0 + \Phi(z,\bar{z},\mu)$ we have

$$R_\varphi X(t) = X(t + \frac{\ell\varphi}{\Omega}) \quad ,$$

so

$$X(t) = R_{\frac{\Omega t}{\ell}} X(0) \quad (\text{hence } R_{\frac{2\pi}{\ell}} X(0) = X(0)) \quad . \tag{3.28}$$

This type of periodic solution is called a "rotating wave". We moreover see that when the "angle" φ rotates by 2π, then we obtain ℓ times the period (ℓ waves).

Exercise. Consider a system invariant under rotations of $2\pi/n$, i.e. assume that F commutes with $R_{2\pi/n}$ (n integer). Assume that (3.20) holds, then show that

$$R_{2\pi/n}\zeta_0 = k\zeta_0$$

where k is an nth root of unity $= e^{2i\pi\frac{\ell}{n}}$. Pose $\ell/n = \ell_0/n_0$ where ℓ_0/n_0 is irreducible and $n = p_0 n_0$.

Show that the bifurcated periodic solution is invariant under $R_{2\pi/p_0}$ (so we break a part of the symmetry). As an example, think of a physical system invariant under translations, having a basic steady spatially periodic solution. Then we can interprete this exercise by saying that we can obtain a bifurcating periodic solution in time, whose spatial period is multiplied by some integer p, if we fix a basic spatial periodicity equal to a multiple n of the period of the basic solution and p is a divisor of n.

3.3 O(2) invariance.

A system (3.1) has a O(2) invariance if it commutes with i) a representation R_φ of the group of plane rotation and ii) with a symmetry S such that (think of $z \to ze^{i\varphi}$ and $z \to \bar{z}$ on the unit circle in $\mathbb{C}$):

$$R_\varphi S = SR_{-\varphi} \quad . \tag{3.29}$$

Hence we have together (3.19) and (3.3) - (3.4) with (3.29).

A physical example with this symmetry can be for instance the case with a cylindrical geometry, with nothing breaking the symmetry $\theta \to -\theta$ for the angle around the axis. Another very common case is when the system is invariant under translations along a direction OZ, and invariant under reflexions $Z \to -Z$. If we impose to the solutions to be Z-periodic with some prescribed period, then we are in this frame.

If we make the assumption that 0 is simple eigenvalue and the other eigenvalues have negative real part then we know that necessarily we have

$$S\xi_0 = \pm\xi_0 \ , \ R_\varphi \xi_0 = \xi_0 \quad .$$

Hence nothing new happens if $S\xi_0 = \xi_0$, except that the steady branch of solutions is O(2) invariant. If $S\xi_0 = -\xi_0$ we obtain as before a pitchfork bifurcation of a branch which is now axisymmetric (invariant under R_φ).

If we make the assumptions (3.20) (i.e. $\pm i\omega_0$ are simple eigenvalues of L), then we have

$$S\zeta_0 = \pm\zeta_0, \quad R_\varphi\zeta_0 = e^{i\ell\varphi}\zeta_0 \text{ for some } \ell \text{ in } \mathbb{Z}.$$

i) If $S\zeta_0 = \zeta_0$ and $\ell = 0$, the bifurcated periodic solution does not break the O(2) symmetry.
ii) If $S\zeta_0 = \zeta_0$ and $\ell \neq 0$, the rotating wave (3.28) is still invariant under S.
iii) If $S\zeta_0 = -\zeta_0$ and $\ell = 0$, the periodic orbit is globally invariant under S, and such that applying S is the same as shifting the time by half of the period.
iv) If $S\zeta_0 = -\zeta_0$ and $\ell \neq 0$, then the rotating wave (3.28) has the property described in iii):

$$X(t) = R_{\frac{\Omega t}{\ell}} X(0) \text{ is such that } SX(t) = R_{\pi/2}X(t) \quad .$$

In fact the most interesting and generic cases in the case of an O(2) invariance occur when the eigenvalues are no longer simple! If we notice that the eigenspaces are invariant under S and R_φ, the simplest generic case is when the eigenspace is 2-dimensional; then we shall assume that S and R_φ act on this invariant subspace in a non-trivial way (S and $R_\varphi \neq \text{Id}$).

Since R_φ is a one parameter group, it has a generator L. Then L satisfies

$$e^{2\pi L} = 1$$

which leads to the fact that the eigenvalues of L are $\pm i\ell$, $\ell \in \mathbb{Z}$. Choosing, as a basis in our two dimensional space, the two eigenvectors ζ_0 and ζ_1 such that ($\ell \neq 0$ since R_φ is not trivial)

$$L\zeta_0 = i\ell\zeta_0 \quad , \quad L\zeta_1 = -i\ell\zeta_1 \quad ,$$

we obtain

$$R_\varphi\zeta_0 = e^{i\ell\varphi}\zeta_0 \quad , \quad R_\varphi\zeta_1 = e^{-i\ell\varphi}\zeta_1 \quad . \tag{3.30}$$

Now, since $R_\varphi S = SR_{-\varphi}$ we can choose

$$\zeta_1 = S\zeta_0 \tag{3.31}$$

since $S\zeta_0$ is an eigenvector of R_φ belonging to the eigenvalue $e^{-i\ell\varphi}$, (we can remark that $S\zeta_0$ is necessarily non colinear to ζ_1 if $\ell \neq 0$).

A new property occurs when the eigenvalue of L is real, since $\zeta_0, \zeta_1 = S\zeta_0, \bar{\zeta}_0, \bar{\zeta}_1$, are all eigenvectors for this eigenvalue.

But, since

$$R_\varphi \bar{\zeta}_0 = e^{-i\ell\varphi}\bar{\zeta}_0 \quad ,$$

we have $\bar{\zeta}_0 = K\zeta_1 = KS\zeta_0$, as well as $\zeta_0 = \bar{K}S\bar{\zeta}_0$, hence $|K|^2 = 1$. Finally multiplying ζ_0 by a suitable number we can arrange things in such a way that

$$\zeta_1 = S\zeta_0 = \bar{\zeta}_0 \quad . \tag{3.32}$$

3.3.1 Steady bifurcation. Let us make the assumptions (3.3), (3.4), (3.16) - (3.18), (3.29), and

> 0 is a double eigenvalue of L, other eigenvalues being of strictly negative real part. Moreover S and R_φ act non-trivially on the eigenspace belonging to 0.

Hence we have the properties (3.30), (3.31), (3.32) here. The decomposition of the solution leads to

$$X = z\zeta_0 + \bar{z}\bar{\zeta}_0 + \Phi(z,\bar{z},\mu) \quad , \tag{3.33}$$

and the dynamics is ruled by a differential equation in $\mathbb{C}$:

$$\frac{dz}{dt} = F(\mu,z,\bar{z}) \quad . \tag{3.34}$$

Here again the commutation relationships with R_φ and S (see eq. (1.42)),

$$\begin{aligned} R_\varphi(z\zeta_0 + \bar{z}\bar{\zeta}_0) &= ze^{i\ell\varphi}\zeta_0 + \overline{(ze^{ie\varphi})}\bar{\zeta}_0 \\ S(z\zeta_0 + \bar{z}\bar{\zeta}_0) &= \bar{z}\zeta_0 + z\bar{\zeta}_0 \quad . \end{aligned} \tag{3.35}$$

Hence we have

$$\begin{aligned} F(\mu, ze^{i\ell\varphi}, \bar{z}e^{-i\ell\varphi}) &= e^{i\ell\varphi}F(\mu,z,\bar{z}) \\ F(\mu,\bar{z},z) &= \overline{F(\mu,z,\bar{z})} \quad . \end{aligned} \tag{3.36}$$

This gives easily the structure of F:

$$F(\mu,z,\bar{z}) = zG(\mu,r) \quad , \tag{3.37}$$

where G is real and even in $r = |z|$.

Let us write more precisely (3.34) in polar coordinates:

$$z = re^{i\theta} \quad ,$$

$$\frac{dr}{dt} = rG(\mu,r) = \xi(\mu)r + a(\mu)r^3 + O(r^5) \quad ,$$

$$\frac{d\theta}{dt} = 0 \quad . \tag{3.38}$$

Hence, we obtain of course a first steady solution $r = 0$, which corresponds to a persisting steady solution of (3.1) having the full O(2) symmetry. Now, from (3.38) we obtain a pitchfork branch of a new steady solutions, depending on an undetermined phase θ_0. These solutions are invariant under $R_{2\pi/\ell}$ and we can obtain all such solutions acting with R_φ. With a suitable choice of the phase, we have a symmetric solution (take $\theta_0 = k\pi$).

The stability of this family of steady solutions is just the same as the one described in section 2.1 for (2.29).

3.3.2 Travelling waves and standing waves. Here we assume that (3,3) - (3.4), (3.16) - (3.18), (3.29) holds and that

> $\pm\, i\omega_0$ are double eigenvalues of L, other eigenvalues being of strictly negative real part. Moreover S and R act non-trivially on the eigenspaces belonging to $\pm\, i\omega_0$. (3.39)

We have then the properties (3.30) and (3.31) for the eigenvectors ζ_0, ζ_1. Let us decompose as usual:

$$X = z_0\zeta_0 + \bar{z}_0\bar{\zeta}_0 + z_1\zeta_1 + \bar{z}_1\bar{\zeta}_1 + \Phi(z_0,\bar{z}_0,z_1,\bar{z}_1,\mu), \tag{3.40}$$

where $y = \Phi$ is the equation of the center manifold which is here 4-dimensional.

The dynamics is ruled by the following system of two complex differential equations:

$$\frac{dz_0}{dt} = F_0(\mu,z_0,\bar{z}_0,z_1,\bar{z}_1)$$

$$\frac{dz_1}{dt} = F_1(\mu,z_0,\bar{z}_0,z_1,\bar{z}_1) \tag{3.41}$$

The actions of R_φ and S are such that

$$R_\varphi(z_0\zeta_0 + z_1\zeta_1) = z_0 e^{i\ell\varphi}\zeta_0 + z_1 e^{-i\ell\varphi}\zeta_1$$
$$S(z_0\zeta_0 + z_1\zeta_1) = z_1\zeta_0 + z_0\zeta_1 \quad , \tag{3.42}$$

and the right hand side of (3.41) commutes with these actions.

This invariance property is not sufficient to put the system (3.41) in normal form. Let us now use the result of section 2.3. For this aim, we note T_t the group generated by L_0. We easily obtain:

$$T_t(z_0\zeta_0 + z_1\zeta_1) = z_0 e^{i\omega_0 t}\zeta_0 + z_1 e^{i\omega_0 t}\zeta_1 \quad , \tag{3.43}$$

since $i\omega_0$ is a double semi-simple eigenvalue of L_0. So it is also easy to check that for any $\Gamma_k \in H_k$, $T_{t*}\Gamma_k$ is $2\pi/\omega_0$ periodic, hence the condition (2.45) is realized. Hence the normal form has to commute with T_t.

Let us show that the normal form of (3.41) commutes then with both T_t and R_φ and S.

We of course know that L_0 and R_φ commute (since L and R_φ commute) so it is easy to check that ${}^k\mathrm{ad}L_0$ and $(R_\varphi)_*$ commute, acting in H_k. So looking at (2.39), we notice that if A_k is invariant under $(R_\varphi)_*$, then Γ_k can be chosen invariant under $(R_\varphi)_*$, since the projection operator P on the kermel of $({}^k\mathrm{ad}L_0)$ commutes with $(R_\varphi)_*$ ($(R_\varphi)_*$ and T_{t*} commute, due to $T_{t*} = \exp({}^k\mathrm{ad}L_0)$). Finally the normal form N_0, N_1 of (3.41) satisfies:

$$N_0[\mu, z_0 e^{i(\ell\varphi + \omega_0 t)}, \bar{z}_0 e^{-i(\ell\varphi + \omega_0 t)}, z_1 e^{i(-\ell\varphi + \omega_0 t)},$$
$$\bar{z}_1 e^{i(\ell\varphi - \omega_0 t)}] = e^{i(\ell\varphi + \omega_0 t)} N_0(\mu, z_0, \bar{z}_0, z_1, \bar{z}_1) \ ,$$
$$N_0(\mu, z_1, \bar{z}_1, z_0, \bar{z}_0) = N_1(\mu, z_0, \bar{z}_0, z_1, \bar{z}_1) \quad . \tag{3.44}$$

Since in (3.44) t and φ are independent, we deduce that N_0 and N_1 are invariant under the group of the 2-torus T^2.

Finally, if we still write the amplitudes z_0, z_1, we have the following system on the center manifold:

$$\frac{dz_0}{dt} = z_0 G_0(\mu, |z_0|^2, |z_1|^2) + O(|z_0| + |z_1|)^N$$
$$\frac{dz_1}{dt} = z_1 G_0(\mu, |z_1|^2, |z_0|^2) + O(|z_0| + |z_1|)^N \quad . \tag{3.45}$$

For generic cases, the dynamics for the full system (3.45) is the same as the one of the truncated system:

$$\frac{dz_0}{dt} = (i\omega_0 + a\mu)z_0 + bz_0|z_0|^2 + cz_0|z_1|^2$$

$$\frac{dz_1}{dt} = (i\omega_0 + a\mu)z_1 + cz_1|z_0|^2 + bz_1|z_1|^2 \quad , \tag{3.46}$$

which gives in polar coordinates: $z_j = r_j e^{i\varphi_j}$:

$$\frac{dr_0}{dt} = r_0(a_r\mu + b_r r_0^2 + c_r r_1^2)$$

$$\frac{dr_1}{dt} = r_1(a_r\mu + c_r r_0^2 + b_r r_1^2)$$

$$\frac{d\varphi_0}{dt} = \omega_0 + a_i\mu + b_i r_0^2 + c_i r_1^2$$

$$\frac{d\varphi_1}{dt} = \omega_0 + a_i\mu + c_i r_0^2 + b_i r_1^2 \tag{3.47}$$

where we note that the (r_0,r_1) system is uncoupled from the angular part. The study of the dynamics of this two dimensional reduced system is easy to handle. Apart from the 0 solution, there are two types of steady solutions: i) $r_0 \neq 0$, $r_1 = 0$ or $r_0 = 0$, $r_1 \neq 0$, and ii) $r_0 = r_1 \neq 0$. These solutions of the (r_0,r_1) system lead to periodic solutions of (3.46) which approach periodic solutions of (3.45).

The type (i) of solutions has the following structure

$$z_0(t) = r_0 e^{i(\Omega_0 t + \varphi_0)} \ , \ z_1(t) = 0 \ , \tag{3.48}$$

where r_0 and Ω_0 are functions of μ easy to obtain (same as for Hopf bifurcation). This type of solution is, as we have already seen, a "rotating wave", since the action of R_φ described in (3.42) is the same on the present variables (z_0,z_1) (commutativity of the change of variable with R_φ) as the original (z_0,z_1) defined by (3.40). In fact these solutions are often called "travelling waves", due to the origin of the O(2) invariance, if it comes from a translational invariance.

These solutions are stable for $\mu > 0$ if $a_r > 0$ and $b_r < 0$, $b_r > c_r$.

The other type (ii) of periodic solutions looks like:

$$z_0(t) = r_0 e^{i(\Omega_1 t + \varphi_0)} \ , \ z_1(t) = r_0 e^{i(\Omega_1 t + \varphi_0)} \quad . \tag{3.49}$$

Now let us note

$$R_{\frac{\Omega_1 t}{\ell}} X^{(0)} = r_0 e^{i\Omega_1 t}\zeta_0 + r_0 e^{-i\Omega_1 t}\bar{\zeta}_0 \quad ,$$

then, the full solution (3.40) of the original system (3.1) takes the form, up to high order terms (arbitrary high!):

$$X(t) = R_{\frac{\Omega_1 t + \varphi_0}{\ell}} X^{(0)} + R_{\frac{-(\Omega_1 t + \varphi_1)}{\ell}} SX^{(0)} + \Phi[R_{\frac{\Omega_1 t + \varphi_0}{\ell}} X^{(0)} + R_{\frac{-(\Omega_1 t + \varphi_1)}{\ell}} SX^{(0)}, \mu] \quad . \tag{3.50}$$

If $\varphi_0 = \varphi_1$ we have $X(t) = SX(t)$, and this family of periodic solutuions has a structure completely different from the one of (3.48), even though on (3.50) it looks like a superposition of two travelling waves running in opposite directions. These are the so called "standing waves". They are stable for $\mu > o$ if $a_r > 0$ and $b_r < c_r$, $b_r + c_r < 0$. Moreover, there are (generically) no other bifurcating flows from the origin when μ crosses 0. It may just happen that the standing waves or the travelling wave, even though they exist, may not be seen since they are unstable (cases $b_r + c_r < 0$, $b_r > 0$ and $b_r < 0$, $b_r + c_r > 0$), [9].

REFERENCES

[1] D. Henry. Geometric theory of semilinear parabolic equations. Lecture Notes in Maths. 840, Springer, 1981.

[2] J. Marsden - M. Mc. Cracken. The Hopf bifurcation and its applications. Appl. Math. Sciences, 19, Springer, 1976.

[3] G. Iooss. Bifurcation of maps and applications. North Holland Math. Studies, 36, 1979.

[4] J. Guckenheimer, P. Holmes. Nonlinear Oscillations, Dynamical Systems, and bifurcations of vector fields. Appl. Math. Sciences, A2, Springer, 1983.

[5] T. Kato. Perturbation theory for linear operators. Springer, 1986.

[6]. C. Elphick, E. Tirapegui, M.E. Brachet, P. Coullet, G. Iooss: A Simple global characterization for normal forms of singular vector fields, preprint Université de Nice NTH 86/6 (1986).

[7] V. Arnold. Chapitres Supplémentaires de la théorie des equations differentialles ordinaires. Mir, Moscow, 1980.

[8] J.J. Duistermaat. Bifurcations of periodic solutions near equilibrium points of Hamiltonian systems. p. 57-105, Lecture Notes in Maths., 1057, 1984.

[9] For more details on how to use symmetries in dynamical systems see the book by M. Golubitsky, D.G. Schaeffer: Singularities and Groups

in Bifurcation Theory, Vol. I, Applied Math. Sciences 51, Springer 1984. For physical examples treated in the spirit of this chapter see for instance P. Chossat and G. Iooss: Primary and Secondary Bifurcations in the Couette-Taylor problem, Japan J. Appl. Math. 2, 37-68 (1985).

STABLE CYCLES WITH COMPLICATED STRUCTURE

J. M. Gambaudo(*)

P. Glendinning(**)

C. Tresser(***)

1. INTRODUCTION

1.1. A nice introductive example: the lunar cycles

The length τ_1 of a lunar month is nearly of 29 days, it follows that a solar year ($\tau_2 \sim 365$ days) contains 12 or 13 full moon nights. In table 1, we have represented the number of full moon nights in a year during the 50 years period from 1950 to 1999 A.C.

1950	1	1960	0	1970	0	1980	1	1990	1
1951	0	1961	1	1971	1	1981	0	1991	0
1952	1	1962	0	1972	0	1982	1	1992	0
1953	0	1963	1	1973	0	1983	0	1993	1
1954	0	1964	0	1974	1	1984	0	1994	0
1955	1	1965	0	1975	0	1985	1	1995	0
1956	0	1966	1	1976	0	1986	0	1996	1
1957	0	1967	0	1977	1	1987	0	1997	0
1958	1	1968	0	1978	0	1988	1	1998	0
1959	0	1969	1	1979	0	1989	0	1999	1

Table I

In this table, the symbol 1 (respectively 0) means that the corresponding year has 13 (respectively 12) full moon nights.

Now, let us consider the ordered sequence $\alpha^{(1)} = (\alpha_n^{(1)})_{n\in\mathbb{N}}$ defined as follows:

For each $n \geq 0$

(*) L.A. 168, Parc Valrose, 06034 Nice Cedex, France.
(**) Math. Department, Warwick University, Coventry, CV47AL, U.K.
(***) L.A. 190, Parc Valrose, 06034 Nice Cedex, France.

E. Tirapegui and D. Villarroel (eds.), Instabilities and Nonequilibrium Structures, 41–62.

i) $\alpha_n^{(1)} \in \{0,1\}$

ii) $\alpha_n^{(1)} = 1$ if the year 1950 + n has 13 full moon nights

$\alpha_n^{(1)} = 0$ if the year 1950 + n has 12 full moon nights

We get:

$$\alpha^{(1)} = 10100100100101001001010010010010100100101001001001$$

and we can make the two observations:

Observation 1: the sequence $\alpha^{(1)}$ does not possess two consecutive "1".

Observation 2: (which makes observation 1 more general) by cutting the sequence $\alpha^{(1)}$ after each "0" we obtain a sequence of blocks with length n_1 and with length $n_1 + 1$ where $n_1 = 1$, (except the last block we cannot construct because of the finite size of ephemeris books !).

Consider now the renormalized sequence $\alpha^{(2)}$ deduced from the sequence $\alpha^{(1)}$ by replacing the blocks with length $n_1 + 1$ by "1" and those with length n_1 by "0", we have:

$$\alpha^{(2)} = 1101010110101101010110101101010$$

As for $\alpha^{(1)}$ we remark that the sequence $\alpha^{(2)}$ does not contain two consecutive "0" and that, by cutting the sequence $\alpha^{(2)}$ after each "0" we get a sequence of blocks with length n_2 and $n_2 + 1$ where $n_2 = 2$.

It is a remarkable and very nice fact that we can now induce on these sequences a renormalization process as shown in table 2.

$\alpha^{(1)}$	10100100100101001001010010010010100100101001001001
$\alpha^{(2)}$	1 10 10 10 1 10 10 1 10 10 10 1 10 10 1 10 10 10
$\alpha^{(3)}$	1 0 0 1 0 1 0 0 1 1 1 0 0
$\alpha^{(4)}$	1 0 1 1 0 1 1 0
$\alpha^{(5)}$	0 1 1

Table II

1.2. Generalization and explanation. Symbolic dynamics of rotations

Let us consider a solid rotation on the one dimensional torus:

$$r_\omega \begin{cases} \Pi^1 & \rightarrow \quad \Pi^1 \\ \Theta & \rightarrow \quad \Theta + \omega \bmod 1 \qquad \text{where} \qquad \omega \in [0,1[; \end{cases}$$

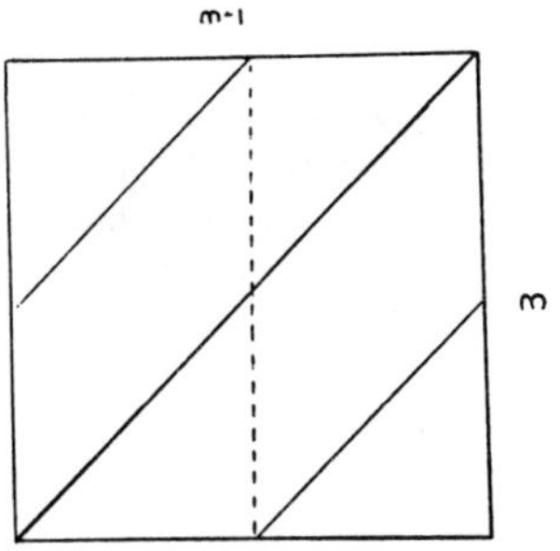

Figure 1.

Now pick a point $x \in [0,1]$ and associate to x a sequence $I_\omega(x) \in \{0,1\}^{\mathbb{N}}$ such that:

$$I_\omega(x) = (x_n)_{n\in\mathbb{N}} \quad \text{where} \quad x_n = 1 \quad \text{if} \quad r_\omega^n(x) \geq 1 - \omega,$$
$$x_n = 0 \quad \text{if} \quad r_\omega^n(x) < 1 - \omega.$$

In the following, such sequences will be called rotation compatible with rotation number ω.

To the natural question: how do these sequences look like?

We can answer in the elegant way:

A sequence $X \in \{0,1\}^{\mathbb{N}}$ is the code $I_\alpha(x)$ of a point $x \in [0,1]$for a rotation r_α if and only if:

i) by cutting X after each "0" we get a sequence of blocks with length n or n + 1, where n satisfies

$$1 - \alpha = 1/(n + \alpha_1) \quad \text{with} \quad \alpha_1 \in [0,1[,$$

excepted perhaps the first block which can be smaller.

ii) The new sequence Y obtained by replacing in X the blocks with length n + 1 by "1" and those with length n by "0" (omitting the first block if smaller) is the code $I_{\alpha_1}(y)$ of a point $y \in [0,1]$ for the rotation r_{α_1}.

A proof of this characterization can be found in many different works [1,2] but we note as a chief reference the paper of C. Series [3].

The relation between rotation compatible sequences and lunar cycles is quite obvious: we consider the solid rotation

$$r_{\alpha_0} : \Pi^1 \to \Pi^1$$

$$x \mapsto x + \alpha_0 \bmod 1 \quad \text{where} \quad \alpha_0 = 1 - \tau_1/\tau_2 \quad .$$

We have $1 - \alpha_0 = 1/(12 + \alpha_1)$ with $\alpha \in [0,1[$, then by cutting a code $I_{\alpha_0}(x)$ we get a sequence of blocks with length 12 or 13 which correspond to the 12 or 13 full moon nights in a year; and, by replacing the blocks with length 13 by "1" and 12 by "0" we obtain a rotation compatible sequence which contains the truncated sequence $\alpha^{(1)}$ of the beginning of this section.

The aim of this paper is to show that rotation compatible sequences arise in the apparently completly different field of dissipative dynamical systems.

In Section 2, we describe experiments where two stable cycles come together to form one large stable cycle (which is essentially a combination of the first two) as a parameter is varied. Small perturbations of such systems can lead to stable cycles or stable aperiodic motions with complicated structures but which possess a natural rotation compatible code.

In Section 3 and 4 we develop the theory of the underlying dynamical system which gives rise to the observed signals in Section 2.

2. EXPERIMENTS WITH COMPLICATED CYCLES

Consider an experiment which produces asymptotically a periodic signal $X_0(t)$, and such that small perturbations give asymptotically the same signal, up to a translation in time. This suggests that the underlying dynamical system has a stable periodic orbit. Now, suppose that for some sufficiently large perturbations of the initial conditions the experiment yields a periodic signal $X_1(t)$ which is not a time translation of $X_0(t)$, but which again corresponds to a stable periodic orbit. This situation can occur, for example, if the underlying dynamical system is invariant under some symmetry.

Although we intend to describe much more general cases, such symmetry conditions arise frequently in physical models and we shall start by describing a typical scenario observed by varying a parameter in an experimental setting corresponding to a symmetric model. Suppose:

(H_1) The output can be chosen so that the signals $X_0(t)$ and $X_1(t)$ corresponding to symmetric initial conditions are not time translations of each other.

(H_2) Consider an initial condition which gives asymptotically the signal $X_i(t)$, $i = 0,1$. If the parameter is varied slowly enough the new signal, $X_i(t)$ has essentially the same profil except that a longer time is spent in an almost stationary phase at the end of each period.

(H_3) As the parameter is varied further, an abrupt change of the signal is observed giving a new signal which is a regular alternation between the profiles X_0 and X_1 (see figure 2).

To simplify we shall say that the signals $X_0(t) = s_0s_0s_0\ldots$ and $X_1(t) = s_1s_1s_1\ldots$ have been "glued together" to give a new signal $s_0s_1s_0s_1s_0\ldots$

Let us now return to the underlying dynamical system and denote by C_0 and C_1 the stable periodic orbits which correspond to the signals $X_0(t)$ and $X_1(t)$ respectively. The change of behavior described above indicates that C_0 and C_1 come closer and closer to the same stationary point until they form a pair of homoclinic curves and are destroyed, giving rise to a new stable periodic orbit which essentially follows successively the loci in phase space of C_0 and C_1.

In this section our aim is to describe the signals which can be generated when two stable cycles come close to the same stationary point in more general systems (i.e. symmetry conditions will be dropped).

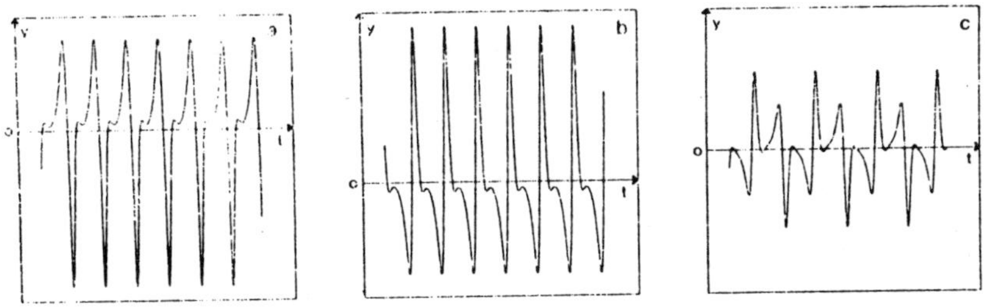

Figure 2. Numerical simulation of the system.

$$\dot{x} = y$$

$$\dot{y} = z$$ which are invariant under the symmetry $(x,y,z) \to (-x,-y,-z)$

$$\dot{z} = -z - 0,5 + \mu x(1 - x^2)$$

By plotting y against time we illustrate a gluing bifurcation:

(a) $\mu = 0.53$ a stable cycle

(b) $\mu = 0.53$ another stable cycle which is the image of (a) under the symmetry.

(c) $\mu = 0.534$ a symmetric signal obtained by "gluing together" the basic signals of (a) and (b).

If a physical system, controlled by two parameters μ_0 and μ_1 is such that:

- for $\mu_0 < 0$ and $\mu_1 < 0$, there are two stable periodic cycles $C_0(\mu_0,\mu_1)$ and $C_1(\mu_0,\mu_1)$ giving respectively the signals $X_0(\mu_0,\mu_1,t) = = s_0s_0...$ and $X_1(\mu_0,\mu_1,t) = s_1s_1...$ (depending on the initial conditions) for the measurement of the same quantity.

- for $i = 0,1$, $\mu_i \to 0^-$, the signal $X_i(\mu_0,\mu_1,t)$ remains essentially unchanged except for longer and longer almost stationary phases between the patterns s_0 or s_1 which reappear periodically (i.e. the cycle $C_i(\mu_0,\mu_1)$ approaches an equilibrium).

- in the region $\mu_0 > 0, \mu_1 > 0$, arbitrarily small μ_0 and μ_1 can be chosen so that there is a stable signal of the form $s_0s_1s_0s_1s_0...$

Then it is likely that the theoretical results of the following section are applicable. This means that for parameter values near $\mu_0 = \mu_1 = 0$

i) there are either 0, 1 or 2 periodic signals which remain close to $C_0(0,0) \cup C_1(0,0)$ and any of these periodic signals is stable,

ii) furthermore, all such periodic orbits can be described by rotation compatible sequences (i.e. the signal has a form

$$\underbrace{s_0 \ldots s_0}_{m_1 \text{ times}} \; \underbrace{s_1 \ldots s_1}_{m_2 \text{ times}} \; \underbrace{s_0 s_0 s_0}_{m_3 \text{ times}} \ldots$$

such that the sequence

$$\underbrace{00\ldots\ldots}_{m_1 \text{ times}} \; \underbrace{11\ldots\ldots}_{m_2 \text{ times}}$$

is a rotation compatible) (figure 3).

iii) if two stable signals $Y_i(\mu_0,\mu_1,t)$, $i = 0,1$ exist at the same parameter values then the rotation numbers, p_i/q_i, written in irreducible form associated with the rotation compatible sequences of Y_i satisfy the condition $|p_0q_1 - p_1q_0| = 1$ (i.e. p_0/q_0 and p_1/q_1 are neighbours in some Farey series) (figure 4),

iv) note also that in some cases there are parameter values for which there exists a stable signal whose rotation compatible sequence has an irrational rotation number so the signal is aperiodic but not chaotic(*).

(*) In particular there exists volume contracting flows in R^3 with an attracting quasi periodic orbit (technically this is the inverse limit of a Cherry flow on a torus with a hole). This is possible since quasi periodic orbits which occur near the double homoclinic loop do not lie on an invariant torus.

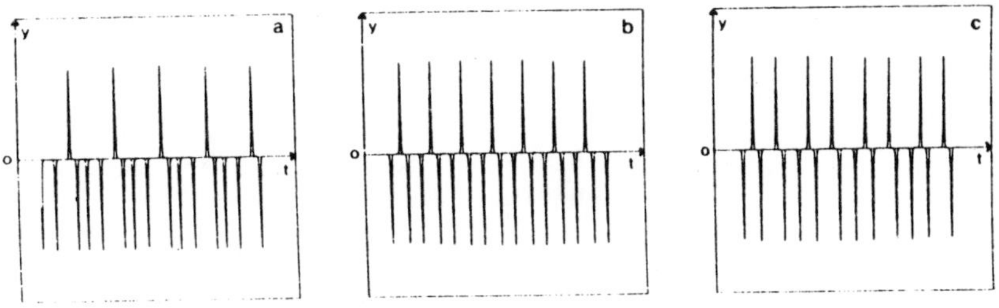

Figure 3. Plots of y against time for the system described in [4].

$$\dot{x} = 1.8(x - y) + \varepsilon x^2$$

$$\dot{y} = - 7.2y + xz + \mu x^3$$

$$\dot{z} = - 2.7z + xy + 0.007z^2$$

showing some stable cycles with $\varepsilon = 0.001$

(a) $\mu = 0.07521$: rotation number 1/4, the symbol is
$S_0S_0S_1S_0S_0S_0S_1S_0S_0S_0S_1S_0S_0S_0S_1S_0S_0S_0S_1\dots$

(b) $\mu = 0.07502$: rotation number 1/3, the symbol is
$S_0S_1S_0S_0S_1S_0S_0S_1S_0S_0S_1\dots$

(c) $\mu = 0.075014$: rotation number 2/5, the symbol is
$S_0S_1S_0S_1S_0S_0S_1S_0S_1S_0S_0S_1S_0S_1S_0S_0S_1S_0S_1S_0\dots$

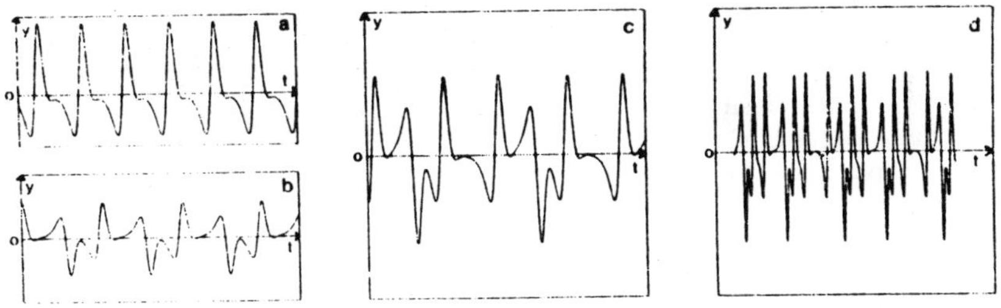

Figure 4. Plots of y against time for the system (which is a small perturbation of the symmetric system of figure 2).

$$\dot{x} = y$$

$$\dot{y} = z + \varepsilon x^2$$

$$\dot{z} = - z + 0.5y + \mu x(1 - x^2)$$

showing some stable cycles:

(a) $\mu = 0.54$, $\varepsilon = 0.005$: a stable cycle with rotation number 1 and symbol $s_1 s_1 s_1 \ldots$

(b) $\mu = 0.54$, $\varepsilon = 0.005$: another stable cycle with rotation number 1/2 which coexists with the cycle shown in (a); the symbol is $s_1 s_0 s_1 s_0 s_1 s_0 s_1 \ldots$

(c) $\mu = 0.556$, $\varepsilon = 0.008$: a stable cycle with rotation number 2/3, the symbol is $s_1 s_0 s_1 s_1 s_0 s_1 s_1 \ldots$

(d) $\mu = 0.556$, $\varepsilon = 0.01$: a stable cycle with rotation number 5/7, the symbol is $s_0 s_1 s_1 s_0 s_1 s_1 s_1 s_0 s_1 s_1 s_0 s_1 s_1 s_1 s_0 s_1 \ldots$

As stated above, these results are not strictly valid. However we shall see in section 3 they can be proved rigorously for sufficiently smooth perturbations of ordinary differential equations with a pair of trajectories bi-asymptotic to the same stationary point 0; providing that the characteristic equation of the linearized flow near 0 has a simple real positive eigenvalue, λ, and all the other eigenvalues are to the left of $-\lambda$ in the complex plane.

The conditions given here above provide strong evidence that the underlying dynamical system which gives rise to the observed signal is indeed of this form.

In particular, the fact that the signal remains stable when it has a long almost stationary phase suggests strongly that the characteristic equation at the stationary point for the underlying dynamical system is of the form given above.

This scenario where two stable cycles come together to form stable cycles with complicated structure is called the "gluing bifurcation". But the very general results we have described can be more explicit if one gives more information about the underlying dynamical system and particularly about the linearized flow near 0, this will be the subject of the section 4.

3. THE GLUING BIFURCATION

3.1. The theoretical problem

Let X_0 be a smooth vector field in $\mathbb{R}^n$, $n \geq 2$, and Φ_0^t the flow defined by X_0. Consider the case where $X_0(0) = 0$ and the spectrum Spec L_0 of the linear part of X_0 at 0 contains a single simple eigenvalue λ_1, to the right of the imaginary axis. It follows that 0 is a saddle point with a (n-1) dimensional stable manifold:

$$W_0^s = \{P \in \mathbb{R}^n \mid \lim_{r \to +\infty} \Phi_0^r (P) = 0\}$$

and a one dimensional unstable manifold:

$$W_0^u = \{P \in \mathbb{R}^n \mid \lim_{r \to -\infty} \Phi_0^r (P) = 0\}$$

Suppose the two branches $W_0^{u,0}$ and $W_0^{u,1}$ of W_0^u are contained in W_0^s, forming two homoclinic loops Γ_0 and Γ_1.

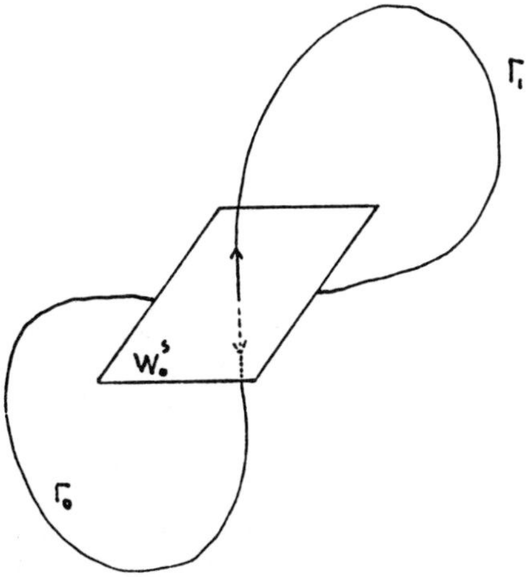

Figure 5.

We shall be concerned in this section with the following type of question: For a small enough neighbourhood $\mathcal{V}$ of X_0 in the space of C^k vector fields in $\mathbb{R}^n$ equipped with the standard C^k topology, and a small enough neighbourhood $\mathcal{U}$ of $\Gamma_0 \cup \Gamma_1$, what sort of dynamical behavior can one find in $\mathcal{U}$ for a vector field X in $\mathcal{V}$?

We only consider the case where:

$$0 < \lambda_1 < \min|\text{Re}\ \mu_k| \equiv -\lambda_2, \quad \mu_k \in \text{Spec}\ L_0 - \{\lambda_1\}$$

As it has been suggested in the previous section, the dynamical behavior of the orbit of a point in $\mathcal{U}$ will be determined by its code: i.e. a sequence $(x_n)_{n \geq 0} \in \{0,1\}^{\mathbb{N}}$ symbolizing the fact that the orbit successively follows $\Gamma_{x_0}, \Gamma_{x_1}, \ldots, \Gamma_{x_n} \ldots$ Of course, if a point Q is on the orbit of a point P, the code of its orbit will be deduced from the code of the orbit of P by a translation.

3.2. Definition and statement of the results

Definition: We shall say that a vector field X_0 in $\mathbb{R}^n$, with $X_0(0) = 0$, is critical if it satisfies the following conditions:

S) X_0 is C^{1+1}, i.e. C^1 with Lipschitz partial derivatives,

L) the linear part L_0 of X_0 at 0 reads

$$L_0 = \lambda_1 x_1 \partial/\partial x_1 + \sum_{1<i,j\leq n} \lambda_{i,j} x_i \partial/\partial x_j$$

with $0 < \lambda_1 < -\lambda_2$ where $\lambda_2 \equiv \max \operatorname{Re} \mu_k$ for

$\mu_k \in \operatorname{Spec} L_0 - \{\lambda_1\}$

G) $\forall \mu_i, \mu_j \in \operatorname{Spec} L_0$, $\mathbb{R}_e(\mu_i) \neq \lambda_1 + \mathbb{R}_e(\mu_j)$

H) X_0 possesses two homoclinic orbits Γ_0 and Γ_1 bi-asymptotic to 0 and bounded away from any other singularity.

In the following, we will only consider the space $C^1(\mathbb{R}^n)$ of C^1 vector fields in $\mathbb{R}^n$, equipped with the C^1 topology.

Remark 1: Condition L implies in particular that the origin is a hyperbolic singularity, then by a theorem of Grobman and Hartmann [5] there exists a continuous change of variables h_{x_0} which linearizes X_0 near 0. Since we will need more regularity, we impose the smoothness condition S and the (generic) non resonance condition G in X_0, so that h_{x_0} can be chosen to be C^1 for X_0 with

$$L_0(P) = Dh_{x_0}[X_0(h_{x_0}^{-1}(P))] \text{ near } 0.$$

Remark 2: As we shall see periodic and quasi periodic orbits arising in this problem are attracting i.e. possess a forward invariant neighbourhood which shrinks to the orbit as $t \to +\infty$.

We can state our results in the following

<u>Theorem</u>: For any critical vector field X_0 there exists a neighbourhood $\mathcal{V}$ of X_0 and a neighbourhood $\mathcal{U}$ of $\Gamma_0 \cup \Gamma_1$ such that, for X in $\mathcal{V}$

1) there can be zero, one or two closed invariant curves not counting 0 in $\mathcal{U}$, if X possesses in $\mathcal{U}$ a closed invariant curve which is a periodic orbit (instead of a homoclinic orbit) then it is attracting, if (and only if) X does not possess in $\mathcal{U}$ a closed invariant curve then X possesses a non periodic attractor.

2) The code of orbits of points on the closed invariant curves (resp on the non periodic attractor) are rotation compatible with rational (resp a unique irrational) rotation number.

3) If X possesses two invariant curves, the rotation numbers p_0/q_0 and p_1/q_1 (written in irreducible form) corresponding to their rotation compatible codes are Farey neighbours:

i.e. $|p_0q_1 - q_0p_1| = 1$

A proof of this theorem can be found in [6]. As for the proof of the characterization of rotation compatible sequence (section 1) it uses an argument of renormalization but in a geometrical way.

Consequences of this result are rather surprising: for example figures 6a and 6b show some situations which cannot occur:

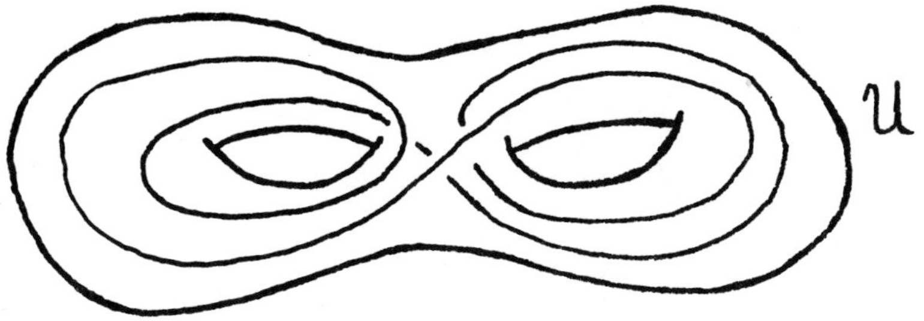

Figure 6a. An orbit with code 11001100

Figure 6b. Two orbits with rotation compatible codes 11 11 and 100 100 100 ... and rotation numbers 1 and 1/3 which are not Farey neighbours.

But this strong and general result which gives in fact necessary conditions on the codes of the orbits for their existences and co-existences, can be made more explicit given more information about the linearized flow near 0 and the geometric configuration of Γ_0 and Γ_1. In some cases, supplementary selection rules can occur (think of the 2 dimensional case !). In the following section we will describe two(*) parameter families of vector fields involving a gluing bifurcation and show similarities and differences between their bifurcation diagrams.

4. BIFURCATION DIAGRAMS

The proofs of the results described in this section can be found in ([7], [8], [9]).

4.1. The similarities

Let X_0 be a critical vector field and X_{μ_0,μ_1} a continuous two parameter family of C^1 vector fields such that $X_{0,0} = X_0$

As we have notice in the previous section, a theorem by Belitskii [10] allows us to do a C^1 change of variables in a neighbourhood V of 0 which linearizes the vector fields family for (μ_0,μ_1) sufficiently small.

Consider now a small cylinder $C(r,h) \subset V$ defined by:

$$C(r,h) = \{P = (x_1,\ldots,x_n) \mid \sum_{1<i\leq n} x_i^2 = r^2 \text{ and } |x_1| \leq h\}$$

For $(\mu_0,\mu_1) = (0,0)$, the first intersections of the two branches of the unstable manifold of the origin, $W_0^{u,0}$ and $W_0^{u,1}$ with the cylinder $C(r,h)$ occur at two points on the hyperplane $x_1 = 0$.

For (μ_0,μ_1) sufficiently small the first intersections of the two branches of the unstable manifold of the origin $W_0^{u,0}$ and $W_0^{u,1}$ occur at points whose first coordinates are respectively $x_1 = \tilde{\mu}_0(\mu_0,\mu_1)$ and $x_1 = -\tilde{\mu}_2(\mu_0,\mu_1)$. Assuming that the map $(\tilde{\mu}_0,\tilde{\mu}_1)$ is a local diffeomorphism, we can then use these two new parameters which have a nice geometrical interpretation (figure 7). Then we have the following global results.

Proposition: For (μ_0,μ_1) sufficiently small and for a small enough neighbourhood $\mathcal{U}$ of $\Gamma_0 \cup \Gamma_1$ the parameter space can be decomposed in 6 regions (Figure 8).

- In region 1 there exists two (stable) periodic orbits in $\mathcal{U}$

  - $C^1(\tilde{\mu}_0,\tilde{\mu}_1)$ such that $\lim_{\tilde{\mu}_1\to 0^-} C^1(\tilde{\mu}_0,\tilde{\mu}_1) = \Gamma_1$ and with the code

11111 ... and

(*) The choice of 2 parameters in obvious: it is the minimum we need to control the position of the two branches of the unstable manifold.

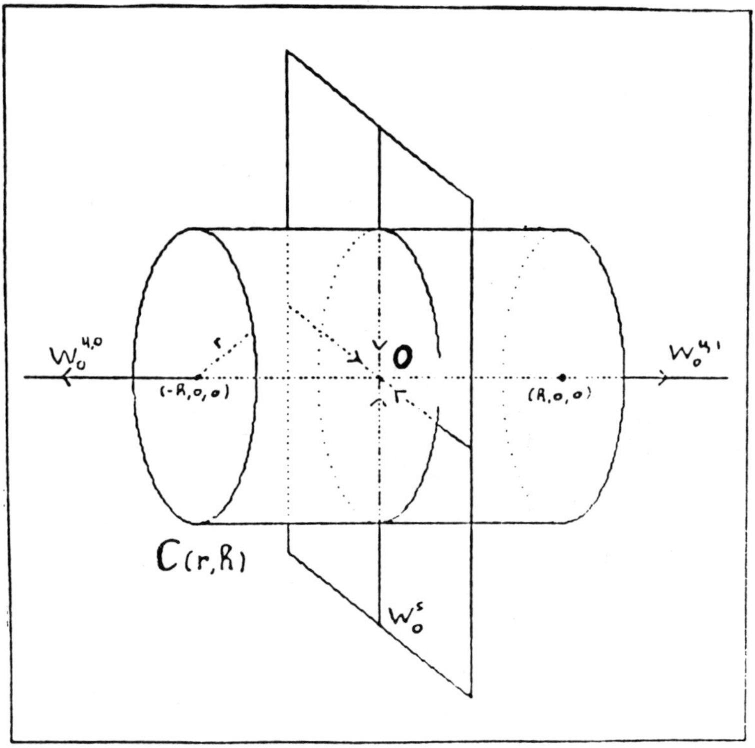

Figure 7.

- $C^0(\tilde{\mu}_0,\tilde{\mu}_1)$ such that $\lim_{\tilde{\mu}_0\to 0^-} C^0(\tilde{\mu}_0,\tilde{\mu}_1) = \Gamma^0$ with the code 000 ...

• In region 2 (resp 3) there is a unique (stable) periodic orbit in $\mathcal{U}$

- $C^1(\tilde{\mu}_0,\tilde{\mu}_1)$

  (resp $C^0(\tilde{\mu}_0,\tilde{\mu}_1)$)

• In region 4 there is a unique (stable) periodic orbit $C^{01}(\tilde{\mu}_0,\tilde{\mu}_1)$ in $\mathcal{U}$ close to $\Gamma_0 \cup \Gamma_1$ for the Hausdorff distance and with the code 010101 ...

• In regions 5 and 6 respectively defined by:

- $C_1\tilde{\mu}_1^\alpha < \tilde{\mu}_0 < C_2\tilde{\mu}_1^\alpha$ and $-C_3\tilde{\mu}_0^\alpha < \tilde{\mu}_1 < C_4\tilde{\mu}_0^\alpha$ where $\alpha = -\lambda_2/\lambda_1$

  and $C_i > 0$.

We cannot say a priori anything about the asymptotic dynamics of the perturbed vector field.

It is in these last two regions that many different things can happen according to the form of the linearized vector field in 0 and the geometry of the vector field in the neighbourhood of the two homoclinic loops $\Gamma_0 \cup \Gamma_1$.

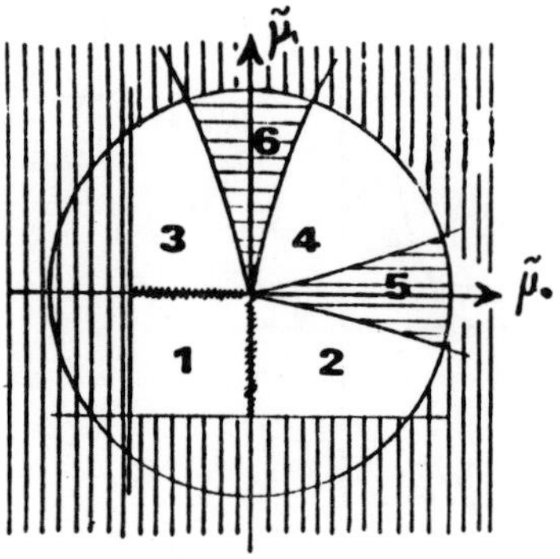

Figure 8.

The way we prove this proposition and the theorem of the previous section is based on the study of the dynamics of a first return map defined as follows:

- thanks to the linearization in V we calculate explicitly the map T^1_{int} (resp T^0_{int}) which associates to a point on the cylinder $C(r,h)$ with coordinate x_1 positive (resp negative) the first intersection of its orbit with the plane $x_1 = h$ (resp $x_1 = -h$),

- the global flow induces also a map T^1_{ext} (resp T^0_{ext}) which associates to a point of $x_1 = h$ (resp $-h$) the first intersection of its orbit with the cylinder,

- then we get a map

$$T : C^*(r,h) \to C(r,h) \text{ where } C^*(r,h) = \{(x_1,\dots,x_n) \in C(r,h) \text{ with } x_1 \neq 0\}$$

which reads:

$$T^1_{ext} \circ T^1_{int} \quad \text{if} \quad x_1 > 0 \quad \text{and} \quad T^0_{ext} \circ T^0_{int} \quad \text{if} \quad x_1 < 0$$

The fact that the linearization in V can be made C^1 and that the eigenvalues of the linearized vector field satisfy condition L (definition in section 3), insures that T is a contraction on the two connected components of $C^*(r,h)$.

The study of such maps is quite general and provides all the results we have given [6].

4.2. The differences.

4.2.1. The complex case. We assume now that the linearized part L_0 of the critical vector field $X_{0,0}$ reads:

$$L_0 = \lambda_1 x_1 \partial/\partial x_1 + (\lambda_2 x_2 - \omega x_3)\partial/\partial x_2 + (\omega x_2 + \lambda_2 x_3)\partial/\partial x_3 + \\ + \sum_{i>3} \left(\sum_{j>3} \lambda_{i,j} x_i \partial/\partial x_j \right)$$

where $0 < \lambda_1 < -\lambda_2$ and $\lambda_2 = \max \operatorname{Re} \mu_k$ for $\mu_k \in \operatorname{Spec} L_0 - \{\lambda_1\}$

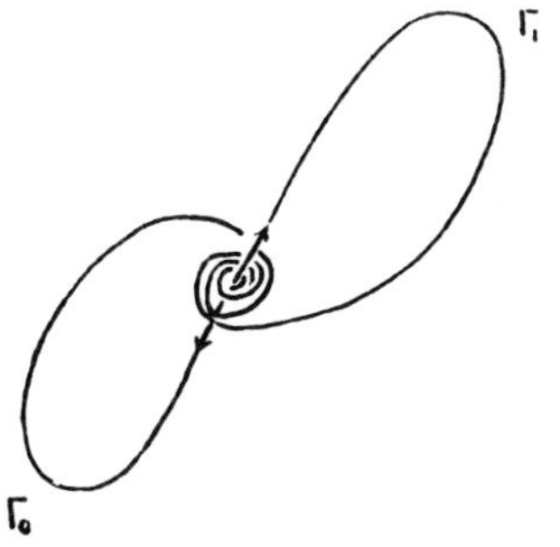

Figure 9. The complex case.

In this case, to understand what happens in regions 5 and 6 we can proceed inductively as follows:

Step 1: There exist in 5 (resp 6):

- a curve (6_1): $\tilde{\mu}_0 \geq 0, \tilde{\mu}_1 = C_1 \tilde{\mu}_0^{-\frac{\lambda_2}{\lambda_1}} \cos((-\omega/\lambda_1)\operatorname{Log}(\tilde{\mu}_0 + \phi_1))$

(resp a curve (5_1), $\tilde{\mu}_1 \geq 0, \tilde{\mu}_0 = C_2 \tilde{\mu}_1^{-\frac{\lambda_2}{\lambda_1}} \cos((-\omega/\lambda_1)\operatorname{Log}(\tilde{\mu}_1 + \phi_2))$ [11]

- a sharp neighbourhood V_{6_1} of (6_1) such that:

If U_{6_1} is a connected open set whose boundary is the union of a segment of the axe $\tilde{\mu}_1 = 0$ and a segment of (6_1), then U_{6_1} contains a non empty open set U_{6_1} whose boundary is the union of a segment of the axis $\tilde{\mu}_1 = 0$ and a segment of the boundary of V_{6_1} (resp. replace 6 by 5 and $\tilde{\mu}_1$ by $\tilde{\mu}_0$).

- The graph G_{6_1} of a function $C^1, \tilde{\mu}_1 = F_{6_1}(\tilde{\mu}_0)$ contained in V_{6_1} such that:

If $(\tilde{\mu}_0, \tilde{\mu}_1) \in G_{6_1}$, the perturbed vector field possesses a homoclinic orbit Γ_{6_1} which starts at 0, follows a small neighbour of Γ_0 then a small neighbour of Γ_1 and goes back to the origin (resp. replace 6 by 5 exchange $\tilde{\mu}_0$ and $\tilde{\mu}_1$ and Γ_0 and Γ_1).

Step 2: we observe that the intersections of G_{5_1}, with $\tilde{\mu}_0 = 0$ (resp. G_{6_1} with $\tilde{\mu}_1 = 0$) correspond again to critical vector fields where new gluing bifurcations occur. The new fact is that, instead of gluing cycles with codes 000 ... and 111 ..., we are gluing cycles with codes 010101 ... and 111 ... (resp. 010101 ... and 000 ...).

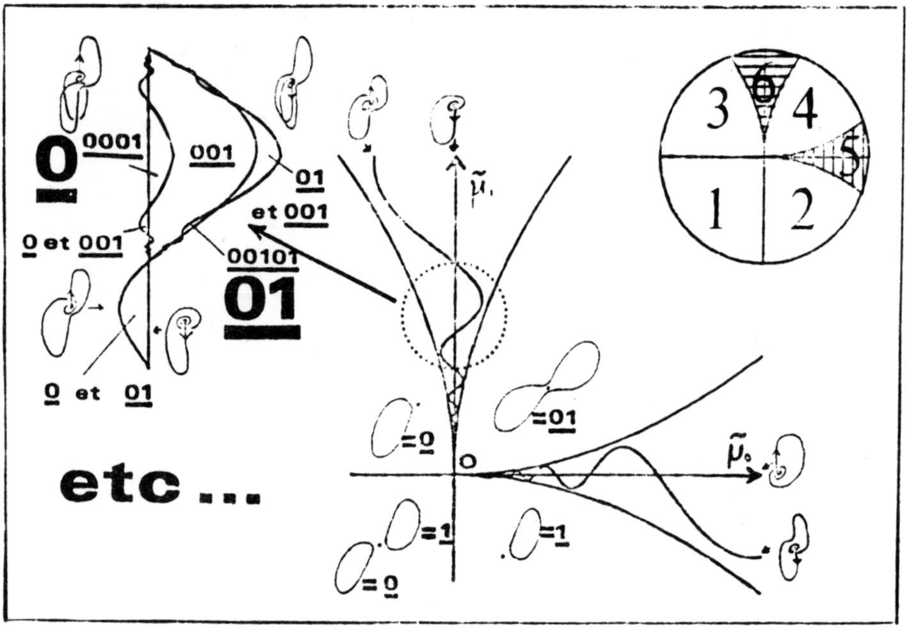

Figure 10.

Step 3: the pair of cycles we have got in step 2 can be now considered as the original cycles, they are gluing themselves together creating a new cycle with code 011011 ... (resp. 100100100...).

We can iterate this construction; after n iterations we get 2^n new types of gluing points in the parameter space corresponding to different pairs of homoclinic orbits (there is of course an infinity of gluing points of each type).

This complicated situation is described in figure 10.

4.2.2. The real case.

α. General aspect. We consider now the case when the linear part of the critical vector field $X_{0,0}$ reads

$$L_0 = \lambda_1 x_1 \partial/\partial x_1 + \lambda_2 x_2 \partial/\partial x_2 + \sum_{i>2} (\sum_{j>2} \lambda_{i,j} x_i \partial/\partial x_j)$$

where $0 < \lambda_1 < -\lambda_2$ and $\lambda_2 = \max \mathrm{Re}\,(\mu_k)$ for $\mu_k \in \mathrm{Spec}\ L_0 - \{\lambda_1\}$.

In this situation, there are (generically) two possibilities depending upon whether the homoclinic orbits approach 0 along the same branch (the butterfly configuration) or along opposite branches (the figure eight configuration) of the x_2 eigenvector (figure 11 a and b)

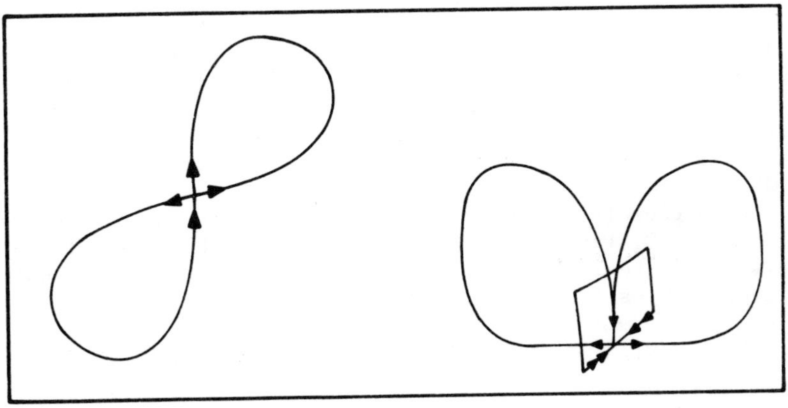

a) "Figure eight" b) "Butterfly"

Figure 11.

Of course we remark that the butterfly configuration is impossible in $\mathbb{R}^2$. In both real cases there are three subcases [12] depending upon whether the two manifolds given by the intersection $W_0^s \cap U$ are orientable or non orientable. These considerations imply that 6 configurations must be considered. A detailed study of the bifurcation diagrams relative to these six configurations in the regions 5 and 6 has been done in

[9], but in the following we will just deal with...

β. A "particular case". We consider a critical vector field in the butterfly configuration; in this particular case we can reduce the study of the first return map T to a neighbourhood N of the point (0,r,0,...0) on the cylinder C(r,h).

Now we can make, as in the geometrical Lorenz model [13], the assumption that there exists a stable foliation for T (figure 12) and that T is orientation preserving.

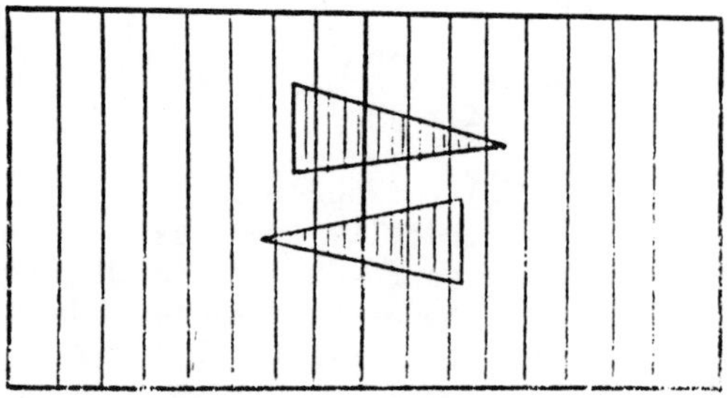

Figure 12. T and its stable foliation.

In fact, this last assumption is not as strong as in the Lorenz case because of the condition $\lambda_1 < -\lambda_2$ (recall that in the Lorenz model we had $-\lambda_2 < \lambda_1$) which insures that T is a contraction on the two connected components of $N \cap C^*(r,h)$ and allows us to construct, thanks to this contraction property, a stable foliation in generic cases [9].

In such a geometrical model it is clear that in the part of region 5 such that $\tilde{\mu}_1 < 0$, (resp region 6 s.t. $\tilde{\mu}_0 < 0$) it happens nothing new compared with the region 2(resp 3). The way, the first return map has been derived insures that is has the form:

$$T_{\tilde{\mu}_0,\tilde{\mu}_1}(x_1,z) = (F_{\tilde{\mu}_0,\tilde{\mu}_1}(x_1),\ G_{\tilde{\mu}_0,\tilde{\mu}_1}(x_1,z)) \text{ with } z = (x_3,..,x_n)$$

where $G_{\tilde{\mu}_0,\tilde{\mu}_1}$ is a contraction and

$F_{\tilde{\mu}_0,\tilde{\mu}_1}$ which determines the dynamics reads:

$$F_{\tilde{\mu}_0,\tilde{\mu}_1}(x) = -\tilde{\mu}_1 + ax^\delta + \text{ h.o.t if } x > 0$$

$$= \tilde{\mu}_0 - bx^\delta + \text{h.o.t. if } x < 0$$

with a and b > 0 and $\delta = -\lambda_2/\lambda_1$.

Here we are just interested in the case $\tilde{\mu}_0 > 0$ and $\tilde{\mu}_1 > 0$. We begin by rescaling our map so that all the important (i.e. recurrent) dynamics is contained in the interval $[0,1]$. This can be done using the change of coordinate $z = (\tilde{\mu}_1 + x)/(\tilde{\mu}_0 + \tilde{\mu}_1)$.

In this new coordinate we get a map $H_{\tilde{\mu}_0,\tilde{\mu}_1}$ satisfying:

1. $H_{\tilde{\mu}_0,\tilde{\mu}_1}$ is piecewise increasing with a single discontinuity at

$$\alpha = \tilde{\mu}_1/(\tilde{\mu}_0 + \tilde{\mu}_1) \text{ and } \lim_{z\to\alpha^-} H_{\tilde{\mu}_0,\tilde{\mu}_1}(z) = 1 \text{ and } \lim_{z\to\alpha^+} H_{\tilde{\mu}_0,\tilde{\mu}_1}(z) = 0,$$

2. $H_{\tilde{\mu}_0,\tilde{\mu}_1}$ is injective,

3. Labelling the orbits of $H_{\tilde{\mu}_0,\tilde{\mu}_1}$, with "0" if $x \in [\,0,\alpha\,[$ and "1" if $x \in [\alpha,1]$, we obtain the same symbol sequence of the corresponding orbit in $\mathcal{U}$ for the vector field X_{μ_0,μ_1}

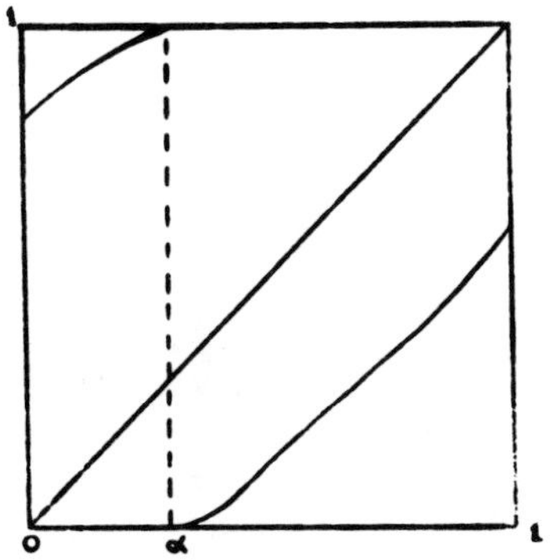

Figure 13. A map $H_{\tilde{\mu}_0,\tilde{\mu}_1}$.

These considerations allow us to consider $H_{\tilde{\mu}_0,\tilde{\mu}_1}$ as an injective map of the circle into itself with a single discontinuity.

Let

$L_{\tilde{\mu}_0,\tilde{\mu}_1}$ a lift of $H_{\tilde{\mu}_0,\tilde{\mu}_1}$ defined on $\mathbb{R}$ as follows:

$$L_{\tilde{\mu}_0,\tilde{\mu}_1}(x) = H_{\tilde{\mu}_0,\tilde{\mu}_1}(x - [x]) + [x] \text{ if } x - [x] < \alpha$$

$$= H_{\tilde{\mu}_0,\tilde{\mu}_1}(x - [x]) + [x] + 1 \text{ if } x - [x] \geq \alpha$$

where [x] denotes the integer part of x, then we know that [13]:

- For all $x \in \mathbb{R}$ $\lim L^n_{\tilde{\mu}_0,\tilde{\mu}_1}(x)/n$ exists and is independent of x.

This limit is called the rotation number of $H_{\tilde{\mu}_0,\tilde{\mu}_1}$ and noted $\rho_{\tilde{\mu}_0,\tilde{\mu}_1}$.

- The code of the itinerary of any point $x \in [0,1]$ is rotation compatible with the rotation number $\rho_{\tilde{\mu}_0,\tilde{\mu}_1}$.

- When $\tilde{\mu}_1$ is fixed and $\tilde{\mu}_0$ is continuously decreasing or when $\tilde{\mu}_0$ is fixed and $\tilde{\mu}_1$ is continuously increasing the lift $L_{\tilde{\mu}_0,\tilde{\mu}_1}$ and then the rotation number are continuously decreasing.

This allows us to complete the bifurcation diagram in regions 5 and 6; the parameter space takes the form of figure 14:

All rotation compatible codes with rotation number between 1/2 and 0, none coexisting.

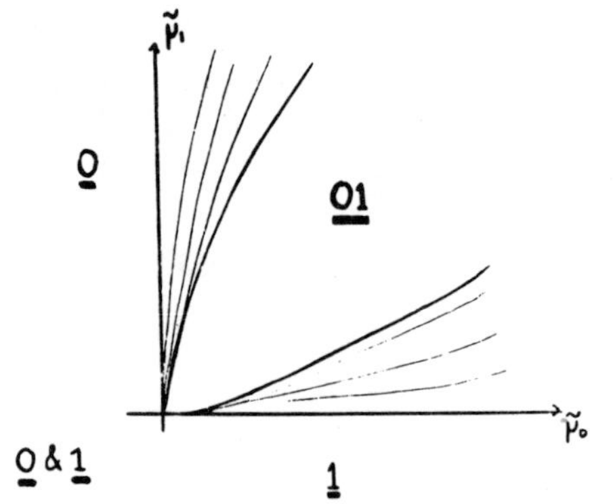

All rotation compatible codes with rotation number between 1 and 1/2, none coexisting.

Figure 14.

Final remark: In such situation, if we look at the flow on the underlying branched manifold (fig. 15a), we can see that the recurrent dynamics occur on a torus with a hole (fig. 15b).

The continuity of the rotation number shows that there are also parameter values with irrational rotation number in any neighbourhood of $(\tilde{\mu}_0,\tilde{\mu}_1) = 0$. It follows that we have for these parameter values a corresponding irrational motion on the punctured torus.

This is a nice and natural example of Cherry flow on the torus [5], i.e. a flow with a saddle, a source and without any other singularities nor periodic orbits (figure 16).

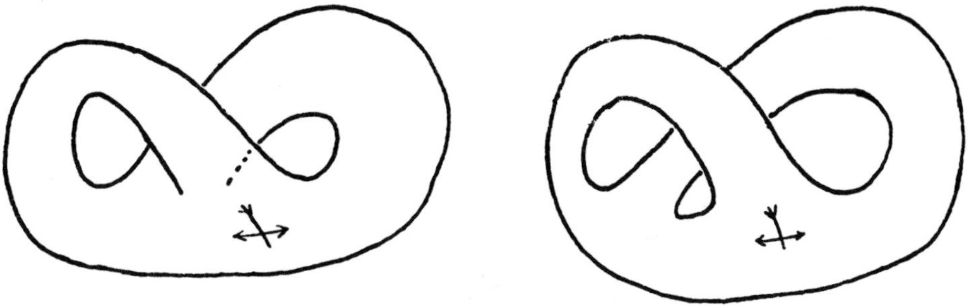

Figure 15.

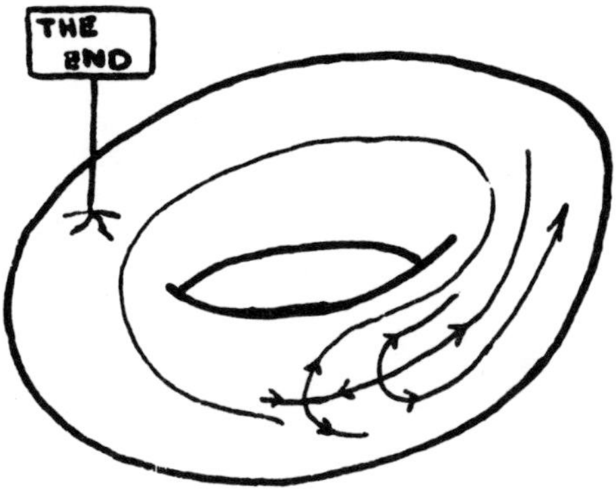

Figure 16. A Cherry flow.

REFERENCES

[1] C. Bernhardt. Proc. London Math. Soc., 3rd. Series 45, part 2 258-280 (1982).
[2] J.M. Gambaudo, O. Lanford III et C. Tresser, Dynamique symbolique des rotations, C.R. Acad. Sc. Paris 299 Série 1, 823-826 (1984).
[3] C. Series. The geometry of Markoff numbers, The Mathematical Intelligencer, Vol. 7, N° 3 (1985).
[4] A. Arneodo, P. Coullet and C. Tresser. A possible new mechanism for the onset of turbulence, Phys. Lett. 81A, 197-201 (1981).
[5] J. Palis and W. de Melo. Geometric theory of dynamical systems,

Springer Verlag, New York (1982).
[6] J.M. Gambaudo, P. Glendinning and C. Tresser. The gluing bifurcation 1, symbolic dynamics of the closed curves, submitted to Com. Math. Phys.
[7] P. Coullet, J.M. Gambaudo et C. Tresser, Une nouvelle bifurcation de codimension 2: le collage de cycles, C.R. Acad. Sc. Paris 299 Série I, 253-256 (1984).
[8] J.M. Gambaudo, P. Glendinning et C. Tresser, Collage de cycles et suites de Farey, C.R. Acad. Sc. Paris 299 Série I, 711-714 (1984).
[9] J.M. Gambaudo, P. Glendinning and C. Tresser. The gluing bifurcation 3, stable foliations and periodic orbits, Preprint.
[10] G.R. Belitskii. Functional equations and conjugacy of local diffeomorphisms of finite smoothness class, Functional Anal. Appl. 7, 268-277 (1973).
[11] P. Glendinning. Bifurcations near homoclinic orbits with symmetry, Phys. Lett. 103A, 163-166 (1984).
[12] V.S. Afraimovich and L.P. Shilnikov. Bifurcation of codimension 1 leading to the appearance of a countable set of tori, Dokl Akad Nauk SSSR 262 (1982).
[13] Guckenheimer and R.F. Williams. Structural stability of Lorenz attractors, Publ. Math. IHES 50, 59-72.
[14] J.M. Gambaudo et C. Tresser. Dynamique régulière ou chaotique. Applications du cercle ou de l'intervalle ayant une discontinuité, C.R. Acad. Sc. Paris 300 Série I, 311-313 (1985).

LARGE SCALE INSTABILITIES OF CELLULAR FLOWS

S. Fauve
Groupe de Physique de Solides
de l'Ecole Normale Supérieure
24 Rue Lhomond, 75231 Paris CEDEX 05
France

1. INTRODUCTION

The close analogy between instabilities in nonlinear systems driven far from equilibrium, and equilibrium phase transitions in condensed-matter physics, is now well documented experimentally as well as theoretically. This idea was fathered by Landau [1] and developped by several authors in the context of hydrodynamics [2], electric circuits, semiconductors, nonlinear optics and chemical instabilities [3].

Chemical or hydrodynamic instabilities often consist of a transition from a motionless state to one varying periodically in time or space. Various examples, widely studied in the past years, are, oscillatory chemical reactions, Rayleigh-Bénard convection, Couette-Taylor flow, waves in shear flows, instabilities of liquid crystals,...[4]. The appearence of periodic structures in these systems driven externally by a forcing homogeneous in space and constant in time, corresponds to a bifurcation, characterized by one or several modes, which become unstable as the control parameters are varied. In the vicinity of the bifurcation, the amplitudes of these critical modes are governed by nonlinear partial differential equations, and describe slow modulations in time or space of the periodic structure envelope [5]. Actually, this large scale description of a periodic structure through its slowly varying envelope, has been used first in nonlinear wave theory, where it has been shown that the motion of a nonlinear wave in a dispersive medium, can be studied through its slowly varying amplitude and phase [6,7].

When the instability saturates nonlinearly and gives rise to a finite amplitude periodic pattern, only its phase remains neutral in the long wavelength limit. Indeed, a spatially uniform modification of the phase corresponds to a shift of the spatial (temporal) periodic pattern, and thus is neutral because of translational invariance in space (time). Correspondingly long wavelength perturbations of the phase behave on a slow time scale; therefore the basic idea is to eliminate adiabatically the fast modes, and to obtain an evolution equation for the phase. This method has been used in order to describe pattern dynamics in oscillatory chemical reactions [8,9,10], or to study the stability of cellular flows which arise in Rayleigh-Bénard convection [11,12,13] or in Couette-Taylor

E. Tirapegui and D. Villarroel (eds.), Instabilities and Nonequilibrium Structures, 63–88.

flow [14,15,16]. These long wavelength modes of periodic patterns trace back to broken symmetries at the instability onset and are thus analogous to Goldstone modes of condensed matter physics [17]. However, in phase unstable regimes, they can lead to turbulence, because there is generally no free energy to minimize for dissipative systems. In these regimes, nonlinear terms of the phase equations have to be considered in order to determine whether the phase instability saturates at long wavelength or cascades to short scales, and thus breaks the long wavelength approximation. It was shown that the form of the nonlinear terms mainly depends on symmetry constraints [18], and consequently we expect that the method of "phase dynamics" provides a universal description of spatio-temporal behaviors of periodic patterns in the long wavelength limit.

2. STATIONARY CELLULAR FLOWS

2.1 Envelope Equations

Rayleigh-Bénard instability provides a canonical example for pattern forming transitions in non equilibrium systems; it occurs in a horizontal layer of fluid heated from below; when the temperature difference accross the layer exceeds a critical value ΔT_c, the buoyancy force overcomes the dissipative effects of viscous shear and thermal conduction, and the motionless state spontaneously breaks up into convective cells. When the temperature is fixed at the boundaries, the buoyancy effects cannot drive the motion on large scales compared to the height of the layer, and small scale motions are inhibited by the dissipative effects; thus the instability occurs at a finite wavenumber k_c, and the growthrate of the unstable modes of wavenumber $k \cong k_c$, reads [19]

$$\eta(k) = \mu - \xi_0^2(k^2 - k_c^2)^2 + \dots \tag{1}$$

where μ measures the distance from criticality ($\mu \sim \Delta T - \Delta T_c$).

A simple one-dimensional model that mimics this instability is [20],

$$u_t = (\mu - (\Delta + k_c^2)^2)\, u - u^3 \tag{2}$$

The instability first occurs for $\mu = 0$ for increasing μ, with a finite critical wavenumber k_c. For $\mu > 0$ there exists a band of unstable modes around k_c (see figure 1). The concept of wave packet can be used to take into account the interaction between these modes in the post bifurcation stage (see for instance reference 5); we write

$$u(x,t) = A(X,T) \exp ik_c x + cc. + \mathcal{U}(x,t,X,T) \tag{3}$$

where $A(X,T)$ is the slowly varying envelope (in x and t) of the roll pattern of wavenumber k_c, and we look for an evolution equation for A in the form

$$A_T = f(A, A_X, A_{XX}, \dots) \tag{4}$$

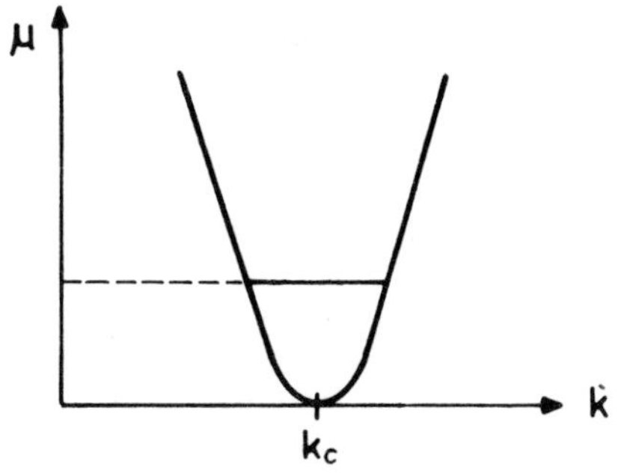

Figure 1. Neutral stability curve.

The main assumption is that, in the vicinity of the bifurcation, the critical modes vary on a time scale much slower than the one of the other modes, and thus contain all the information about the asymptotic time dependence of $u(x,t)$ near the instability onset. We next assume that $\mathcal{U}$ and f can be expanded in Taylor series in $A, A_X, A_{XX}, \ldots$ Solvability conditions then give the evolution equation (4). The form of this equation can be found by symmetry arguments:

- Translational invariance in space implies that if $u(x)$ is a solution of (2), $u(x+\phi)$ is also a solution; from (3), this implies the invariance of the evolution equation under the transformation

$$A \to A \exp ik_c\phi$$

Therefore the only possible nonlinear term, up to the third order in A is $|A|^2A$.

- Space reflection invariance implies the invariance of the evolution equation under the transformation

$$X \to -X,\ A \to \bar{A}\ .$$

Therefore the coefficient of $|A|^2A$ should be real. The linear part of the amplitude equation can be obtained by the Fourier transformation of equation (1). The amplitude equation thus reads at leading order

$$A_T = \mu A + A_{XX} + \varepsilon|A|^2A, \quad (\varepsilon = \pm 1) \tag{5}$$

where we have simplified the coefficients by appropriate scalings of amplitude and space. A similar equation has been obtained first for the convection problem by Newell and Whitehead [21], and Segel [22].

2.2. The Eckhaus instability

We consider the evolution equation (5) for a supercritical bifurcation ($\varepsilon = -1$). The solutions of the form

$$A_0(X) = Q \exp iqx, \quad Q^2 = \mu - q^2, \quad |q| < \sqrt{\mu} \tag{6}$$

represent perfectly periodic patterns of wavenumber $k_c + q$. To study their stability, we write

$$A(X,T) = A_0(X) + a(X,T)\exp iqx \tag{7}$$

and find from (5)

$$a_T = -2Q^2 \, \mathcal{R}(a) + 2iq\,a_X + a_{XX} - Q(a^2 + 2|a|^2) - |a|^2 a \tag{8}$$

Writing

$$a = R + i\phi \tag{9}$$

we get from (8)

$$R_T = -2Q^2 R - 2q\phi_X + R_{XX} + \dots \tag{10.a}$$

$$\phi_T = 2qR_X + \phi_{XX} + \dots \tag{10.b}$$

Equations (10) read in Fourier space

$$s \begin{pmatrix} R_k \\ \phi_k \end{pmatrix} = \begin{pmatrix} -2Q^2 - k^2 & -2iqk \\ 2iqk & -k^2 \end{pmatrix} \begin{pmatrix} R_k \\ \phi_k \end{pmatrix}$$

The eigenvalues are given by

$$s_{\pm}(k) = -(Q^2 + k^2) \pm (Q^4 + 4q^2k^2)^{\frac{1}{2}}$$

In the long wavelength limit ($k \to 0$), we have

$$s_-(k) \cong -2Q^2$$

The corresponding mode in damped; it is an amplitude mode. The second eigenvalue vanishes in the long wavelength limit,

$$s_+(k) = -k^2(1 - 2q^2/Q^2) - 2q^4k^4/Q^6 + o(k^4) \tag{11}$$

This is due to the translational invariance of the problem. Indeed, adding a perturbation $a = i\phi$ in (7) corresponds for small ϕ to the transformation $qx \to qx + \phi/Q$ in the solution (6), and therefore to a global phase shift of the pattern. In the complex plane this corresponds to a displacement along the circle of radius $|A_o| = Q$ (see figure 2).

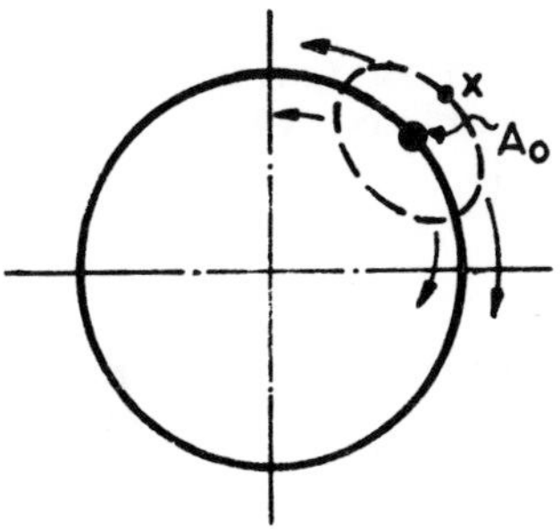

Figure 2. Ensemble of periodic solutions with different phases, in the complex plane.

Thus the translational symmetry implies that the phase of the periodic structure is a mode with a vanishing growth rate in the long wavelength limit. We thus expect that this mode can be easily destabilized. We see from (11) that this indeed occurs for $Q^2 = 2q^2$ or $\mu = 3q^2$. This is the Eckhaus instability that limits the range of stable wavenumbers around k_c (see figure 3).

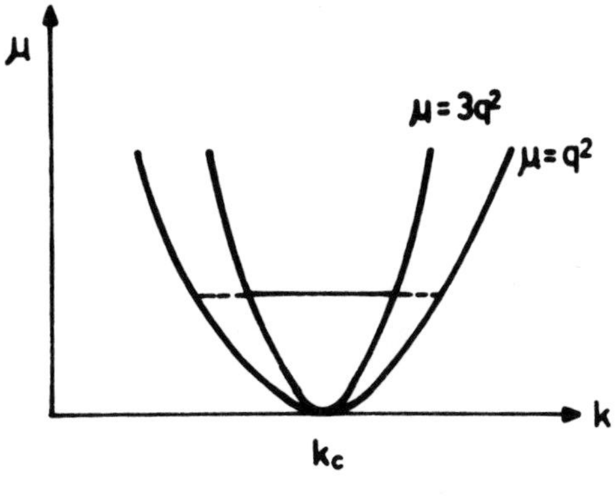

Figure 3. The Eckhaus instability.

The mechanism of this instability can be understood as follows: let us consider an initial condition shown in figure 2; the dotted line represents a perturbation of a homogeneous solution A_o, with R and ϕ varying periodically in space. The first term of the righthandside of (10.a) tends to contract the dotted circle along the circle of radius $Q(R \rightarrow 0)$, but an antagonistic effect is generated by the term $2qR_X$ in (10.b), that consists of a stretching which tends to spread the phase along the circle of radius Q (arrows in figure 2). When q is large enough, this effect becomes important and the phase instability occurs. In the long wavelength limit, R is damped on a fast time scale compared to the one of ϕ, which is nearly marginal. Therefore R follows adiabatically the gradients of the phase, and can be eliminated from equations (10) to get an evolution equation for the phase. Its linear part is the Fourier transform of (11)

$$\phi_T = (1 - 2q^2/Q^2)\phi_{XX} - 2q^4/Q^6\phi_{XXXX} + \ldots \qquad (12)$$

The form of the nonlinear terms can be determinated by symmetry considerations. Translational invariance in space implies the invariance of the phase equation under the transformation

$$\phi \to \phi + \text{constant}$$

Therefore ϕ cannot appear explicitely in the phase equation, and the leading order nonlinear term is ϕ_X^2. However, this term is also forbidden because of the reflection symmetry which implies the invariance of the phase equation under the transformation

$$X \to -X, \ \phi \to -\phi$$

We thus get at leading order an equation of the form

$$\phi_T = D(\mu,q)\phi_{XX} - \kappa\phi_{XXXX} + g\phi_X\phi_{XX} \qquad (13)$$

We next notice that $\phi = hX$ is a particular solution of (13); from (6) and (9) we see that it represents an homogeneous solution of wavenumber $q + h/Q$. Its linear stability is governed by the dispersion relation

$$s = -D(q + h/Q)k^2 + o(k^2) \qquad (14)$$

We can also compute s by linearisation of (13) near $\phi = hX$; we have

$$s = -[D(q) + hg]k^2 + o(k^2) \qquad (15)$$

We get from (14) and (15)

$$g = 1/Q(\partial D/\partial q)$$

This implies that the nonlinearity in the phase equation is associated with local changes in the phase diffusion coefficient [18].

The nonlinear term in (13) does not saturate the instability as we can see by computing the amplitude equation for an unstable mode of wavenumber k_0; we write

$$\phi(X,T) = A(T)\exp ik_0X + cc. \ldots$$

This mode is marginally stable if

$$L.\phi = D_0\phi_{XX} - \kappa\phi_{XXXX} = 0$$

or

$$D_0 = -\kappa k_0{}^2$$

We consider the weakly unstable regime $D < D_0$, and expand

$$D = D_0 - \varepsilon D_1 - \varepsilon^2 D_2 + \ldots$$

$$\phi = \varepsilon\phi_1 + \varepsilon^2\phi_2 + \ldots$$

We choose the time scale $\partial_T = \varepsilon^2\partial_\tau$ in order to have nonlinear and the time dependent terms at the same order in the equation for A. At order ε, we find the linear problem, and

$$\phi_1 = A(T)\exp ik_0X + cc.$$

At order ε^2, we get

$$L.\phi_2 = -D_1k_0^2A\exp ik_0X - igk_0^3A^2\exp 2ik_0X + cc.$$

and the solvability condition implies $D_1 = 0$. The amplitude equation is obtained at order ε^3

$$L.\phi_3 = (-D_2k_0^2A + A_\tau - (g^2k_0^2/6\kappa)|A|^2A)\exp ik_0X + cc.$$

$$+ \text{ non resonant terms}$$

The solvability condition gives

$$A_\tau = D_2k_0^2A + (g^2k_0^2/6\kappa)|A|^2A$$

and the Eckhaus instability is thus subcritical. We can write the evolution equation (13) under the form of a conservation equation

$$\phi_T = \partial_X(D\phi_X - \kappa\phi_{XXX} + g/2\ \phi_X^2)$$

and see that the spatial average of ϕ is a conserved quantity for periodic boundary conditions along the x-axis. Therefore an unstable periodic pattern cannot evolve to a stable one, by continuously changing its wavelength. Thus, in the unstable regime, changes in wavelength occur by nucleation or annihilation of a pair of rolls; so the instability does not saturate at long wavelength, and therefore the description of the unstable regime should incorporate the fast modes that we have eliminated.

2.3. The zig-zag instability

We consider now the dynamics of the torsion mode of the roll pattern described by equation (6), and look for long wavelength perturbations of the phase along the y-axis (see figure 4).

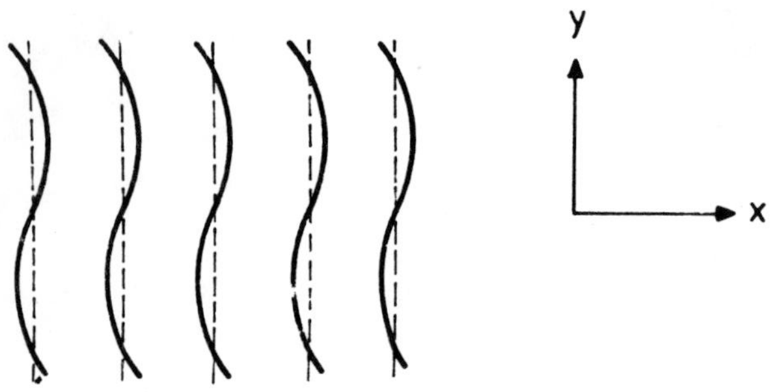

Figure 4. The zig-zag instability.

We have thus to consider solutions of (2) under the form

$$u(x,y,t) = A(X,Y,T)\exp ik_c x + cc. + \mathcal{U}(x,t,X,Y,T) \quad . \tag{16}$$

We have, as in section 2.2, a family of stationary solutions

$$A_o(X) = Q\exp iqx, \; Q^2 = \mu - q^2, \; |q| < \sqrt{\mu} \tag{17}$$

We consider perturbations of the form

$$A(X,Y,T) = A_o(X) + a(Y,T)\exp iqx \tag{18}$$

with

$$a = R + i\phi \tag{19}$$

The form of the non linear evolution equation for ϕ can be determined by the following symmetry arguments:

- invariance under the transformation

$\phi \to \phi + \text{constant}$ (translational invariance)

- invariance under the two transformations

$X \to -X, \; \phi \to -\phi$

$Y \to -Y$ (space reflection symmetry)

We have thus at leading order

$$\phi_T = D_\perp \phi_{YY} - \kappa_\perp \phi_{YYYY} + g_\perp (\phi_Y^3)_Y \tag{20}$$

$D_\perp, \kappa_\perp$ are given by the linear stability analysis of the roll pattern;

$g_\perp$ can be computed as follows: we first notice that $\phi = hY$ is a particular solution of (20). We see from (16), (18) and (19) that for h small, it corresponds to a slight rotation of the pattern, of an angle $h/Q(k_c + q) \cong h/Qk_c$. The rotated pattern has a modified wavenumber $k_c + q + h^2/2Q^2k_c$, and with the same argument as in section 2.2, we obtain

$$g_\perp = 1/6k_c q^2 (\partial D_\perp/\partial q) \quad .$$

The zig-zag instability occurs for decreasing $D_\perp$, when $D_\perp$ vanishes. The nonlinear terms renormalizes the diffusivity which become $D_\perp + 3g_\perp\phi_y^2$, and saturates the instability if $g_\perp$ is positive. This can be checked by computing the amplitude equation for an unstable mode

$$\phi(Y,T) = A(\tau) \exp iky + cc. + \ldots$$

We obtain

$$A_\tau = -D_\perp k^2 A - 3g_\perp k^4 |A|^2 A$$

Consequently the zig-zag instability is supercritical if $g_\perp$ is positive, and in that case the evolution equation (20) for the phase also describes the unstable regime. The subcriticality of the Eckhaus instability thus appears as a singular behavior of one-dimensional cellular structures.

2.4. Mean flow effect

The long wavelength instabilities described above in the framework of phase dynamics consist of stationary instabilities. However, stationary cellular flows also undergo bifurcations to time dependent states; a well known example is the oscillatory instability of convection rolls, that consists of a transverse oscillation of the rolls which propagates along their axis. The instability is connected with the appearance of vertical vorticity [23,24]. Its critical wavenumber vanishes in the case of stress-free top and bottom boundary conditions [23], and is finite for rigid boundary conditions [25,26]. Just as the Eckhaus and zig-zag instabilities trace back to translational invariance, the oscillatory instability is related to translational and Galilean invariances [27]. Therefore its description requires two coupled phase variables, and a propagative behavior results [28].

Let us consider the Oberbeck-Boussinesq equations that govern the convective motions in a fluid layer of height d, of infinite extension in the horizontal plane, heated from below.

$$\nabla \cdot \vec{v} = 0 \tag{21.a}$$

$$\partial_t \vec{v} + \vec{v} \cdot \nabla \vec{v} = -\nabla p + P\, \Delta\vec{v} + RP\theta\hat{z} \tag{21.b}$$

$$\partial_t \theta + \vec{v} \cdot \nabla\, \theta = \vec{v} \cdot \hat{z} + \Delta\theta \tag{21.c}$$

where $\underset{\sim}{v} = (u,v,w)$ is the velocity vector, θ is the temperature deviation from the linear profile, and p represents the pressure disturbances; $\hat{z}$

is the vertical unit vector opposite to gravity. Two dimensionless numbers characterize the problem; the Rayleigh number R, and the Prandtl number P

$$R = g\alpha d^3 \Delta T/\nu\kappa, \quad P = \nu/\kappa$$

where g is the acceleration of gravity and α is the isobaric expansion coefficient.

Two different kinds of boundary conditions are usually considered for the above set of equations. In the case of stress-free boundary conditions, one requires

$$w = \partial_z u = \partial_z v = 0 \quad \text{at} \quad z = 0,1 \tag{22.a}$$

whereas for rigid boundaries one has

$$\vec{v} = 0 \quad \text{at} \quad z = 0,1 \tag{22.b}$$

If the heat conductivity of the boundaries is large compared to the one of the fluid, one has

$$\theta = 0 \quad \text{at} \quad z = 0,1 \tag{22.c}$$

At the convection onset, the only possible stable steady solution of equations (21) with the boundary conditions (22) consists of two-dimensional rolls [29], of the form

$$U_\circ(x,z) = [u_\circ(x,z), 0, w_\circ(x,z), p_\circ(x,z), \theta_\circ(x,z)] \tag{23}$$

Translational symmetry in the horizontal plane implies that $U_\circ(x + \phi, z)$ is also a solution, and consequently $\partial_x U_\circ$ is a neutral mode of the roll pattern described by (23).

With stress-free boundary conditions another neutral mode exists, and consists of a uniform horizontal velocity ψ along the x-axis. We thus look for a perturbation of the basic pattern under the form

$$U(x,z,t) = U_\circ[x + \phi(t), z] - \psi(t)1 \tag{24}$$

where $1 = (1,0,0,0,0)$

We obtain from (21) and (24)

$$\phi_t \partial_x U_\circ - \psi_t = \psi \partial_x U_\circ$$

or

$$\partial_t \begin{pmatrix} \phi \\ \psi \end{pmatrix} = \begin{pmatrix} 0 & 1 \\ 0 & 0 \end{pmatrix} \begin{pmatrix} \phi \\ \psi \end{pmatrix} \tag{25}$$

The existence of two phase modes ϕ and ψ is related to translational and Galilean invariances; their linear coupling can be understood as follows: the advection of the pattern at a constant speed ψ along the x-axis leads to a spatial phase ϕ that increases linearly in time. This makes the phase dynamics second order in time, and thus one can expect a propagative behavior for long wavelength disturbances of the periodic pattern. For low Prandtl number fluids, a propagative torsion mode of the roll pattern is amplified close to the convection onset, and gives rise to the oscillatory instability [23]; the roll pattern is shifted perpendicular to its axis, periodically along the axis and in time.

We thus look for a perturbation of the basic pattern under the form

$$U(x,y,z,t) = U_o[x+\phi(Y,T),z] - \psi(Y,T)1 + \mathcal{U}(x,z,Y,T) \qquad (26)$$

where $\phi(Y,T)$ represents a long wavelength modulation of the spatial phase along the roll axis, and $\psi(Y,T)$ is a slowly varying horizontal velocity; the corresponding vertical vorticity mode has been first considered by Busse in his stability analysis of two-dimensional convective rolls in a low Prandtl number fluid [23]. It was shown by Siggia and Zippelius [24] that this mode also modifies at leading order the evolution equation which governs the rolls amplitude in the vicinity of the convection onset [21,22].

In the limit of zero wavenumber, equation (26) reduces to (24), and the evolution of $\phi(T)$ and $\psi(T)$ is governed by equation (25), which represents a codimension two singularity [30,31]. The structure of the Galilee group implies that the eigenmode $\partial_x U_o$ has a double zero eigenvalue. The first part of expression (26) consists of a perturbation of the basic state $U_o(x,z)$ along the generalized eigenspace $\{\partial_x U_o, 1\}$, whereas $\mathcal{U}(x,z,Y,T)$ stands for corrections perpendicular to this eigenspace. In the long wavelength limit, the eigenmodes are almost neutral; therefore we assume that all the other modes behave on time scales much faster than the one of ϕ and ψ, and thus can be eliminated to give evolution equations for ϕ and ψ. Because of the structure of the codimension-two singularity described by equation (25), we can look these evolution equations under the form

$$\phi_T = \psi \qquad (27.a)$$

$$\psi_T = f(\phi,\psi,\partial_Y) \qquad (27.b)$$

We next assume that f and $\mathcal{U}$ can be expanded in multiple Taylor series of ϕ, ψ, and their gradients. The compatibility between (21), (26) and (27) gives evolution equations for ϕ and ψ at each order in the gradient expansion. Their form is determined by symmetry constraints:

- Translational and Galilean invariances imply the invariance of the equations under the transformations

$$\phi \rightarrow \phi + \text{constant}, \quad \psi \rightarrow \psi + \text{constant} ,$$

thus f depends on ϕ and ψ only through their Y-derivatives.

- Space reflection symmetry implies the invariance of the phase equations under the transformations

$$x \to -x,\ \phi \to -\phi,\ \psi \to -\psi$$

and

$$Y \to -Y$$

It follows from the first constraint that f involves no quadratic nonlinearities. The second one implies that the number of Y-derivatives is even in each term of f. Thus we obtain up to the fourth order in ∂_Y and keeping only the leading order nonlinear terms

$$\phi_T = \psi \tag{28.a}$$

$$\begin{aligned}\psi_T &= a\phi_{YY} + b\psi_{YY} - c\phi_{YYYY} - d\psi_{YYYY}\\ &\quad + (g_1'\phi_Y^2 + g_2'\phi_Y\psi_Y + g_3'\psi_Y^2)\phi_{YY}\\ &\quad + (g_4'\phi_Y^2 + g_5'\phi_Y\psi_Y + g_6'\psi_Y^2)\psi_{YY}\end{aligned} \tag{28.b}$$

The large scale velocity field $\psi(Y,T)$ is driven by inhomogeneities of the roll pattern, and in turn tends to convect the pattern. However we cannot admit a global advection of the pattern along the x-axis. Therefore the average of $\psi(Y,T)$ must stay zero at the oscillatory instability onset. Thus we have

$$\bar{\psi}_T = \frac{1}{2L}\partial_T \int_{-L}^{L} \psi(Y,T)\,dY = 0$$

For periodic boundary conditions along the Y-axis, this implies $g_4' = g_2'/2$, $g_3' = g_5'/2$. Thus we obtain

$$\phi_T = \psi \tag{28.a}$$

$$\begin{aligned}\psi_T &= \partial_Y[a\phi_Y + b\psi_Y - c\phi_{YYY} - d\psi_{YYY}\\ &\quad + (g_1'/3)\phi_Y^3 + (g_2'/2)\phi_Y^2\psi_Y + g_3'\phi_Y\psi_Y^2 + g_4'\psi_Y^3]\end{aligned} \tag{28.b}$$

We next observe that the two last terms are non resonant and can be eliminated at this order by the following change of variables

$$\phi = \tilde{\phi} + (g_3'/6)(\tilde{\phi}_Y^3)_Y + (g_4'/6)(\tilde{\phi}_Y^2\tilde{\psi}_Y)_Y$$

$$\psi = \tilde{\psi} + (g_3'/2)(\tilde{\phi}_Y^2\tilde{\psi}_Y)_Y + (g_4'/3)(\tilde{\phi}_Y\tilde{\psi}_Y^2)_Y$$

Finally we write $\theta = \tilde{\phi}_Y$, and $\chi = \tilde{\psi}_Y$, the phase gradients, and obtain

$$\theta_T = \chi \tag{29.a}$$

$$\chi_T = \partial_{YY}[a\theta + b\chi - c\theta_{YY} - d\chi_{YY} + g_1\phi^3 + g_2\theta^2\chi] \qquad (29.b)$$

where $g_1 = g_1'/3$, and $g_2 = g_2'/2$.

Equations (29) describe the stability of the roll pattern with respect to long wavelength perturbations, i.e. θ and χ varying like $\exp(\eta T \pm ikY)$ with $k \to 0$. In this limit the dispersion relation is

$$\eta^2 + bk^2\eta + ak^2 = o(k^2) \qquad (30)$$

where the coefficients depend on the Rayleigh number R, the Prandtl number P, and the wavenumber α of the basic roll pattern. The oscillatory instability occurs when $b(R,P,\alpha)$ vanishes i.e. on the critical surface $R = R_o(P,\alpha)$ of the R-P-α space. Its frequency at onset is

$$\omega_o = k\sqrt{a(R_o,P,\alpha)}$$

The coefficients a and b have been computed by Busse (see reference 23, relations 3.11 and 3.17).

The nonlinear terms of equation (29.b) simply renormalize the damping and the frequency of the instability, which is a supercritical Hopf bifurcation if $g_2 > 0$. We can compute the coefficients of the nonlinear terms on noting that $\theta = p$ is a particular solution of (29), which corresponds to a slight rotation of the roll pattern. We consider

$$\theta = p + \varepsilon \exp(\eta T + ikY)$$

and linearize equations (29) in ε. We obtain

$$\eta^2 + [b(\alpha) + g_2p^2]k^2\eta + [a(\alpha) + 3g_1p^2]k^2 = o(k^2) \qquad (31)$$

The wavenumber of the slightly rotated pattern described by $\theta = p$, is

$$\alpha' \cong \alpha(1 + p^2/2) \qquad (32)$$

We can also write the dispersion relation which gives the growthrate of the slighty rotated pattern, under the form

$$\eta^2 + b(\alpha + \alpha p^2/2)k^2\eta + a(\alpha + \alpha p^2/2)k^2 = o(k^2) \qquad (33)$$

We obtain from (31) and (32)

$$g_1 = \alpha/6\ \partial a/\partial\alpha \qquad (34.a)$$

$$g_2 = \alpha/2\ \partial b/\partial\alpha \qquad (34.b)$$

Thus the nonlinear effects can be understood as follows: the wavy disturbances slightly modify the local wavenumber (see equation 32), and in turn, this changes locally the propagation speed and the damping rate of the disturbances.

In most experimental realizations of thermal convection, the top and bottom boundaries are rigid. This externally breaks the Galilean invariance and a large scale Z-invariant horizontal velocity is not allowed by the boundary conditions. In other words, the vertical vorticity modes are damped, with a damping rate $\nu = Pk^2$, for the first Fourier mode in Z, with a vanishing horizontal wavenumber. For low Prandtl number fluids, this damping is small and we extend phenomenologically equations (28) by incorporating the damping term $-\nu\psi$ in the right hand side of equation (28.b). We do not consider the nonlinear renormalization of this damping by a term of the form $\psi\phi_Y^2$ because we expect its coefficient to vanish since ν does not depend on α[32]. Thus we obtain for the phase gradients θ and χ

$$\theta_T = \chi \tag{35.a}$$

$$\chi_T = -\nu\chi + \partial_{YY}[a\theta + b\chi - c\theta_{YY} - d\chi_{YY} + g_1\theta^3 + g_2\theta^2\chi] \tag{35.b}$$

The growthrate η of the oscillatory instability obeys the dispersion relation

$$\eta^2 + (\nu + bk^2 + dk^4)\eta + ak^2 + ck^4 = o(k^4) \tag{36}$$

Stability of short wavelength requires $d > 0$. The oscillatory instability occurs for

$$b_o = -2(\nu d)^{\frac{1}{2}}$$

with a finite wavenumber

$$k_o = (\nu/d)^{\frac{1}{4}}$$

and a frequency at onset

$$\omega_o = [a(\nu/d)^{\frac{1}{4}} + c(\nu/d)^{\frac{1}{2}}]^{\frac{1}{2}}$$

The real and imaginary parts of the growthrate as a function of the wavenumber are given below (figure 5); the form of this curve is in agreement with the one computed numerically for the oscillatory instability onset (see reference[33], figure 3). We observe that the oscillatory instability comes from the interaction between the translation mode ($\eta(0) = 0$), and the damped Galilean mode ($\eta(0) = -\nu$).

In the vicinity of the instability onset, we write

$$\theta(Y,T) = A(\zeta,\tau)\exp i(\omega_o T + k_o Y)$$
$$+ B(\zeta,\tau)\exp i(\omega_o T - k_o Y) + cc. + C(\zeta,\tau) + \dots$$

and derive the evolution equations for A,B,C. We obtain at leading order

$$A_\tau = (k_o^2/2)(b_o - b)A + \omega_1 A_\zeta + (\xi_o^2 - i\omega_2)A_{\zeta\zeta}$$
$$- (\beta - i\gamma)[3|A|^2 + 6|B|^2 + 3C^2]A + \dots \tag{37.a}$$

$$B_\tau = (k_o^2/2)(b_o - b)B - \omega_1 B_\zeta + (\xi_o^2 - i\omega_2)B_{\zeta\zeta}$$
$$- (\beta - i\gamma)[6|A|^2 + 3|B|^2 + 3C^2]B + \ldots \quad (37.b)$$

$$C_\tau = (a/\nu)C_{\zeta\zeta} + \ldots \quad (37.c)$$

with

$$\omega_1 = (k_o/\omega_o)(a + 2ck_o^2)$$

$$\omega_2 = (1/2\omega_o)[(a + 6ck_o^2) - (k_o/\omega_o)^2(a + 2ck_o^2)^2]$$

$$\xi_o^2 = 2(\nu d)^{\frac{1}{2}}$$

$$\beta = -(k_o^2 g_2)/2$$

$$\gamma = (k_o^2 g_1)/2\omega_o$$

The oscillatory instability is supercritical if g_2 is positive, and consists of travelling waves (see section 3.3); this last result is in agreement with direct numerical simulations [34,35], and with experimental observations [36].

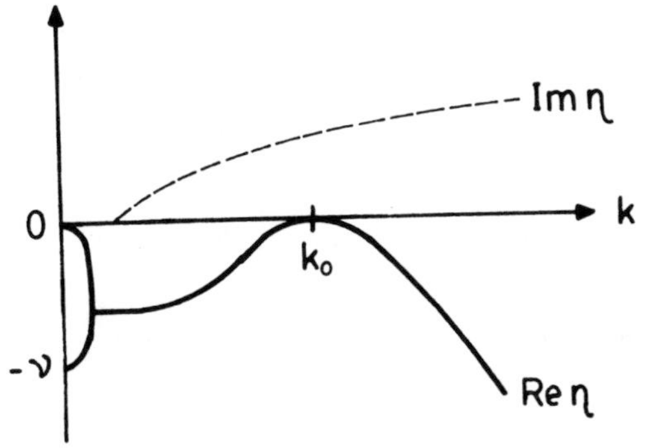

Figure 5. Growth-rate of the oscillatory instability at onset.

2.5. Phase dynamics of two-dimensional patterns.

Hydrodynamic instabilities also generates two-dimensional patterns, that consists of squares (convection between top and bottom boundaries of low thermal conductivity), or hexagons (Marangoni instability). The phase dynamics of these patterns involves two coupled phases ϕ and ψ, associated with the broken translational invariances along the x and y-axis. Symmetry arguments are less straightforward to use than for one-dimensional patterns, but, still, they constrain the form of the evolution equations for the phases.

3. NONLINEAR DISPERSIVE WAVES

3.1. The nonlinear Schrodinger equation

In this section we consider spatio-temporal cellular structures, i.e. waves. We begin with conservative systems for wich nonlinear wave equations arise in many different physical situations [6]. Famous examples are, the Korteweg-de Vries equation (KdV)

$$u_t + uu_x + u_x + u_{xxx} = 0 \tag{1}$$

which is an approximation for waves in shallow water, ($u(x,t)$ is the surface elevation), and the Sine-Gordon equation (SG)

$$u_{tt} - u_{xx} + \omega_o^2 \sin u = 0 \tag{2}$$

which is found in nonlinear optics where it models the propagation of pulses in resonant media, in condensed-matter physics, where it describes charge-density waves in periodic pinning potentials or propagation along Josephson transmission lines [37].

These equations admit travelling solutions of the form $u(\omega t - kx)$ where the frequency ω depends on the wavenumber k, as for dispersive waves, but also on the amplitude of u. This effect was shown first by Stokes for gravity waves [38]. In the small amplitude limit, we can study the stability of these travelling solutions with a method introduced by Benney and Newell [7]. We write

$$u(x,t) = A(X,T) \exp i(\omega t - kx) + cc. + \mathcal{U}(x,t,X,T) \tag{3}$$

where $A(X,T)$ is the slowly varying envelope of the wave of frequency ω and wavenumber k, and we look for an evolution equation for A in the form

$$A_T = f(A, A_X, A_{XX}, \ldots) \tag{4}$$

At leading order we obtain the dispersion relation between ω and k;

$$\omega = k - k^3 \qquad \text{(KdV)}$$

$$\omega^2 = \omega_o^2 + k^2 \qquad \text{(SG)}$$

Solvability conditions at higher orders give the amplitude equation

$$A_T = -cA_X + i\alpha A_{XX} - i\beta |A|^2 A \tag{5}$$

where

$$c = \partial\omega/\partial k = \begin{cases} 1 - 3k^2 & \text{(KdV)} \\ k/\omega & \text{(SG)} \end{cases}$$

$$\alpha = -1/2\partial^2\omega/\partial k^2 = \begin{cases} 3k & \text{(KdV)} \\ 1/2\omega(k^2/\omega^2 - 1) & \text{(SG)} \end{cases}$$

$$\beta = \begin{cases} -1/6k & \text{(KdV)} \\ \omega_o^2/4\omega & \text{(SG)} \end{cases}$$

In the reference frame that moves with the group velocity c of the wave-packet, equation (5) simplifies, and we obtain the nonlinear Schrodinger equation

$$A_T = i\alpha A_{XX} - i\beta|A|^2A \tag{6}$$

This equation governs the envelope of nonlinear dispersive waves. Its universality can be understood with symmetry arguments.

- Translational invariance in space and time of (1) and (2) implies that (4) is invariant under the transformation $A \to A\exp i\theta$. Therefore the leading order nonlinear term is $|A|^2A$.

- The invariance of (1) and (2) under the transformation $x \to -x$, $t \to -t$, (conservative systems with space reflection symmetry), implies the invariance of (4) under the transformation $X \to -X$, $T \to -T$, $A \to \bar{A}$. Therefore the coefficients of the terms with an odd number of x-derivatives are real, and the ones of the terms with an even number of x-derivatives are pure imaginary. Note also that the linear part of equation (5) is simply the Fourier transform of

$$\omega = K\partial\omega/\partial k + 1/2K^2\partial^2\omega/\partial k^2 + \dots$$

where K is the wavenumber associated with X.

3.2. The Benjamin-Feir instability [39]

Equation (6) admits a family of travelling wave solutions

$$A_o(\Omega T - KX) = Q\exp i(\Omega T - KX) \tag{7}$$

with

$$\Omega = -\alpha K^2 - \beta Q^2 \tag{8}$$

These solutions describe travelling wave solutions of equations (1) and

(2), with a frequency $\omega + \Omega$ and a wavenumber $k + K$. To study their stability we write

$$A(X,T) = A_o(\Omega T - KX) + a(X,T)\exp i(\Omega T - KX) \qquad (9)$$

and find from (6)

$$a_T = 2\alpha K a_X + i\alpha a_{XX} - 2i\beta Q^2 \mathcal{R}(a) - i\beta Q(a^2 + 2|a|^2) - i\beta|a|^2 a \qquad (10)$$

We write

$$a = R + i\phi$$

and linearize (10) in R and ϕ; we obtain

$$\phi_T = -2\beta Q^2 R + 2\alpha K\phi_X + \alpha R_{XX} \qquad (11.a)$$

$$R_T = 2\alpha K R_X - \alpha\phi_{XX} \qquad (11.b)$$

For R and ϕ varying like $\exp(\eta T - iqX)$, the dispersion relation is

$$\eta^2 + 4i\alpha Kq\eta + (2\alpha\beta Q^2 - 4\alpha^2K^2)q^2 + \alpha^2q^4 = 0$$

and we obtain for the growthrate η

$$\eta = -2i\alpha Kq \pm (-2\alpha\beta Q^2q^2 - \alpha^2q^4)^{\frac{1}{2}}$$

Therefore monochromatic travelling waves are unstable for

$$\alpha\beta < 0$$

This is the Benjamin-Feir instability [39]. It occurs at long wavelength

$$0 < q^2 < 2Q^2|\beta/\alpha|$$

and consists of a resonant amplification of side band modes [40]. The waves breaks into pulses the solutions of the nonlinear Schrodinger equation.

3.3. Waves in dissipative systems

Hydrodynamic instabilities often lead to time-dependent cellular patterns. The Couette flow between counter-rotating cylinders [42], or thermal convection in the presence of a salinity gradient [43], can undergo a Hopf bifurcation, with critical modes of finite spatial wavenumber. Thus the flow consists of a periodic pattern in space, with oscillatory behavior in time. Depending on the coefficient of the nonlinear terms in the amplitude equations, one may get at the instability onset, either propagating waves, or standing waves [44].

Let us consider a one space dimensional system described by a set of scalar fields $U(x,t) = (U_1(x,t), U_2(x,t), \ldots, U_n(x,t))$ which is invariant by space relfection $x \to -x$, and by translations in space and time, $x \to x + \Sigma$, $t \to t + \Theta$. We suppose that this system undergoes a bifurcation at $k = k_o$, with an imaginary growthrate $i\omega$, and thus write the growthrate of the mode $\Phi_k(x) = \Phi_k \exp ikx$,

$$\eta_k \cong \mu - \xi_o(k^2 - k_o^2)^2 + i[\omega + \omega_1(k^2 - k_o^2)] + \ldots \tag{12}$$

The instability occurs at $\mu = 0$, and for $\mu > 0$ one expects a behavior characterized by rapid space and time scales $2\pi/k_o$ and $2\pi/\omega$ respectively. We look for U in the form

$$U(x,t) = A(X,T)\exp i(\omega t - k_o x)\Phi_k + B(X,T)\exp i(\omega t + k_o x)\Phi_k + cc. + \mathcal{U}(x,t,X,T) \tag{13}$$

The fields A and B describe respectively the amplitudes of the waves propagating to the right and the left, and obey amplitude equations, the form of which can be found, as in the previous section, by symmetry considerations:

- Translational invariance in space and time, which implies the invariance of the amplitude equations under the transformations

$$A \to A\exp -i\Sigma, \quad B \to B\exp i\Sigma$$

$$A \to A\exp i\Theta, \quad B \to B\exp i\Theta$$

- Space reflection symmetry, which implies the invariance under the transformation

$$X \to -X, \quad A \to B, \quad B \to A$$

This leads at lowest order to the equations

$$A_T = \mu A - A_X + (1 + i\alpha)A_{XX} - (\varepsilon + i\beta)|A|^2A - (\gamma + i\delta)|B|^2A \tag{14.a}$$

$$B_T = \mu B + B_X + (1 + i\alpha)B_{XX} - (\gamma + i\delta)|A|^2B - (\varepsilon + i\beta)|B|^2B \tag{14.b}$$

where we have simplified the coefficients by appropriate scaling, and $\varepsilon = \pm 1$. We note that although symmetry arguments allow $(1 + i\eta)A_X$ and $(\pm 1 + i\alpha)A_{XX}$, stability requirements lead to (14). We assume that we are in the supercritical situation, $\varepsilon = 1$ and $\gamma > -1$; the spatially homogeneous solutions

$$A_o(T) = Q_A \exp i\Omega_A T, \quad B_o(T) = Q_B \exp i\Omega_B T,$$

describe either:

- propagating waves corresponding to

$$Q_A^2 = 0;\ Q_B^2 = \mu,\ \Omega_B = -\beta\mu \qquad (15.a)$$

or

$$Q_A^2 = \mu,\ \Omega_A = -\beta\mu;\ Q_B^2 = 0 \qquad (15.b)$$

and are stable with respect to spatially homogeneous perturbations when $\gamma > 1$;

- standing waves which correspond to

$$Q_A^2 = Q_B^2 = \mu/(1+\gamma),\ \Omega_A = \Omega_B = -\mu(\beta+\delta)/(1+\gamma) \qquad (16)$$

which are stable when $-1 < \gamma < 1$.

3.3.1 Large scale instability of propagating waves. The system (14) reduces to one equation, say for A; in the reference frame which moves with the group velocity of the wave packet, we obtain an equation of the form

$$A_T = \mu A + (1 + i\alpha)A_{XX} - (1 + i\beta)|A|^2A \qquad (17)$$

To investigate the stability of the homogeneous solution (15.a) we write

$$A(X,T) = [Q + a(X,T)] \exp i\Omega T$$

and obtain from (17)

$$a_T = -2(1 + i\beta)Q^2\mathcal{R}(a) + (1 + i\alpha)a_{XX}$$
$$- (1 + i\beta)Q[(a^2 + 2|a|^2) - |a|^2a] \qquad (18)$$

We write

$$a = R + i\phi$$

The linear part of (18) reads in Fourier space

$$s \begin{pmatrix} R_k \\ \phi_k \end{pmatrix} = \begin{pmatrix} -2Q^2 - k^2 & \alpha k^2 \\ -2\beta Q^2 - \alpha k^2 & -k^2 \end{pmatrix} \begin{pmatrix} R_k \\ \phi_k \end{pmatrix} \qquad (19)$$

The eigenvalues are given by

$$s_\pm(k) = -(Q^2+k^2) \pm [(Q^2+k^2)^2 - 2Q(1+\alpha\beta)k^2 - (1+\alpha^2)k^4]^{1/2}$$

There is one amplitude mode which corresponds to $\mathcal{R}(s_-) < 0$, and is damped. The other eigenvalue s_+ corresponds to a phase mode; we have in the long wavelength limit

$$s_+(k) = -(1 + \alpha\beta)k^2 - \alpha^2(1 + \beta^2)k^4/2Q^2 + o(k^4) \qquad (20)$$

An instability occurs at zero wavenumber when $1 + \alpha\beta$ vanishes; it corresponds to a large scale spatial desynchronization of the travelling wave corresponding to (15.a). The linear part of the evolution equation for the phase is given by the Fourier transform of (20). Symmetry considerations (translational invariance in space and time) constrain the form of the nonlinear terms. We have at leading order an evolution equation of the form [10]

$$\phi_T = D\phi_{XX} - \kappa\phi_{XXXX} + g\phi_X{}^2 \qquad (21)$$

with

$$D = 1 + \alpha\beta \; .$$

We can compute g on noting that (17) has a family of travelling wave solutions

$$A_q = Q_q \exp i(\Omega_q T - qX) \qquad (22)$$

with

$$Q^2 = \mu - q^2, \; \Omega_q = -(\alpha - \beta)q^2 - \beta\mu$$

We next look for a phase equation under the form

$$\phi_T = c\phi_X + D\phi_{XX} - \kappa\phi_{XXXX} + g\phi_X^2 \qquad (23)$$

which has the particular solution, $\phi = \rho t - rx$, if $\rho = -cr + gr^2$; we notice that this solution corresponds to a travelling wave of frequency $\omega - (\alpha - \beta)q^2 - \beta\mu + \rho/Q$, and of wavenumber $q + r/Q$; therefore we have

$$\Omega_q(q + r/Q) = \Omega_q(q) + \rho/Q$$

We get from the expression of ρ by identifying terms in r and r^2,

$$c = -\partial\Omega_q/\partial q; \; g = 1/2Q\partial^2\Omega_q/\partial q^2$$

For $q = 0$, we find

$$c = 0; \; g = (\beta - \alpha)/Q$$

In order to determine the effect of the nonlinear term in (21) we consider a weakly unstable mode

$$\phi(X,T) = F(\tau)\exp ikx + cc.$$

with

$$D_o = -\kappa k^2$$

and look for an evolution equation for F. We thus expand ϕ and D

$$\phi = \varepsilon\phi_1 + \varepsilon^2\phi_2 + \ldots$$

$$D = D_o - \varepsilon D_1 - \varepsilon^2 D_2 + \ldots$$

and solve (21) at each order in ε. At order ε we have to take into account a mode, constant in space, that is marginal and forced by F through nonlinear interactions. Thus we take

$$\phi_1 = F(\tau_2)\exp ikx + cc. + G(\tau_1)$$

The solvability condition at order ε^2 determines the evolution equation for G

$$G_{\tau_1} = 2gk^2|F|^2$$

At order ε^3, we find the evolution equation for F

$$F_{\tau_2} = D_2k^2F - g^2/3\kappa|F|^2F$$

In the unstable regime ($D_2 > 0$), we obtain

$$|F|^2 = 3\kappa D_2k^2/g^2;\ G = 6\kappa D_2k^4t/g\ ,$$

which shows that a change in the wave oscillation frequency occurs ($G \sim t$) because of the phase instability. The instability is saturated by the nonlinear term in equation (21), which therefore governs the long wavelength dynamics of the phase in the weakly unstable regime.

3.3.2 Large scale instability of standing waves. In order to study the stability of standing waves (16), it is useful to remark that equations (14) admit a more general class of solutions than (16),

$$A_o(X,T) = Q_A\exp(-ipX + i\Omega_AT) \qquad (24.a)$$

$$B_o(X,T) = Q_B\exp(iqX + i\Omega_BT) \qquad (24.b)$$

with

$$Q_A^2 = [\mu(1-\gamma) - p^2 + \gamma q^2]/(1-\gamma^2)$$

$$Q_B^2 = [\mu(1-\gamma) + \gamma p^2 - q^2]/(1-\gamma^2)$$

$$\Omega_A = -\mu(\beta+\delta)/(1+\gamma) + p - [\alpha + (\delta\gamma-\beta)/(1-\gamma^2)]p^2 +$$
$$+ (\delta-\beta\gamma)/(1-\gamma^2)q^2$$

$$\Omega_B = -\mu(\beta+\delta)/(1+\gamma) + q + (\delta - \beta\gamma)/(1-\gamma^2)p^2$$
$$- [\alpha + (\delta\gamma - \beta)/(1-\gamma^2)]q^2$$

To study the stability of this solution, we write,

$$A(X,T) = A_o(X,T) + a(X,T)\exp(-ipX + i\Omega_A T)$$

$$B(X,T) = B_o(X,T) + b(X,T)\exp(iqX + i\Omega_B T)$$

In the limit of homogeneous perturbations i.e. a and b independent of X, one finds two types of modes; the former associated with the real part of a and b are stable, at least for p and q small enough. The latter, which are marginal, correspond to phase perturbations and are associated to the imaginary parts ϕ and ψ of a and b. The elimination of the amplitude modes leads to the phase equations which are of the form

$$\phi_T = a_1\phi_X + b_1\psi_X + a_2\phi_{XX} + b_2\psi_{XX} + \ldots$$
$$+ g\phi_X^2 + h\phi_X\psi_X + k\psi_X^2 + \ldots \qquad (25.a)$$

$$\psi_T = -b_1'\phi_X - a_1'\psi_X + b_2'\phi_{XX} + a_2'\psi_{XX} + \ldots$$
$$+ k'\phi_X^2 + h'\phi_X\psi_X + g'\psi_X^2 + \ldots \qquad (25.b)$$

where the coefficients are functions of p and q. These equations admit the solutions

$$\phi = -rx + \Omega_\phi T$$

$$\psi = sx + \Omega_\psi T$$

with

$$\Omega_\phi = -a_1 r + b_1 s + gr^2 - hrs + ks^2$$

$$\Omega_\psi = b_1' r - a_1' s + k'r^2 - h'rs + g's^2$$

which can be interpreted in the limit $r \to 0$, $s \to 0$, as modifications of the wavenumbers of the initial solutions: $p \to p + r/Q_A$, $q + s/Q_B$. Thus we have

$$\Omega_A(p + r/Q_A, q + s/Q_B) \cong \Omega_A(p,q) + \Omega_\phi \qquad (26.a)$$

$$\Omega_B(p + r/Q_A, q + s/Q_B) \cong \Omega_B(p,q) + \Omega_\psi \qquad (26.b)$$

In order to compute the coefficients (a_1, a_1', b_1, b_1', g, h, l, g', h', l') we expand the left hand side of (26) in r and s and identify the terms in r, s, r^2, rs, s^2. In the case p = q = 0, i.e. when one considers the homogeneous solution (16), the phase equations become

$$\phi_T = -\phi_X + a_2\phi_{XX} + b_2\psi_{XX} + \dots + g\phi_X^2 + h\phi_X\psi_X + k\psi_X^2 + \dots \tag{27.a}$$

$$\psi_T = \psi_X + b_2\phi_{XX} + a_2\psi_{XX} + \dots + k\phi_X^2 + h\phi_X\psi_X + g\psi_X^2 + \dots \tag{27.b}$$

with

$$g = 1/2Q_A(\partial^2\Omega_A/\partial p^2)_{p=q=0} = -[(1+\gamma)/\mu]^{\frac{1}{2}}[\alpha + (\gamma\delta - \beta)/(1-\gamma^2)]$$

$$h = 1/2Q_B(\partial^2\Omega_A/\partial p\partial q)_{p=q=0} = 0$$

$$k = Q_A/2Q_B^2(\partial^2\Omega_B/\partial q^2)_{p=q=0} = [(1+\gamma)/\mu]^{\frac{1}{2}}(\gamma\delta - \beta)/(1-\gamma^2)$$

$$a_2 = 1 + (\alpha\beta/1-\gamma^2) - (\alpha\gamma\delta/1-\gamma^2)$$

$$b_2 = \alpha(\delta - \beta\gamma)/(1 - \gamma^2)$$

Equations (27) are invariant by $X \to -X$, $\phi \to \psi$, $\psi \to \phi$, which reflects the space reflection invariance of the initial system, which is not broken by the choice of the homogeneous solution (16). The stability of this solution is determined by the dispersion relation

$$s_\pm(k) = \pm ik + s_2k^2 + \dots \tag{28}$$

where

$$s_2 = -1 - \alpha(\beta - \gamma\delta)/(1 - \gamma^2)$$

A bifurcation occurs when s_2 changes sign and it represents the appearance of a low frequency which modulates the standing wave envelope. This bifurcation occurs at zero spatial wavenumber if the coefficient s_4 of k^4 in (28) is positive. In fact s_2 and s'_4 can vanish simultaneously. This defines a codimension-two surface in parameter space in the vicinity of which the bifurcation can occur first either at zero or finite spatial frequency. It is immediate to interprete the phases ϕ and ψ in term of the spatial and temporal phases,

$$\Sigma = (\psi - \phi)/2, \quad \Theta = (\psi + \phi)/2 \ ,$$

of the oscillating pattern. In these new variables, equations (27) become

$$\Theta_T = \Sigma_X + a\Theta_{XX} + \dots + u(\Theta_X^2 + \Sigma_X^2) + \dots \tag{29.a}$$

$$\Sigma_T = \Theta_X + a\Sigma_{XX} + \dots + w\Theta_X\Sigma_X + \dots \tag{29.b}$$

These equations describe the phase dynamics of a time periodic pattern with respect to long wavelength perturbations. The only hypothesis are,

translational invariances in space and time, and the reflection invariance of the basic pattern which implies the invariance of (29) under the transformation, $X \to -X$, $\Theta \to \Theta$, $\Sigma \to -\Sigma$.
Let us note finally that when the phase instability occurs at a finite wavelength and is nonlinearly saturated, we obtain a quasi-periodic pattern that involves a larger spatial scale than the one of the initial standing wave. If this new pattern itself undergoes a phase instability, one can thus describe a cascade mechanism toward larger and larger scales.

ACKNOWLEDGMENTS

The first part of this review was presented as lectures at the Geophysical Fluid Dynamics summer program, at Woods Hole in 1985; I thank the Woods Hole Oceanographic Institution for support. I have benefitted from discussions with P. Coullet, M.E. Brachet and E. Tirapegui with whom I have worked on several topics reviewed here.

REFERENCES

[1] L. Landau, "On the theory of phase transitions", "On the problem of turbulence", Collected Papers of L. Landau, Gordon and Breach (1967).
[2] For a review, see P.H. Coullet and E.A. Spiegel, SIAM J. Appl. Math. 43, 775-821 (1983).
[3] For a review, see A. Nitzan, P. Ortoleva, J. Deutch and J. Ross, J. Chem. Phys. 61, 1056-1074 (1974).
[4] For a review, see J.P. Gollub and H.L. Swinney, Hydrodynamic Stability and the Transition to Turbulence, Topics in applied physics 45, Springer Verlag (1981) or J.E. Wesfreid and S. Zaleski, Cellular Structures in Instabilities. Lecture notes in physics, Springer Verlag (1984).
[5] A.C. Newell, Lectures in Applied Math. 15, 157-163 (1974).
[6] G.B. Witham, Linear and Nonlinear Waves, Wiley (1974).
[7] D.J. Benney and A.C. Newell, J. Math. Phys. 46, 133-139 (1967).
[8] P. Ortoleva and J. Ross, J. Chem. Phys. 58, 5673-5680 (1973).
[9] L.N. Howard and N. Kopell, Stud. Appl. Math. 56, 95-145 (1977).
[10] Y. Kuramoto and T. Tsuzuki, Prog. Theor. Phys. 55, 356-369 (1976).
[11] Y. Pomeau and P. Manneville, J. Physique Lettres 40, 609-612 (1979)
[12] M.C. Cross, Phys. Rev. A27, 490-498 (1983).
[13] M.C. Cross and A.C. Newell, Physica 10D, 229-328 (1984).
[14] P. Tabeling, J. Physique Lettres 44, 665-672 (1983).
[15] H. Brand and M.C. Cross, Phys. Rev. A27, 1237-1239 (1983).
[16] H. Brand, Prog. Theor. Phys. 71, 1096-1099 (1984).
[17] P.W. Anderson, Basic Notions of Codensed Matter Physics, Frontiers in Physics, Benjamin (1984).
[18] Y. Kuramoto, Prog. Theor. Phys. 71, 1182-1196 (1984).
[19] S. Chandrasekhar, Hydrodynamic and Hydromagnetic Stability, Clarendon Press (1961).

[20] J. Swift and P.C. Hohenberg, Phys. Rev. A15, 319-328 (1979).
[21] A.C. Newell and J.A. Whitehead, J. Fluid Mech. 38, 279-303 (1969).
[22] L.A. Segel, J. Fluid Mech. 38, 203-224 (1969).
[23] F. Busse, J. Fluid Mech. 52, 97-112 (1972).
[24] E.D. Siggia and A. Zippelius, Phys. Rev. Lett. 47, 835-838 (1981).
[25] R.M. Clever and F.H. Busse, J. Fluid. Mech. 65, 625-645 (1974).
[26] B.F. Edwards and A.L. Fetter, Phys. Fluids 27, 2795-2802 (1984).
[27] S. Fauve, Instabilités convectives et équations d'amplitude, Ecole de Goutelas, Observatoire de Meudon (1983).
[28] P. Coullet and S. Fauve, Phys. Rev. Lett. 55, 2857-2859 (1985).
[29] A. Schlüter, D. Lortz, and F. Busse, J. Fluid Mech. 23, 129-144 (1965).
[30] V.I. Arnold, Supplementary Chapters to the Theory of Ordinary differential Equations, Nauka (1978).
[31] J. Guckenheimer and P. Holmes, Nonlinear Oscillations, Dynamical System and Bifurcation of Vector Fields, Springer Verlag (1984).
[32] Note that in this model, the large scale horizontal velocity field should be of the form $Q(z)\ \psi(Y,T)$, where $Q(z)$ satisfies the boundary conditions.
[33] E.W. Bolton, F.H. Busse and R.M. Clever, Preprint (1985).
[34] F.B. Lipps, J. Fluid Mech. 75, 113-148 (1976).
[35] J.B. McLaughlin and S.A. Orszag, J. Fluid. Mech. 122 (1982).
[36] G.E. Willis and J.W. Deardorff, J. Fluid Mech. 44, 661-672 (1970).
[37] A.R. Bishop, J.A. Krumhansl and S.E. Trullinger, Physica 1D, 1-44 (1980).
[38] G.B. Stokes, Trans. Camb. Phil. Soc. 8,441, 197 (1847).
[39] T.B. Benjamin and J.E. Feir, J. Fluid Mech. 27, 417-430 (1967).
[40] J.T. Stuart and R.C. Di Prima, Proc. Roy. Soc. A362, 27-41 (1978).
[41] R.C. Di Prima and R.N. Grannick, IUTAM Symp. on Instabilities of Continuous Systems, H. Leipholtz ed. Springer Verlag (1971).
[42] C.S. Bretherton and E.A. Spiegel, Phys. Lett. 96A, 152-156 (1983).
[43] Y. Demay and G. Iooss, J. Mécanique (numéro spécial 1984).

PHYSICAL INTERPRETATION OF OPTICAL BIFURCATIONS

H. M. Nussenzveig
Departamento de Física
Pontifícia Universidade Católica
Cx. P. 38071, Rio de Janeiro, RJ
Brasil

ABSTRACT. It is shown that the bifurcations found in the standard model for optical bistability in a homogeneously broadened unidirectional ring laser, including the bistability threshold and switching points, self-pulsing and the Ikeda instability, can all be understood in terms of a common physical mechanism, the generation and amplification of sidebands by parametric processes. The bifurcation threshold follows from the usual self-oscillation condition in feedback systems. The basic nonlinear processes involved are the production of combination tones and phase-matched forward four-wave mixing. The relevant nonlinear constitutive parameters of the medium are determined. Results previously given in the literature are reviewed and extended to more general situations.

1. INTRODUCTION

It is now well known that a great variety of nonlinear dissipative systems undergo a sequence of bifurcations, possibly ending in chaos, when a control parameter is varied. Through the study of such systems, one is beginning to develop an extension of statistical mechanics to situations that are far from thermodynamic equilibrium. Bifurcations may be regarded as extensions of the concept of phase transitions in this analogy.

In most treatments of these problems, the powerful mathematical techniques of bifurcation theory and of dynamical systems theory are employed. This has the advantage of drawing attention to universal features, that do not depend on the specific nature of the system.

However, an important disadvantage of this approach is the loss of physical insight: one would like to know what physical properties are relevant, what is the actual mechanism involved. Physical insight into the mechanism responsible for bifurcations in a given system may be helpful in the search for answers to some basic yet unsolved questions: What determines the choice of a specific path to chaos? What is the role of quantum effects?

To investigate the physical mechanism and interpretation of bifurcations, we choose as model system a unidirectional homogeneously broad-

E. Tirapegui and D. Villarroel (eds.), Instabilities and Nonequilibrium Structures, 89–115.

ened passive ring laser, driven by near-resonant incident laser radiation. An introductory treatment of the bifurcations that take place in this system, including optical bistability, self-pulsing, the Ikeda instability and a period-doubling route to chaos, has been given in a previous lecture set [1], to which frequent reference will be made.

The ring laser model is relatively simple, rather well understood and not too unrealistic, in contrast with other models that have been investigated, e.g., in fluid dynamics. We discuss the physical interpretation of bifurcations in this model in terms of the nonlinear interactions among its constituents, namely, the laser field and a collection of two-level atoms. We show that the basic mechanism can be reduced to well-known effects in nonlinear oscillations and nonlinear optics.

The results are related to the traditional linear stability analysis around equilibrium [1], but the approach is different, and it may suggest new developments. The main ideas are due to Mollow [2,3], Sargent [4], Hillman et al [5], Firth et al [6], and Bar-Joseph and Silberberg [7].

2. GENERATION OF COMBINATION TONES BY TWO-LEVEL ATOMS

The generation of combination tones is a well-known phenomenon in nonlinear oscillations. If we consider, e.g., a weak quadratic nonlinearity and a two-frequency driving term,

$$\ddot{x} + \omega_0^2 x + \varepsilon x^2 = X_1\cos(\omega_1 t) + X_2\cos(\omega_2 t),$$

we find, with a trial solution of the unperturbed form $x = x_1\cos(\omega_1 t) + x_2\cos(\omega_2 t)$, that the quadratic perturbation generates (besides dc components and harmonics $2\omega_1$, $2\omega_2$) the combination tones $\omega_1 \pm \omega_2$.

The nonlinear systems in our ring laser model are the two-level atoms driven by the laser field. We therefore begin by discussing what happens to a single two-level atom driven by a strong, near-resonant coherent field [2].

Let H_A denote the atomic Hamiltonian and $|\pm\rangle$ the state vectors associated with the two levels, with

$$H_A|\pm\rangle = \pm\hbar\omega_0/2|\pm\rangle. \tag{2.1}$$

A pure state $|\psi\rangle$ may also be represented by the density operator $\rho = |\psi\rangle\langle\psi|$, with $\mathrm{Tr}\rho = 1$. For a driving field

$$\vec{E} = E\exp(-i\omega t)\vec{i}\,, \tag{2.2}$$

where, as usual, it is understood that one must take the real part, the total (semiclassical) Hamiltonian, in dipole approximation, is $H = H_A - e\vec{x}\cdot\vec{E}$.

We introduce the dimensionless variables

$$\rho_{+-} = p\exp(-i\omega t), \quad p = u + iv, \tag{2.3}$$

$$\frac{1}{2}(\rho_{++} - \rho_{--}) \equiv w, \tag{2.4}$$

where p is proportional to the atomic polarization and w to the atomic inversion, with equilibrium value $w_{eq} = -1/2$. Let

$$\mu = e\langle +|x|-\rangle \tag{2.5}$$

be the transition dipole moment (taken as real).

Near resonance, i.e., for $|\Delta\omega| \ll \omega$, where

$$\Delta\omega = \omega - \omega_0 \tag{2.6}$$

is the atomic detuning, we may employ the rotating-wave approximation, and Schrödinger's equation yields [1]

$$\dot{p} = i\Delta\omega p - \frac{i\mu}{\hbar} wE, \tag{2.7}$$

$$\dot{w} = \frac{i\mu}{2\hbar}(p^*E - pE^*). \tag{2.8}$$

These equations can be rewritten as a vector precession equation,

$$\dot{\vec{s}} = \vec{\Omega} \times \vec{s}, \tag{2.9}$$

where

$$\vec{s} \equiv (u,v,w), \qquad \vec{s}^{\,2} = 1/4 \tag{2.10}$$

$$\vec{\Omega} \equiv (\Omega_{or}, \Omega_{oi}, \Delta\omega) \tag{2.11}$$

are called "pseudospin" and "pseudofield", respectively. Here,

$$\Omega_{or} + i\Omega_{oi} \equiv \mu E/\hbar \tag{2.12}$$

and

$$\Omega_0 = \mu|E|/\hbar \tag{2.13}$$

is the resonant Rabi frequency.

A medium formed by identical two-level atoms with number density N acts as a source for the electromagnetic field through the medium polarization

$$P = 2N\mu p. \tag{2.14}$$

As a parenthetical remark to provide insight into (2.9), let us consider for a moment a quite different situation, in which E is not the driving field, but rather the field generated by the spontaneous decay - after some initial excitation - of the atoms in a resonant ($\Delta\omega = 0$) point-like sample, where all atoms see the same field. Maxwell's equations then yield

$$\dot{E} = 2\pi i\omega P. \tag{2.15}$$

Taking the initial value of E to be real, the solution evolves in the $u = 0$ plane, where we can set

$$v = \frac{1}{2}\sin\theta(t), \qquad w = -\frac{1}{2}\cos\theta(t), \tag{2.16}$$

and (2.9) shows that $\dot{\theta} = \Omega_0 = \mu E/\hbar$, so that (2.14) and (2.15) yield

$$\ddot{\theta} + (2\pi N\mu^2\omega/\hbar)\sin\theta = 0, \tag{2.17}$$

the equation of motion of a well-known nonlinear oscillatory system, the simple pendulum. The pseudospin is also called the Bloch pendulum vector; the angle θ measures the degree of excitation of the two-level atoms.

Going back to the model, we obtain the Bloch equations [1] by adding phenomenological damping terms to (2.7) - (2.8):

$$\dot{p} = i\Delta\omega p - \frac{i\mu}{\hbar} wE - \gamma_\perp p, \tag{2.18}$$

$$\dot{w} = \frac{i\mu}{2\hbar}(p^*E - pE^*) - \gamma_\parallel (w + \frac{1}{2}), \tag{2.19}$$

where $\gamma_\perp$ and $\gamma_\parallel$ are the transverse and longitudinal widths, respectively, with $\gamma_\perp \geq \gamma_\parallel/2$. Only homogeneous broadening is considered.

If the driving field is weak, so that $\Omega_0 \ll \gamma_\parallel, \gamma_\perp$, the Bloch pendulum gets damped back towards equilibrium before it can swing far out, so that it performs small oscillations around $\theta = 0$, corresponding to the linear regime. A strong driving field, capable of producing significant nonlinear effects, must satisfy the condition

$$\Omega_0 \gg \gamma_\parallel, \gamma_\perp. \tag{2.20}$$

The <u>stationary solution</u> of the Bloch equations is obtained by setting $\dot{p} = \dot{w} = 0$, which yields

$$\bar{w} = -\frac{1/2}{1+\dfrac{\Omega_0^2}{\gamma_\parallel \gamma_\perp (1+\Delta^2)}}, \quad \bar{p} = -\frac{i\mu\bar{w}E}{\hbar\gamma_\perp(1+i\Delta)}, \tag{2.21}$$

where Δ is the dimensionless atomic detuning,

$$\Delta \equiv -\Delta\omega/\gamma_\perp = (\omega_0 - \omega)/\gamma_\perp . \tag{2.22}$$

Note that inversion never takes place: in the saturation limit of a very strong field ($\Omega_0 \to \infty$), $\bar{w}$ approaches zero from below (equal population in upper and lower level).

The transient solution of the Bloch equations for a constant driving field E and for given initial values $p(0)$, $w(0)$ can also be found [2,8], by applying the Laplace transform; e.g., for $p(t)$, one gets

$$p(t) = g_{pp}(t)p(0) + g_{pp^*}(t)p^*(0) + g_{pw}(t)w(0), \tag{2.23}$$

where the Laplace transforms of the response functions,

$$G_{ij}(\lambda) = \int_0^\infty g_{ij}(t)\exp(-\lambda t)dt, \quad (i,j = p,p^*,w), \tag{2.24}$$

contain a common denominator

$$f(\lambda) = (\lambda+\gamma_\parallel)(\lambda+\xi)(\lambda+\xi^*) + \Omega_0^2(\lambda+\gamma_\perp), \tag{2.25}$$

with

$$\xi \equiv \gamma_\perp(1 - i\Delta). \tag{2.26}$$

The poles of the response function (zeros of $f(\lambda)$) define the atomic natural modes [1]. In the strong-field regime (2.20), we find from (2.25) that the poles are approximately given by

$$\lambda_0 \approx -\gamma_\perp, \quad \lambda_\pm \approx -\frac{1}{2}(\gamma_\parallel + \gamma_\perp) \pm i\Omega_0. \tag{2.27}$$

The transient response, which describes the decay of deviations from the stationary solution (2.21), is related, through the regression theorem [9], with the fluorescence spectrum of the strongly driven atoms. The roots $\lambda_\pm$ lead to the well-known [10] satellite peaks, shifted by $\pm\Omega_0$ (AC Stark shift).

After this preparatory discussion, let us consider an atom, driven by a strong field of the form (2.2), which is in the stationary state (2.21), and let us perturb it by a weak signal field at a neighboring frequency $\omega+\nu$ $(|\nu| \ll \omega)$,

$$E\,\exp(-i\omega t) \to [E + E'\exp(-i\nu t)]\exp(-i\omega t), \quad |E'| \ll |E|. \tag{2.28}$$

Because of the nonlinearity, the atom responds by generating also a combination tone at the frequency

$$\omega - \nu = 2\omega - (\omega + \nu), \tag{2.29}$$

so that the perturbed values of p and w are of the form

$$\bar{p} \to (\bar{p} + \delta p_0) + \delta p_+ \exp(-i\nu t) + (\delta p_-)^* \exp(i\nu t), \tag{2.30}$$

$$\bar{w} \to (\bar{w} + \delta w_0) + \delta w\,\exp(-i\nu t) + (\delta w)^* \exp(i\nu t), \tag{2.31}$$

where δp_0, $\delta p_\pm$, δw_0 and $\delta w_\pm$ are obtained by substituting (2.28), (2.30) and (2.31) back into the Bloch equations (2.18) - (2.19) and taking into account that $\bar{p}$, $\bar{w}$ are the stationary solutions.

The result [2] for $\delta p_\pm$, to lowest order in E', is

$$\delta p_+ = -\frac{i\mu\bar{w}}{\hbar f(-i\nu)}\left[(\gamma_{\parallel} - i\nu)(\xi - i\nu) + \frac{i\Omega_0^2\nu}{2\xi}\right]E', \tag{2.32}$$

$$\delta p_- = -\frac{i\mu\bar{w}}{\hbar f(-i\nu)}\,\frac{(\gamma_\perp - \frac{i\nu}{2})}{\xi}\left(\frac{\mu E^*}{\hbar}\right)^2 E', \tag{2.33}$$

where $f(\lambda)$ is defined by (2.25).

According to (2.1) and (2.4),

$$< H_A > = \hbar\omega_0 w, \tag{2.34}$$

so that (2.19) can be interpreted as an energy balance relation. The derivation that leads to (2.32) - (2.33) also yields

$$\hbar\omega_0\gamma_{\parallel}\,\delta w_0 = \delta W_0 + W', \tag{2.35}$$

where (Im denoting the imaginary part)

$$\delta W_0 = \mu\omega_0 \mathrm{Im}(E^*\delta p_0), \tag{2.36}$$

$$W' = \mu\omega_0 \mathrm{Im}(E'^*\delta p_+). \tag{2.37}$$

The l.h.s. of (2.35) is the decrease in the rate of atomic energy dissipation as a consequence of the perturbation E'; δW_0 is the corresponding increase in the rate of energy absorption from the strong field E, and W' is the rate of energy absorption from the signal field E'.

At exact atomic resonance $(\Delta = 0)$, in the strong-field regime (2.20), we get from (2.32) and (2.37)

$$W'(\nu) \approx -\frac{\hbar\omega_0\bar{w}}{2\gamma_\perp}\left(\frac{\mu|E'|}{\hbar}\right)^2\Omega_0^2 \frac{(\nu^2-\Omega_0^2)(\nu^2-2\gamma_\parallel\gamma_\perp^3/\Omega_0^2)}{(\nu^2+\gamma_\perp^2)[(\nu^2-\Omega_0^2)^2+(\gamma_\parallel+\gamma_\perp)^2\nu^2]} \quad . \quad (2.38)$$

The absorption line-shape function $W'(\nu)$ is plotted in Fig.1 for a specific choice of the parameters.

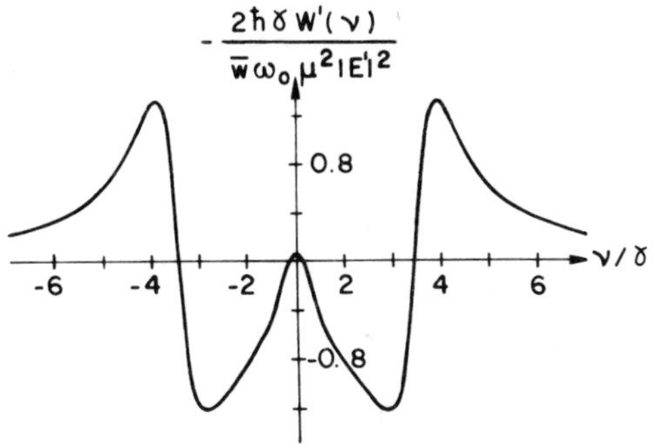

Figure 1. Plot of the absorption line shape (2.38) for $\gamma_\parallel = \gamma_\perp = \gamma$ and $\Omega_0 = 5\gamma$ (after [2]).

According to (2.21), $\bar{w} < 0$, i.e., there is never any atomic inversion in the stationary state. In spite of this, (2.38) shows that $W'(\nu)$ becomes negative, representing stimulated emission, rather than absorption, for ν^2 in the interval

$$2\gamma_\parallel\gamma_\perp^3/\Omega_0^2 < \nu^2 < \Omega_0^2, \quad (2.39)$$

where the l.h.s. is $\ll \Omega_0^2$ (cf. (2.20)).

Therefore, if the signal frequency $\omega + \nu$ is such that ν falls within this range, i.e., within one Rabi frequency from resonance (except for a tiny interval around $\nu = 0$), the signal field is amplified, rather than attenuated, by the strongly driven atom.

The strong driving field E plays the role of a pump field. The effect of the signal field on the rate of energy absorption from the pump field can be computed from (2.36). Under the above conditions, one finds that [2]

$$\delta W_0 \approx -W', \quad (2.40)$$

so that the amplification of the signal field is the result of an energy transfer from the pump field, which undergoes increased attenuation.

These results have been experimentally verified by Wu et al [11]. They employed an atomic beam of Na that was simultaneously excited by a fixed-frequency driving field and by a weak tunable probe field. The evolution of the absorption line shape as the driving field strength increased was observed, ranging from a single absorption peak through the three-peaked AC Stark spectrum to the stimulated emission dips illustrated in Fig. 1.

3. THE MOLLOW INSTABILITY

We now consider a medium formed by N identical two-level atoms per unit volume, through which a strong coherent pump wave propagates [3], with electric field

$$\vec{E}(z,t) = E_0 \exp[i(kz - \omega t)]\vec{i}, \quad k = \omega/c, \tag{3.1}$$

where we have assumed N to be small enough, so that the refractive index ≈ 1.

In order to find out whether (3.1) is a stable solution, we perturb it by a weak (signal) field at a neighboring frequency $\omega + \nu$,

$$\vec{E}'(\vec{r},t) = E'(\vec{r})\exp[-i(\omega+\nu)t]\vec{i}, \quad |\nu| \ll \omega, \quad |E'| \ll |E|. \tag{3.2}$$

Comparing (3.1) and (3.2) with (2.2) and (2.28), we see that, in order to apply the results of Sect. 2, it suffices to make the substitutions: $E \to E_0 \exp(ikz)$, $E' \to E'(\vec{r})$.

Taking into account (2.3), (2.14), (2.32) and (2.33), we find that the medium polarization induced by (3.2) is given by

$$\vec{P}' = P'(\vec{r},t)\vec{i}, \tag{3.3}$$

$$P'(\vec{r},t) = \chi_+(\nu)E'(\vec{r})\exp[-i(\omega+\nu)t] + \chi_-^*(\nu)E'^*(\vec{r})\exp(2ikz)\exp[-i(\omega-\nu)t], \tag{3.4}$$

$$\chi_+(\nu) = -\frac{2iN\mu^2\overline{w}}{\hbar f(-i\nu)}\left[(\gamma_{\parallel} - i\nu)(\xi - i\nu) + \frac{i\Omega_0^2\nu}{2\xi}\right], \tag{3.5}$$

$$\chi_-(\nu) = -\frac{2iN\mu^2\overline{w}}{\hbar f(-i\nu)}\frac{(\gamma_\perp - i\nu/2)}{\xi}\left(\frac{\mu E_0^*}{\hbar}\right)^2, \tag{3.6}$$

where $\Omega_0 = \mu|E_0|/\hbar$.

Since the medium is rarefied, with refractive index ≈ 1, (3.2) implies

$$E'(\vec{r}) \approx E'\exp(i\vec{k}'\cdot\vec{r}), \quad |\vec{k}'| = (\omega+\nu)/c. \tag{3.7}$$

Replacing this in (3.4), we see that, in order for the second term to contain the correct propagation factor, we must have

$$\exp(-i\vec{k}'\cdot\vec{r} + 2ikz) = \exp[i(\omega - \nu)z/c], \tag{3.8}$$

which, in view of (3.7), is only possible if

$$\vec{k}'\cdot\vec{r} = |\vec{k}'|z = (\omega + \nu)z/c. \tag{3.9}$$

Thus, in order for the signal field at $\omega + \nu$, in the presence of the pump field at ω, to generate an "idler" field at $\omega - \nu$, all three fields must propagate in the same direction (phase matching condition). For such fields, travelling in the positive z direction, Maxwell's equations yield

$$\left(\frac{\partial}{\partial z} + \frac{1}{c}\frac{\partial}{\partial t}\right)E(z,t) = -\frac{2\pi}{c}\frac{\partial}{\partial t}P(z,t). \tag{3.10}$$

Let

$$\tilde{E}(z,\tilde{\omega}) \equiv \int_{-\infty}^{\infty} \exp\left[i\tilde{\omega}(t - \frac{z}{c})\right]E(z,t)dt, \tag{3.11}$$

with a similar definition for $\tilde{P}(z,\tilde{\omega})$. Then, for $\tilde{\omega}$ close to ω, (3.10) becomes

$$\frac{\partial\tilde{E}}{\partial z}(z,\tilde{\omega}) = 2\pi i\frac{\tilde{\omega}}{c}\tilde{P}(z,\tilde{\omega}) \approx 2\pi ik\tilde{P}(z,\tilde{\omega}). \tag{3.12}$$

From (3.4) - (3.9), we find that $\tilde{P}(z,\tilde{\omega})$ at $\tilde{\omega} = \omega + \nu$ arises from $\tilde{E}(z,\tilde{\omega})$ at $\tilde{\omega} = \omega + \nu$ (cf. (3.2)) as well as at $\tilde{\omega} = \omega - \nu$, namely,

$$\tilde{P}(z,\omega + \nu) = \chi_+(\nu)\tilde{E}(z,\omega + \nu) + \chi_-^*(-\nu)\tilde{E}^*(z,\omega - \nu), \tag{3.13}$$

where the second term represents the phase-matched combination tone.

Substituting (3.13) into (3.12) at $\tilde{\omega} = \omega + \nu$, and repeating this procedure at $\tilde{\omega} = \omega - \nu$, we find the coupled pair of equations

$$\frac{\partial}{\partial z}\tilde{E}(z,\omega + \nu) = a(\nu)\tilde{E}(z,\omega + \nu) + b^*(-\nu)\tilde{E}^*(z,\omega - \nu), \tag{3.14}$$

$$\frac{\partial}{\partial z}\tilde{E}^*(z,\omega - \nu) = b(\nu)\tilde{E}(z,\omega + \nu) + a^*(-\nu)\tilde{E}^*(z,\omega - \nu), \tag{3.15}$$

where

$$a(\nu) = 2\pi ik\chi_+(\nu), \quad b(\nu) = -2\pi ik\chi_-(\nu). \tag{3.16}$$

For the remainder of this Section, we specialize these results to an exactly resonant pump field, $\Delta = 0$, taking E_0 to be real in (3.1). Under these conditions,

$$a^*(-\nu) = a(\nu), \quad b^*(-\nu) = b(\nu), \tag{3.17}$$

and (3.14) - (3.15) are analogous to the equations [12] for traveling-wave parametric amplification, yielding the solution

$$\tilde{E}(z,\omega+\nu) = \exp(az)[\tilde{E}(0,\omega+\nu)\cosh(bz) + \tilde{E}^*(0,\omega-\nu)\sinh(bz)],$$

(3.18)

$$\tilde{E}^*(z,\omega-\nu) = \exp(az)[\tilde{E}(0,\omega+\nu)\sinh(bz) + \tilde{E}^*(0,\omega-\nu)\cosh(bz)],$$

which may be rewritten in terms of linear combinations of $\exp\left[(a\pm b)z\right]$.

From (3.16), (3.5) and (3.6) we find that

$$\mathrm{Re}[a(\nu) - b(\nu)] < 0 \quad \text{for all } \nu. \tag{3.19}$$

Therefore, in order to obtain amplification, we must have

$$\mathrm{Re}\left[a(\nu) + b(\nu)\right] > 0, \tag{3.20}$$

$$\tilde{E}(0,\omega+\nu) + \tilde{E}^*(0,\omega-\nu) \neq 0. \tag{3.21}$$

A familiar example of parametric amplification is a pendulum of periodically variable length [13], described by Mathieu's equation

$$\ddot{\theta} + \omega_0^2[1 + h\cos(2\omega t)]\,\theta = 0. \tag{3.22}$$

When $h \neq 0$, the oscillatory solution found for $h = 0$ becomes unstable, leading to an exponentially growing amplitude for a range of values of h around $\omega = \omega_0$. This is the well-known phenomenon of parametric resonance, illustrated by a child pumping on a swing. If the swing is given pushes in the right direction at the oscillation extrema (frequency $2\omega_0$), the amplitude keeps increasing. It can also be pushed every other time $(\omega = \omega_0/2)$, every third time $(\omega = \omega_0/3)$, etc., with less efficient results (subharmonic resonances).

In our ring laser model, the parametric interaction between signal and idler takes place through the pump field. Note that the last term in (3.4), according to (3.6), is proportional to E_0^2, which is associated with the requency 2ω. The parametric process involved is the generation of combination tones. Thus, according to Sect. 2, we expect to get amplification for $\nu^2 < \Omega_0^2$.

Indeed, in the strong-field regime (2.20), we find from (3.16), (3.5) and (3.6) that

$$\mathrm{Re}[a(\nu) + b(\nu)] \approx -\frac{4\pi k N \mu^2 \overline{w}\Omega_0^2(\Omega_0^2 - \nu^2)}{\hbar\gamma_\perp[(\Omega^2 - \nu^2)^2 + (\gamma_\parallel + \gamma_\perp)^2\nu^2]}, \tag{3.23}$$

so that (3.20) is equivalent to

$$-\Omega_0 < \nu < \Omega_0. \tag{3.24}$$

Within this interval, i.e., within one Rabi frequency from resonance, signal and idler reinforce each other through their parametric interaction with the pump, leading to exponential growth of both of them in (3.18). Note that the medium behaves like an amplifier even though it is not inverted ($\bar{w} < 0$). Actually, of course, the exponential growth will be checked by higher-order terms in E', not included in the above analysis (pump depletion has not been taken into account).

We conclude that a strong coherent field traveling through a resonant medium is subject to a frequency instability [3]: it tends to generate sidebands at symmetrically placed frequencies $\omega \pm \nu$, traveling in the same direction, in response to fluctuations within the interval (3.24). Initial fluctuations within this interval, arising from spontaneous emission by the strongly pumped atoms, are always present.

4. SIDEBAND GENERATION IN A KERR MEDIUM

The discussion of sideband generation following (3.16) in the previous Section was specialized to an exactly resonant pump field, $\Delta = 0$, corresponding to a purely absorptive medium (cf. (4.3) below). We now go over to the opposite extreme situation, by taking the dispersive limit

$$\Delta^2 >> 1 + (\Omega_0^2/\gamma_{\parallel}\gamma_{\perp}). \tag{4.1}$$

We also assume fast transverse relaxation, i.e.,

$$\gamma_{\perp} >> |\nu|. \tag{4.2}$$

The nonlinear complex refractive index of the Bloch medium [1]

$$n'(\omega) = 1 + \frac{2\pi N \mu^2 (\Delta + i)}{\hbar\gamma_{\perp}(1+\Delta^2+\Omega_0^2/\gamma_{\parallel}\gamma_{\perp})} \tag{4.3}$$

may be approximated, according to (4.1), by

$$n'(\omega) \approx n_0'(\omega) + n_2|E|^2, \tag{4.4}$$

where $n'(\omega)$ is the complex linear refractive index, given by

$$n_0'(\omega) - 1 = \frac{2\pi N \mu^2 (\Delta + i)}{\hbar\gamma_{\perp}(1 + \Delta^2)} = n_0(\omega) - 1 + i\,\frac{\alpha_0(\omega)}{2k}, \tag{4.5}$$

with n_0 representing the real index and α_0 the absorption coefficient, and

$$n_2(\omega) = -\frac{2\pi N \mu^4 \Delta}{\hbar^3 \gamma_{\parallel} \gamma_{\perp}^2 (1+\Delta^2)^2} \approx -\frac{2\pi N \mu^4}{\hbar^3 \gamma_{\parallel} \gamma_{\perp}^2 \Delta^3} . \tag{4.6}$$

Thus, the Bloch medium behaves like a Kerr medium, with quadratic coefficient of refractive index n_2.

It follows from (2.21) and (4.1) that

$$\bar{w} \approx -\frac{1}{2} + \frac{\Omega_0^2}{2\gamma_{\parallel} \gamma_{\perp} \Delta^2} , \tag{4.7}$$

so that the degree of atomic excitation is small. Also. from (2.25),

$$\frac{1}{f(-i\nu)} \approx \frac{1}{(\gamma_{\parallel} -i\nu)(\xi-i\nu)(\xi^*-i\nu)} - \frac{\Omega_0^2(\gamma_{\perp} -i\nu)}{(\gamma_{\parallel} -i\nu)^2(\xi-i\nu)^2(\xi^*-i\nu)^2} ; \tag{4.8}$$

Substituting these results in (3.5) and (3.6), and taking into account (4.2), we find

$$2\pi\chi_+(\nu) \approx n_0' - 1 + n_2\left(1 + \frac{1}{1-i\sigma}\right)|E_0|^2 , \tag{4.9}$$

$$2\pi\chi_-(\nu) \approx \frac{n_2}{(1-i\sigma)} (E_0^*)^2 , \tag{4.10}$$

where n_0' and n_2 are given by (4.5) and (4.6), respectively, and

$$\sigma \equiv \nu/\gamma_{\parallel} . \tag{4.11}$$

Omitting the z-dependence in (3.13), we see from (4.9) and (4.10) that it may be rewritten, under the present conditions, as

$$\tilde{P}(\omega+\nu) = \chi_0'(\omega+\nu)\tilde{E}(\omega+\nu) + \chi^{(3)}(-\omega-\nu; -\omega,\omega,\omega+\nu)E_0^*(\omega)E_0(\omega)\tilde{E}(\omega+\nu)$$

$$+ \chi^{(3)}(-\omega-\nu;\omega,\omega,-\omega+\nu)E_0(\omega)E_0(\omega)\tilde{E}^*(\omega-\nu) , \tag{4.12}$$

where $\chi_0'(\omega) = [n_0'(\omega) - 1]/2\pi$ is the linear susceptibility, and

$$2\pi\chi^{(3)}(-\omega-\nu;-\omega,\omega,\omega+\nu) = n_2 + \frac{n_2}{1-i\sigma} , \tag{4.13}$$

$$2\pi\chi^{(3)}(-\omega-\nu;\omega,\omega,-\omega+\nu) = \frac{n_2}{1-i\sigma} . \tag{4.14}$$

The nonlinear third-order susceptibility tensor $\chi^{(3)}_{ijk\ell}(-\omega_4;\omega_1,\omega_2,\omega_3)$ is defined, in general, by

$$P_i^{(3)}(\omega_4) = \chi^{(3)}_{ijk\ell}(-\omega_4;\omega_1,\omega_2,\omega_3)E_j(\omega_1)E_k(\omega_2)E_\ell(\omega_3), \tag{4.15}$$

where $\omega_4 = \omega_1 + \omega_2 + \omega_3$ and $E_j(-\omega) = E_j^*(\omega)$. In the present case, since the polarizations coincide, all indices are equal, so that they have been omitted.

We see, therefore, that (4.12) contains, besides the linear susceptibility at $\omega + \nu$, the lowest-order nonlinear optical processes that contribute in a Kerr medium, namely phase-matched forward four-wave mixing processes: (4.13) represents a pump-induced modulation of the refractive index at $\omega + \nu$, whereas (4.14) is associated with a process of the stimulated Rayleigh scattering type [14]. While (3.5) - (3.6) are valid to all orders in the pump field strength, (4.12) results from an expansion in which only these lowest-order nonlinear contributions are kept.

One can give an equivalent interpretation of (4.12) in terms of a time-dependent nonlinear refractive index n_{NL}, slowly-varying in the scale of ω^{-1}, which obeys a Debye relaxation equation [7]

$$\frac{1}{\gamma_{||}}\frac{dn_{NL}}{dt} + n_{NL} = n_2 < E^2(t) > , \tag{4.16}$$

where E is the real field and the angular brackets denote a time average over one or several periods $2\pi/\omega$, so that the r.h.s. is a slowly-varying intensity envelope. The solution of (4.16) is

$$n_{NL}(t) = \gamma_{||} n_2 \int_{-\infty}^{t} \exp[\gamma_{||}(t' - t)] < E^2(t') > dt'. \tag{4.17}$$

In the present case,

$$E = \mathrm{Re}\{E_0(\omega)\exp(-i\omega\tau) + \tilde{E}(\omega+\nu)\exp[-i(\omega+\nu)\tau] + $$
$$+ \tilde{E}(\omega-\nu)\exp[-i(\omega-\nu)\tau]\}, \qquad \tau \equiv t - z/c. \tag{4.18}$$

Substituting this in (4.17) and remembering that $|\tilde{E}(\omega+\nu)| << |E_0(\omega)|$, we get

$$n_{NL} = n_2|E_0(\omega)|^2 + \frac{n_2}{(1-i\sigma)}[E_0^*(\omega)\tilde{E}(\omega+\nu) + E_0(\omega)\tilde{E}^*(\omega-\nu)]\exp(-i\nu\tau)$$
$$+ \frac{n_2}{(1+i\sigma)}[E_0(\omega)\tilde{E}^*(\omega+\nu) + E_0^*(\omega)\tilde{E}(\omega-\nu)]\exp(i\nu\tau). \tag{4.19}$$

The first term represents a uniform bias of refractive index produced by the pump field. The other terms may be called "holographic", because they arise from a phase grating resulting from interference between the pump and the sidebands.

The nonlinear contribution to the polarization is

$$P_{NL} = n_{NL}E(\tau)/2\pi, \tag{4.20}$$

where $E(\tau)$ is the expression within curly brackets in (4.18). Replacing n_{NL} by (4.19) and taking the Fourier component of frequency $\omega+\nu$, we recover the nonlinear terms in (4.12).

Thus, in (4.13), the first term is a uniform bias term, whereas the second one, as well as (4.14), are holographic terms. The phase shift in these terms is a typical relaxation effect [14, 15]. In this interpretation, the parametric interaction is mediated by the time-dependent nonlinear refractive index. The parametric process described by (4.14) may be thought of as transfer from one sideband to the other through diffraction by the holographic grating.

Substituting (4.9) and (4.10) into (3.14) - (3.16) and reestablishing the z-dependence, we find

$$\frac{\partial}{\partial z}\tilde{E}(z,\omega+\nu) = ik\{[(n_0'-1) + n_2|E_0(z,\omega)|^2]\tilde{E}(z,\omega+\nu)$$

$$+ \frac{n_2}{(1-i\sigma)}[|E_0(z,\omega)|^2\tilde{E}(z,\omega+\nu) + E_0^2(z,\omega)\tilde{E}^*(z,\omega-\nu)]\}, \tag{4.21}$$

$$\frac{\partial}{\partial z}\tilde{E}(z,\omega-\nu) = ik\{[(n_0'-1) + n_2|E_0(z,\omega)|^2]\tilde{E}(z,\omega-\nu)$$

$$+ \frac{n_2}{(1+i\sigma)}[|E_0(z,\omega)|^2\tilde{E}(z,\omega-\nu) + E_0^2(z,\omega)\tilde{E}^*(z,\omega+\nu)]\}. \tag{4.22}$$

Similarly, for the pump field, we find

$$\frac{\partial}{\partial z}E_0(z,\omega) = ik[(n_0'-1) + n_2|E_0(z,\omega)|^2]E_0(z,\omega), \tag{4.23}$$

neglecting terms of relative order $|\tilde{E}(z,\omega\pm\nu)/E_0(z,\omega)|^2$. This amounts to neglecting pump depletion in the parametric process. Setting

$$E_0(z,\omega) = \exp[i(n_0'-1)kz]\rho\exp(i\phi), \tag{4.24}$$

we find from (4.23) that

$$\partial\rho/\partial z = 0, \quad \partial\phi/\partial z = n_2k\rho^2\exp(-\alpha_0 z), \tag{4.25}$$

so that the solution of (4.23) is

$$E_0(z,\omega) = E_0(0,\omega)\exp[ik(n_0'-1)z + in_2kh^2(z)E_0^2(0,\omega)], \quad (4.26)$$

where we have assumed $E_0(0,\omega)$ to be real, for simplicity, and

$$h(z) \equiv \{[1 - \exp(-\alpha_0 z)]/\alpha_0\}^{1/2}. \quad (4.27)$$

Setting

$$\tilde{E}(z,\omega\pm\nu) \equiv \exp\{ik[(n_0'-1)z + n_2E_0^2(0,\omega)h^2(z)]\}a_\pm(z), \quad (4.28)$$

we get, from (4.21) - (4.22) and (4.26),

$$\frac{\partial a_+}{\partial z} = i\,\frac{n_2k}{(1-i\sigma)}\,E_0^2(0,\omega)\,[a_+(z) + a_-^*(z)]\exp(-\alpha_0 z), \quad (4.29)$$

$$\frac{\partial a_-}{\partial z} = i\,\frac{n_2k}{(1+i\sigma)}\,E_0^2(0,\omega)\,[a_+^*(z) + a_-(z)]\exp(-\alpha_0 z). \quad (4.30)$$

It follows that

$$a_+(z) + a_-^*(z) = a_+(0) + a_-^*(0) = \tilde{E}(0,\omega+\nu) + \tilde{E}^*(0,\omega-\nu) = \text{const.}, \quad (4.31)$$

so that the solution of (4.29) - (4.30) is

$$a_+(z) = a_+(0) + \frac{in_2k}{(1-i\sigma)}\,E_0^2(0,\omega)\,[a_+(0) + a_-^*(0)]h^2(z) \quad (4.32)$$

$$a_-(z) = a_-(0) + \frac{in_2k}{(1+i\sigma)}\,E_0^2(0,\omega)\,[a_+^*(0) + a_-(0)]h^2(z). \quad (4.33)$$

For propagation over distances z much smaller than the absorption length α_0^{-1}, we have

$$h^2(z) \approx z \qquad (\alpha_0 z \ll 1), \quad (4.34)$$

and (4.32) - (4.33) reduce to the results of Bar-Joseph and Silberberg [7]. In this situation, the amplitudes vary linearly with the distance, rather than exponentially, as in (3.18).

The relative gain $G_\pm(z) = |a_\pm(z)/a_\pm(0)|^2$ for the signal or idler depends on their initial phase relationship. Taking the initial amplitudes to be the same, with

$$a_-(0) = a_+(0)\exp(i\psi), \quad (4.35)$$

one finds from (4.32) - (4.34) that $G_+(z) = 1$ when $\psi = \pi$. Indeed, in this case, we see from (4.18) that the sidebands produce only frequency modulation, so that $\langle E^2(t) \rangle$ = constant in (4.16), corresponding just to a refractive index bias, with no holographic coupling between the sidebands (cf. also (3.21)). On the other hand, if $\psi = 0$, we have amplitude modulation, leading to amplification of one sideband for $\sigma > 0$ and of the other one for $\sigma < 0$, with peaks at $\sigma = \nu/\gamma_{\parallel} = \pm 1$. For intermediate values of ψ, the peak amplification shifts between $\sigma = 0$ and $\sigma = \pm 1$ [7].

Four-wave parametric interactions of this or of similar type have been theoretically discussed by several authors [5 - 7,16], and experimental observations have been reported [17]. The results can also be interpreted in terms of population pulsations [4].

5. APPLICATION TO OPTICAL BIFURCATIONS

We now apply the above results to the sequence of optical bifurcations that take place in a homogeneously broadened unidirectional ring laser, driven by a near-resonant laser beam, as the incident beam intensity is gradually raised [1]. The system is shown in Fig. 2, where mirrors 1 and 2 have reflectivity R and transmissivity T(R + T = 1), and mirrors 3

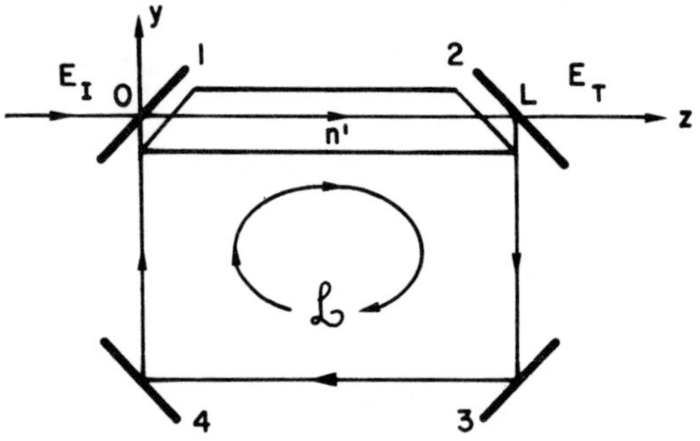

Figure 2. The ring cavity.

and 4 are totally reflecting. We consider only a "good" cavity, for which $T \ll 1$. The sample length is L and the cavity roundtrip length is $\mathcal{L} = L + \ell$. Neglecting transverse effects, the electric field along the z axis is of the form

$$\vec{E} = E(z,t)\exp[i(kz - \omega t)]\vec{i}, \tag{5.1}$$

where E(z,t) is the slowly-varying envelope amplitude. The corresponding incident and transmitted amplitudes at the entrance and exit of the sample are $E_I(t)$ and $E_T(t)$, respectively.

The system is described by the Maxwell-Bloch equations, subject to the boundary conditions [1]

$$E_T(t) = \sqrt{T}\, E(L,t), \tag{5.2}$$

$$E(0,t) = \sqrt{T}\, E_I(t) + R \exp(ik\mathcal{L})E(L,t - \tfrac{\ell}{c}). \tag{5.3}$$

In the absence of the sample, the cavity behaves like a Fabry-Perot, with complex natural mode [1] wave numbers (poles of the transfer function E_T/E_I)

$$\tilde{k}_m \approx (2m\pi/\mathcal{L}) - i/c\tau_c = k_m - i\gamma_c/c \tag{5.4}$$

where m is an integer (mode number), and

$$\tau_c = t_c/T \gg t_c \equiv \mathcal{L}/c \tag{5.5}$$

is the mode lifetime, much greater than the cavity roundtrip time t_c. For a "good" cavity, we also have $\tau_c^{-1} = \gamma_c \ll \gamma_{\parallel}, \gamma_{\perp}$. The Fabry-Perot transmissivity shows very sharp transmission resonances peaked around $k = k_m$, with great sensitivity to the roundtrip phase shift near the peaks.

5.1. Bifurcations in Purely Absorptive Optical Bistability

We begin with the purely absorptive case, in which the frequency ω of the driving field is tuned both to atomic resonance and to cavity resonance with a given mode m:

$$\Delta = 0, \quad k = \omega/c = k_m, \tag{5.6}$$

so that $\exp(ik\mathcal{L}) = 1$ in (5.3).

The simplest situation corresponds to the "mean field" regime, in which not only $T \ll 1$, but also $\alpha_0 L \ll 1$ (where α_0 is the linear absorption coefficient of the active medium), with a given value of the <u>cooperation parameter</u> [1]

$$C = \alpha_0 L/4T. \tag{5.7}$$

Under these conditions, one may employ a single-mode approximation for the steady-state field, replacing position-dependent quantities F(z,t) by their "mean-field approximation"

$$F(z,t) \to F(t) \equiv \langle F(z,t) \rangle \equiv \frac{1}{L}\int_0^L F(z,t)\,dz. \tag{5.8}$$

5.1.1. First Bifurcation: Optical Bistability Threshold. Since $\nu = 0$ falls within the range (3.24) of the Mollow instability, we expect to find differential gain (small-signal amplification) for sufficiently strong incident fields. This optical transistor effect indeed takes place [1] and is a precursor of optical bistability.

The combination of gain with the feedback loop condition (5.3) may be expected to lead to self-oscillation. When this happens for fluctuations at $\nu = 0$, we have a marginal (neutral) mode; the presence of such a zero eigenvalue signals a bifurcation point [18].

With $\nu = 0$ and $\tilde{E}(0,\omega) = \tilde{E}^*(0,\omega)$, we get from (3.18)

$$\tilde{E}(L,\omega) = \exp\{[a(0) + b(0)]L\}\tilde{E}(0,\omega). \tag{5.9}$$

Comparing (4.18) with (5.1), we note that E_0 represents the steady-state solution (satisfying the boundary conditions), which is perturbed by the small signal $\tilde{E}$. Therefore, taking into account (5.6), the boundary condition (5.3) becomes

$$\tilde{E}(0,\omega) = R\exp(ik\mathcal{L})\tilde{E}(L,\omega) = (1 - T)\tilde{E}(L,\omega). \tag{5.10}$$

Since $T \ll 1$, we have $1 - T \approx \exp(-T)$, so that (5.9) and (5.10) yield

$$1 = R\exp(ik\mathcal{L})\exp\{[a(0) + b(0)]L\} \approx \exp\{[a(0) + b(0)]L - T\}. \tag{5.11}$$

This is the well-known self-oscillation condition that the overall gain around the feedback loop must be equal to unity. The loss due to transmission through the mirrors is offset by the nonlinear parametric amplification gain.

From (2.21), (3.5), (3.6) and (3.16), we get

$$a(0) + b(0) = -\frac{2\pi k\mathcal{N}\mu^2}{\hbar\gamma_\perp}\frac{(1 - x^2)}{(1 + x^2)^2}, \tag{5.12}$$

where

$$x \equiv \Omega_0/\sqrt{\gamma_\parallel\gamma_\perp} = \mu\langle\bar{E}\rangle/\hbar\sqrt{\gamma_\parallel\gamma_\perp} \tag{5.13}$$

and $\langle\bar{E}\rangle$ is the steady-state mean field (5.8) in the sample.

Substituting (5.12) into (5.11), and taking into account (4.5) and (5.7), we get

$$T + \frac{\alpha_0 L}{2}\frac{(1 - x^2)}{(1 + x^2)^2} = T\left[1 + \frac{2C(1 - x^2)}{(1 + x^2)^2}\right] = 0. \tag{5.14}$$

Solving for x^2, we find the roots (cf. [1])

$$x_\pm^2 = C - 1 \pm \sqrt{C(C - 4)}. \tag{5.15}$$

Therefore, the threshold for bistability, where $x_+ \equiv x_-$ (analogue of a critical point) is [1]

$$C = 4. \tag{5.16}$$

5.1.2. Second Bifurcation: Switching Points. For $C > 4$, equation (5.15) yields the switching points in the bistable steady-state response curve, where a stable and an unstable branch bifurcate. At these points, the system must jump to another stable branch (analogue of a first order phase transition). The zero eigenvalue leads to critical slowing-down [1].

5.1.3. Third Bifurcation: Self-Pulsing. We now discuss bifurcations with $\nu \neq 0$, in which, besides the resonant mode (5.6), two side modes $\omega \pm \nu$ are excited [1,19]. Applying the boundary condition (5.3), where $\tilde{E}(L,\omega \pm \nu)$ is given by (3.18), and taking (4.18) into account, we arrive at the self oscillation conditions

$$\tilde{E}(0,\omega+\nu) = R \exp[a(\nu)L + i\nu t_c]\{\cosh[b(\nu)L]\tilde{E}(0,\omega+\nu)$$

$$+ \sinh[b(\nu)L]\tilde{E}^*(0,\omega-\nu)\}, \tag{5.17}$$

$$\tilde{E}^*(0,\omega-\nu) = R \exp[a(\nu)L + i\nu t_c]\{\sinh[b(\nu)L]\tilde{E}(0,\omega+\nu)$$

$$+ \cosh[b(\nu)L]\tilde{E}^*(0,\omega-\nu)\}. \tag{5.18}$$

In order for these equations to have a nontrivial solution, the determinant of the system must vanish. Taking into account (3.19) and (3.21), this requires $\tilde{E}(0,\omega-\nu) = \tilde{E}^*(0,\omega+\nu)$ (real field) and

$$R \exp\{[a(\nu) + b(\nu)]L + i\nu t_c\} = 1, \tag{5.19}$$

which is the generalization of (5.11) and has the same simple physical interpretation.

Equating modulus and phase of both sides of (5.19), we get

$$R \exp\{\mathrm{Re}[a(\nu) + b(\nu)]L\} = 1 \tag{5.20}$$

and

$$\mathrm{Im}[a(\nu) + b(\nu)]L + \nu t_c \equiv 0 \pmod{2\pi}. \tag{5.21}$$

Taking into account (3.5), (3.6) and (3.16), we find that (5.20) is equivalent to

$$1+\frac{2C\gamma_\perp\{\gamma_\parallel(1-x^2)[\gamma_\parallel\gamma_\perp(1+x^2)-\nu^2]+\nu^2(\gamma_\parallel+\gamma_\perp)\}}{(1+x^2)\{[\gamma_\parallel\gamma_\perp(1+x^2)-\nu^2]^2+\nu^2(\gamma_\parallel+\gamma_\perp)^2\}}=0, \qquad (5.22)$$

which is indeed the condition defining the self-pulsing instability threshold (cf. [1], eqs. (8.37) - (8.38) and Sect. 13.1). Note that it reduces to (5.14) for $\nu=0$.

While (5.20) may be compared with the threshold condition in semi-classical laser theory [20], (5.21) plays the role of the frequency-determining equation. In a weak-field approximation $x^2 \ll 1$, where x is defined by (5.13), we find from (3.5), (3.6) and (3.16) that

$$\mathrm{Im}[a(\nu)+b(\nu)] \approx -\frac{2\pi N\mu^2 k\nu}{\hbar(\gamma_\perp^2+\nu^2)} = k[n_0(\omega+\nu)-1], \qquad (5.23)$$

where n_0 is the linear real refractive index of the Bloch medium, given by (4.5). Thus, (5.21), in this approximation, becomes

$$\frac{\nu}{c}\mathcal{L} + [n_0(\omega+\nu)-1]kL \equiv 0 \quad (\mathrm{mod}\ 2\pi), \qquad (5.24)$$

which, together with (5.6), is equivalent to the requirement that $\omega \pm \nu$ are the frequencies of cavity modes, taking into account that these frequencies are modified by the presence of the active medium (mode pulling). While (5.24) accounts for linear pulling effects, the true frequency of self-oscillation must be obtained from (5.21), which includes nonlinear effects, and it is not just a multiple of $2\pi/t_c$, although it is close to one for a sufficiently rarefied medium.

The above results may be applied to the adjacent side modes $\nu \approx \nu_1 = 2\pi/t_c$ as well as to higher-order side modes $\nu_n \approx n\nu_1$. For $\nu \approx \nu_1$, we get self-pulsing with period t_c, arising from beats between the pump mode m and modes $m \pm 1$.

One usually has $x^2 \gg 1$ in the self-pulsing range, so that the approximation (3.23) may be employed. Thus, the Mollow instability condition (3.24) applies for $\nu \approx \nu_1$, yielding instability boundaries in the (ν_1, x) plane. For a more detailed discussion, we refer to [1].

5.2. Bifurcations in the Dispersive Limit

We now go over from purely absorptive optical bistability to the dispersive limit (4.1), in which the Kerr medium approximation holds. We also assume fast transverse relaxation (cf. (4.2)), and we take propagation effects within the sample into account, so that the mean-field approximation (5.8) no longer holds. We further assume that the quadratic coefficient of refractive index n_2 gives rise to a large ($\gg 1$) nonlinear contribution to the roundtrip phase shift.

Under these circumstances, the stationary solution of the Maxwell-

Debye equations leads to a multistable response curve [21], that arises from scanning through several Fabry-Perot resonances through the variation of the nonlinear phase shift with the incident intensity [1]. However, linear instability analysis reveals that not only branches of the response curve with negative slope, but also large portions of branches with positive slope are unstable (Ikeda instability).

It was found by Ikeda [21], through numerical integration of the Maxwell-Debye equations, that, as one crosses the instability threshold on a positive-slope branch, there occurs a bifurcation to a self-pulsing solution with period $2t_c$. As one increases the incident intensity, this is followed by a sequence of period-doubling bifurcations (periods $4t_c$, $8t_c$,...) that lead to a chaotic regime. We now discuss the physical interpretation of these new bifurcations that take place within the Ikeda instability domain [6,7].

5.2.1. Fourth Bifurcation: The Ikeda Instability. To obtain the stationary solution in the present situation, it suffices to apply boundary condition (5.3), in the form

$$E_0(0,\omega) = \sqrt{T}\, E_I + R \exp(ik\mathcal{L})E_0(L,\omega), \tag{5.25}$$

and to get $E_0(L,\omega)$ from (4.26). Taking $n_2 > 0$, for definiteness, and introducing the notations [1]

$$\sqrt{n_2 k}\; h(L)E_0(0,\omega) \equiv \mathcal{E}\quad , \tag{5.26}$$

$$\sqrt{n_2 k}\; h(L)\sqrt{T}\, E_I = \sqrt{T}\;\; \mathcal{E}_I \equiv A, \tag{5.27}$$

$$R \exp(-\alpha_0 L/2) \equiv B, \tag{5.28}$$

$$k\mathcal{L} + (n_0 - 1)kL \equiv \delta_0 \pmod{2\pi} \tag{5.29}$$

where δ_0 is the linear roundtrip phase shift, we find

$$\mathcal{E} = A + B\mathcal{E}\exp[i(\delta_0 + |\mathcal{E}|^2)], \tag{5.30}$$

so that $|\mathcal{E}|^2$ is the nonlinear contribution to the phase shift. From (5.30) we get the stationary state equation

$$|\mathcal{E}|^2[1 - 2B\cos(\delta_0 + |\mathcal{E}|^2) + B^2] = A^2, \tag{5.31}$$

which leads to the multistable features mentioned above [1,21].

For the sidebands $\tilde{E}(z,\omega \pm \nu)$, boundary condition (5.3) becomes

$$\tilde{E}(0,\omega\pm\nu) = R \exp[i(k \pm \tfrac{\nu}{c})\mathcal{L}]\tilde{E}(L,\omega\pm\nu)$$

$$= R\exp(ik\mathcal{L} \pm i\nu t_c)\tilde{E}(L,\omega\pm\nu), \tag{5.32}$$

where $\tilde{E}(L,\omega \pm \nu)$ is obtained from (4.28), (4.32) and (4.33). In terms of the above notations, this leads to the coupled pair of conditions

$$\tilde{E}(0,\omega+\nu) = B \exp[i(\delta_0+|\mathcal{E}|^2+\nu t_c)]\{\tilde{E}(0,\omega+\nu) + \\ + \frac{i|\mathcal{E}|^2}{(1-i\sigma)}[\tilde{E}(0,\omega+\nu) + \tilde{E}^*(0,\omega-\nu)]\}, \quad (5.33)$$

$$\tilde{E}^*(0,\omega-\nu) = B \exp[-i(\delta_0+|\mathcal{E}|^2-\nu t_c)]\{\tilde{E}^*(0,\omega-\nu) - \\ - \frac{i|\mathcal{E}|^2}{(1-i\sigma)}[\tilde{E}(0,\omega+\nu) + \tilde{E}^*(0,\omega-\nu)]\}. \quad (5.34)$$

Again, the determinant of this linear system must vanish. This yields the self-oscillation condition

$$1 - 2B\mathcal{S}\exp(i\nu t_c) + B\exp(2i\nu t_c) = 0 \quad (5.35)$$

where [22]

$$\mathcal{S} = \cos(\delta_0 + |\mathcal{E}|^2) - \frac{|\mathcal{E}|^2}{(1-i\sigma)}\sin(\delta_0 + |\mathcal{E}|^2). \quad (5.36)$$

This condition coincides with Ikeda's characteristic equation [21], obtained from linear stability analysis, taking

$$\delta\mathcal{E}(t) = \delta\mathcal{E}(0)\exp(\lambda t), \quad (5.37)$$

where one must take

$$\lambda = -i\nu. \quad (5.38)$$

Therefore, the physical mechanism for the Ikeda instability is the creation of sidebands by the above process.

What is ν? In order to find out, we assume with Ikeda that one also has fast longitudinal relaxation, i.e.,

$$\gamma_{\parallel}^{-1} \ll t_c, \quad (5.39)$$

from which it follows that

$$|\sigma| = |\nu/\gamma_{\parallel}| \ll |\nu t_c|. \quad (5.40)$$

As it will turn out that νt_c is of order unity, we can approximate $1 - i\sigma \approx 1$ in (5.36). This renders $\mathcal{S}$ a known (ν-independent) quantity, so that (5.35) becomes a quadratic equation in $\exp(i\nu t_c)$. The roots of this equation are given by

$$\nu_n t_c = 2n\pi - i\ln[(S + \sqrt{S^2 - 1})/B] \qquad (n = 0, \pm 1, \ldots). \tag{5.41}$$

In order for ν_n to be real, we must have either

$$S = (1 + B^2)/2B, \qquad \nu'_n = 2n\pi/t_c \tag{5.42}$$

or

$$S = -(1 + B^2)/2B, \qquad \nu_n = (2n + 1)\pi/t_c. \tag{5.43}$$

From (5.31), we get the slope of the response curve,

$$G = d(|\mathcal{E}|^2)/dA^2 = (1 + B^2 - 2BS)^{-1}. \tag{5.44}$$

Therefore, the bifurcation points (5.42), where $G \to \infty$, are the analogues of the switching points found in (5.15) for optical bistability, where a stable and an unstable branch bifurcate and the response curve has a vertical tangent. On the other hand, the bifurcation points (5.43) are associated with positive slope, so that they correspond to the Ikeda instability.

The threshold frequency associated with Ikeda's self-pulsing instability is thus given by

$$\nu_0 = \pi/t_c. \tag{5.45}$$

This explains why the Ikeda oscillation arises with period $2t_c$. According to (5.43), $\nu_n = (2n + 1)\nu_0$, so that odd-order harmonics should be dominant in the power spectrum. This agrees with the results found by Ikeda et al [23]. The present approximation, with its neglect of σ in (5.36), represents the limit $\gamma_\parallel \to \infty$, in which all modes become simultaneously unstable. For finite $\gamma_\parallel$, the number of unstable modes is finite. From (5.35) and (5.33) - (5.34), we also find that $|\tilde{E}(0,\omega + \nu)/\tilde{E}(0,\omega - \nu)| \approx 1$, so that the two sidebands are excited with approximately equal amplitudes.

For a high-finesse cavity, with (cf. (5.28))

$$B = 1 - \varepsilon, \qquad 0 < \varepsilon \ll 1, \tag{5.46}$$

the lowest possible threshold intensity for the Ikeda instability, according to (5.36) and (5.43), is obtained for [7]

$$\delta_0 \geq \pi - \sqrt{3}\,\varepsilon, \tag{5.47}$$

i.e., for nearly antiresonant cavity detuning [24]. The corresponding threshold intensity is given by [22]

$$|E|^2_{th} = 2\varepsilon/\sqrt{3}. \tag{5.48}$$

As was discussed above in connection with (5.21), the self-oscillation conditions (5.33) - (5.34) identify the frequencies $\omega \pm \nu_0$ of two adjacent cavity modes (cf. (5.45) and (5.47)), taking into account both linear and nonlinear dispersion, as well as nonlinear parametric coupling effects. These frequencies differ from their low-pump-intensity values in the linear regime, becoming gradually shifted as $|E|^2$ increases (Fig. 3). This effect has been named "transphasing" [6].

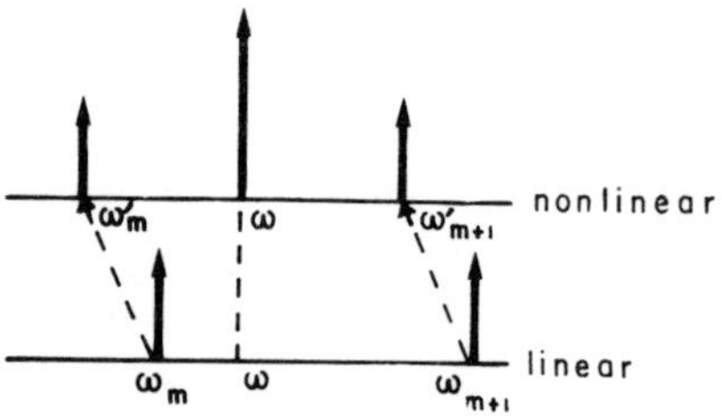

Figure 3. Mode spectrum in the linear (low pump intensity) and nonlinear regimes, showing the effect of transphasing (after [7]).

The results (5.45) and (5.47) mean, therefore, that the threshold intensity for the Ikeda instability is reached when the "transphased" mode frequencies are such that the incident laser frequency ω falls halfway between two consecutive modes, with sufficient gain for self-oscillation.

5.2.2. Fifth and Higher Bifurcations: Period Doubling. Beyond the threshold for the Ikeda instability, one finds a cascade of period-doubling bifurcations as the input intensity keeps increasing. It has been suggested by Firth et al [6] and by Bar-Joseph and Silberberg [7] that transphasing and sideband generation are also responsible for these bifurcations.

Immediately above the threshold for the Ikeda instability, the operating regime becomes the $2t_c$ periodic self-pulsation. Thus, the properties of the medium are no longer steady. However, they repeat themselves at intervals of $2t_c$, corresponding to two cavity roundtrips. In this sense, the effective cavity length is doubled, and for this cavity with a time-periodic medium one may think in terms of a new "Ikeda mode", falling halfway between a pair of original cavity modes.

The transphasing process tunes the frequencies of the new cavity modes as the pump intensity is increased. By a mechanism similar to that for the Ikeda instability, we get a new bifurcation when a shifted mode is brought halfway between the pump and "Ikeda" modes, for high enough gain. Therefore, the new threshold frequency is $\nu_0/2 = \pi/2t_c$, corresponding to a period of $4t_c$ (first period-doubling bifurcation).

From this point on, the argument just repeats itself at each stage, yielding the whole sequence of period-doubling bifurcations. The threshold intensities get closer and closer together, because the incremental nonlinear frequency shift needed keeps getting smaller and smaller. One can also see the origin of the scaling features.

5.2.3. Mechanism of Optical Chaos. As was mentioned following (5.5), the field in a "good" Fabry-Perot cavity is strongly dependent on the total roundtrip phase shift $\delta = \delta_0 + \delta_{NL}$, where $\delta_{NL} = |E|^2$ (cf. (5.30)) is the nonlinear (Kerr) contribution. The chaotic regime occurs for atomic relaxation times $\gamma_\perp^{-1}$, $\gamma_\parallel^{-1}$ much smaller than the cavity roundtrip time t_c.

Thus, as the field goes round the feedback loop, the medium refractive index has already had time to relax to the value determined by the field on the previous trip. Let E_j be the value of E on the jth trip (this is analogous to a stroboscopic map or a Poincaré section). The retarded feedback leads, therefore, to a strong dependence of E_j on $\delta_{NL}(j-1) = |E_{j-1}|^2$, i.e., to an erratic behavior of E_j.

6. CONCLUSION

We have shown that the Mollow instability (Sect. 3) and its Kerr medium counterpart (Sect. 4) are the basic physical mechanisms underlying all optical bifurcations found in a driven (homogeneously broadened) unidirectional ring laser. In all cases, sidebands (possibly degenerate, for $\nu = 0$) are generated and amplified by parametric processes, and the bifurcation threshold corresponds to the standard self-oscillation condition in a feedback system.

The results of this "dual-sideband approach" for the bifurcation points and instability boundaries coincide with those obtained by linear stability analysis [1]. It is readily seen that the two procedures are equivalent [25]. Indeed, the characteristic equation found in linear stability analysis for the eigenvalues λ_n has real coefficients, so that complex eigenvalues appear in complex conjugate pairs. At a bifurcation point, where $\mathrm{Re}\lambda_n = 0$, they therefore define a pair of frequencies $\pm\nu$, corresponding to the sidebands.

It also follows that one may extend the dual-sideband approach to the discussion of linear stability, by analytically extending the self-oscillation condition (equivalent to the characteristic equation) to complex values of ν or λ (cf. (5.38)), and employing it to determine the complex eigenvalues λ_n and the sign of their real part [4,25].

The advantage of the present approach is the physical insight it provides, by dealing with the fluctuations around the stationary solutions as actual signals, allowing us to bring out the nonlinear physical processes involved in their generation and propagation, and to determine the nonlinear constitutive parameters of the medium that are relevant for the bifurcation.

Similar techniques have been applied to the discussion of laser instabilities [26], both in homogeneously broadened and in inhomogeneously broadened lasers [4,5,27]. A quantized field treatment of sidemode build-up has also been presented [28].

This work was supported by FINEP and CNPq.

REFERENCES

[1] H. M. Nussenzveig, in Nonlinear Phenomena in Physics, ed. F. Claro (Springer-Verlag, Berlin, 1985), p. 162.
[2] B. R. Mollow, Phys. Rev. A5, 2217 (1972). Cf. also S. Haroche and F. Hartmann, Phys. Rev. A6, 1280 (1972).
[3] B. R. Mollow, Phys. Rev. A7, 1319 (1973). Cf. also S. L. McCall, Phys. Rev. A9, 1515 (1974).
[4] M. Sargent III, Phys. Rep. 43, 223 (1978); S. Hendow and M. Sargent III, Opt. Commun. 40, 385 (1982); 43, 59 (1982); J. Opt. Soc. Am. B2, 84 (1985).
[5] L. W. Hillman, R. W. Boyd and C. R. Stroud, Jr., Opt. Lett. 7, 426 (1982); L. W. Hillman, J. Krasinski, K. Koch and C. R. Stroud, Jr., J. Opt. Soc. Am. B2, 211 (1985).
[6] W. J. Firth, Opt. Commun. 39 343 (1981); W. J. Firth, E. Abraham and E. M. Wright, Appl. Phys. B28, 170 (1982); W. J. Firth, E. M. Wright and E. J. D. Cummins, in Optical Bistability 2, eds. C. M. Bowden, H. M. Gibbs and S. L. McCall (Plenum, N. York, 1984), p. 111.
[7] I. Bar-Joseph and Y. Silberberg, Opt. Commun. 48, 53 (1983); Y. Silberberg and I. Bar-Joseph, in Optical Bistability 2, eds. C. M. Bowden, H. M. Gibbs and S. L. McCall (Plenum, N. York 1984), p. 61; J. Opt. Soc. Am. B1, 662 (1984).
[8] H. C. Torrey, Phys. Rev. 76, 1059 (1949).
[9] M. Lax, Phys. Rev. 129, 2342 (1963); 157, 213 (1967).
[10] B. R. Mollow, Phys. Rev. 188, 1969 (1969).
[11] F. Y. Wu, S. Ezekiel, M. Ducloy and B. R. Mollow, Phys. Rev. Lett. 38, 1077 (1977).
[12] J. A. Armstrong, N. Bloembergen, J. Ducuing and P. S. Pershan, Phys. Rev. 127, 1918 (1962).
[13] L. D. Landau and E. M. Lifshitz, Mechanics (Addison-Wesley, Reading, Mass., 1960).
[14] N. Bloembergen, Am. J. Phys. 35, 989 (1967). D. C. Hanna, M. A. Yuratich and D. Cotter, Nonlinear Optics of Free Atoms and Molecules (Springer-Verlag, Berlin, 1979).
[15] N. Bloembergen and P. Lallemand, Phys. Rev. Lett. 16, 81 (1966).
[16] R. Y. Chiao, P. L. Kelley and E. Garmire, Phys. Rev. Lett. 17, 1158 (1966); R. W. Boyd, M. G. Raymer, P. Narum and D. J. Harter, Phys. Rev. A24, 411 (1981).
[17] R. L. Carman, R. Y. Chiao and P. L. Kelley, Phys. Rev. Lett. 17, 1281 (1966); A. C. Tam, Phys. Rev. A19, 1971 (1979); D. J. Harter, P. Narum, M. G. Raymer and R. W. Boyd, Phys. Rev. Lett. 46, 1192 (1981).

[18] P. Coullet, in Nonlinear Phenomena in Physics, ed. F. Claro (Springer-Verlag, Berlin, 1985), p. 16.
[19] M. Gronchi, V. Benza, L. A. Lugiato, P. Meystre and M. Sargent III, Phys. Rev. A24, 1419 (1981).
[20] W. E. Lamb, Jr., Phys. Rev. 134, A1429 (1964).
[21] K. Ikeda, Opt. Commun. 30, 257 (1979); K. Ikeda, H. Daido and O. Akimoto, Phys. Rev. Lett. 45, 709 (1980).
[22] H. J. Carmichael, R. R. Snapp and W. C. Schieve, Phys. Rev. A26, 3408 (1982); J. . Englund, R. R. Snapp and W. C. Schieve, in Progress in Optics XXI, ed E. Wolf (Elsevier, Amsterdam, 1984), p. 355.
[23] K. Ikeda, K. Kondo and O. Akimoto, Phys. Rev. Lett. 49, 1467 (1982).
[24] H. J. Carmichael, Phys. Rev. Lett. 52, 1292 (1984); Opt. Commun. 53, 122 (1985).
[25] N. B. Abraham, L. A. Lugiato, P. Mandel, L. M. Narducci and D. Bandy, J. Opt. Soc. Am. B2, 35 (1985).
[26] N. B. Abrham, L. A. Lugiato and L. M. Narducci, J. Opt. Soc. Am. B2, 7 (1985).
[27] L. W. Casperson, in Laser Physics, eds. J. Harvey and D. Walls (Springer-Verlag, Berlin 1983), p. 88.
[28] M. Sargent III, M. S. Zubairy and F. de Martini, Opt. Lett. 8, 76 (1983); M. Sargent III, D. A. Holm and M. S. Zubairy, Phys. Rev. A31, 3112 (1985); S. Stenholm, D. A. Holm and M. Sargent III, Phys. Rev. A31, 3124 (1985).

THE INTERACTION OF SOUND AND VORTICITY

Fernando Lund
Departamento de Física
Facultad de Ciencias Físicas y Matemáticas
Universidad de Chile
Casilla 487/3, Santiago, Chile

ABSTRACT. This is a set of two lectures on vortex dynamics in a slightly compressible, adiabatic, inviscid, fluid. The first, of an introductory nature, is a summary of well established facts and an overview of the current approaches to finding the sound radiated by vortices in unsteady motion. The second, a description of recent work done in collaboration with N. Zabusky (Lund and Zabusky, 1986) concerning both the flow generated by a vortex filament in prescribed motion as well as the response of such a vortex to an external flow.

1. GENERAL PROPERTIES OF VORTEX DYNAMICS IN THE INCOMPRESSIBLE APPROXIMATION. FAR FIELD EXPRESSIONS FOR AERODYNAMICALLY GENERATED SOUND

1.1. Generalities

A fluid of density ρ, velocity $\vec{v}$ and pressure p obeys the well known [1], [2] equations of mass and momentum conservation

$$\frac{\partial \rho}{\partial t} + \nabla \cdot (\rho \vec{v}) = 0 \tag{1.1}$$

$$\frac{\partial \vec{v}}{\partial t} + (\vec{v} \cdot \nabla)\vec{v} = - \frac{1}{\rho} \nabla p \tag{1.2}$$

when viscous effects can be neglected, for instance, when the time scale of interest is short compared to the dissipative time scale, and no external force is present. Moreover, if the flow is adiabatic, the right hand side of (1.2) is equal to the specific enthalphy H:

$$\frac{1}{\rho} \nabla p = \nabla H, \tag{1.3}$$

and the equation of energy conservation does not add new information. Equations (1.1) and (1.2) supplemented by an equation of state relating ρ and p, for instance $S(\rho,p)$=const for adiabatic flow with S the entropy,

E. Tirapegui and D. Villarroel (eds.), Instabilities and Nonequilibrium Structures, 117–140.

form a closed set. In the sequel we shall confine ourselves to this case. The problems encountered when this restriction is lifted have been nicely described by Chu and Kovasznay [3].

When (1.3) holds, a convenient equation involving only the velocity $\vec{v}$ is obtained by taking the curl of (1.2):

$$\frac{D\vec{\omega}}{Dt} = (\vec{\omega} \cdot \nabla)\vec{v} - (\nabla \cdot \vec{v})\vec{\omega} \tag{1.4}$$

where we have introduced the vorticity $\vec{\omega} \equiv \nabla \wedge \vec{v}$ and the convective derivative $D/Dt \equiv \partial/\partial t + \vec{v} \cdot \nabla$. Since $\nabla \cdot \vec{\omega} \equiv 0$, the vorticity is a solenoidal field like the magnetic field in electromagnetism and vortex lines, i.e. curves that are everywhere tangent to $\vec{\omega}$, are closed. An interesting and useful result that goes back to Helmholtz is that for an incompressible flow ($\nabla \cdot \vec{v} = 0$) vortex lines moves as material lines. Indeed, take two neighboring points, $\vec{r}_1$ and $\vec{r}_2 = \vec{r}_1 + \delta\vec{\ell}$ on such a line. Moving with the fluid as they do, they satisfy

$$\vec{r}_1(t + Dt) = \vec{r}_1(t) + \vec{v}(r_1, t)Dt$$

$$\vec{r}_2(t + Dt) = \vec{r}_2(t) + \vec{v}(r_2, t)Dt = \vec{r}_2(t) + [\vec{v}(r_1, t) + (\delta\vec{\ell} \cdot \nabla)\vec{v}]Dt.$$

So the material line element $\delta\vec{\ell}$ changes by

$$D(\delta\vec{\ell}) = (\delta\vec{\ell} \cdot \nabla)\vec{v}Dt \tag{1.5}$$

in the time interval Dt, which is the same as (1.4) when $\nabla \cdot \vec{v} = 0$. In two dimensions the vorticity is orthogonal to the velocity and (1.4) is further simplified to

$$\frac{D\vec{\omega}}{Dt} = 0 \quad ; \quad (2\text{-D}).$$

More generally, a compressible inviscid fluid satisfies

$$\frac{D}{Dt}\left(\frac{\vec{\omega}}{\rho}\right) = \left[\left(\frac{\vec{\omega}}{\rho}\right) \cdot \nabla\right]\vec{v}.$$

Another useful result is that of conservation of circulation (Kelvin's Theorem). The circulation Γ of the velocity field $\vec{v}$ along a closed curve C is

$$\Gamma = \oint_C \vec{v} \cdot d\vec{\ell} = \iint \vec{\omega} \cdot d\vec{s}$$

where the last expression is over any surface bounded by the curve C. Kelvin's circulation theorem states that $d\Gamma/dt = 0$ whenever C moves with the fluid (i.e. as a material line), when the latter is assumed to be compressible and adiabatic. To see this, consider

$$\Gamma(t+Dt) - \Gamma(t) = \oint_{C(t+Dt)} d\vec{\ell}\cdot\vec{v}\Big|_{t+Dt} - \oint_{C(t)} d\vec{\ell}\cdot\vec{v}\Big|_{t}$$

using (1.5)

$$= \oint_{C(t)} [d\vec{\ell} + \{(d\vec{\ell}\cdot\nabla)\vec{v}\}Dt]\cdot(v + \frac{D\vec{v}}{Dt}Dt)\Big|_{t} - \oint_{C(t)} d\vec{\ell}\cdot\vec{v}\Big|_{t}$$

using (1.2) - (1.3)

$$= Dt \oint_{C(t)} d\vec{\ell}\cdot\nabla(\frac{1}{2}v^2 + H)$$

$$= 0.$$

The last equality holds whenever velocity and enthalpy are single valued.

1.2. Vortex Dynamics in the Incompressible Approximation

For an incompressible fluid, considerable progress can be made whenever vorticity is confined to a small fraction of the fluid volume because it is then enough to find the evolution of the vorticity field. The velocity is found by integration of $\vec{\omega} = \nabla\wedge\vec{v}$ with appropriate boundary conditions and the pressure is found by integrating (1.2), the density ρ being of course a constant. One has that $\nabla\cdot\vec{v} = 0$ implies the existence of a vector potential that might be taken divergenceless: $\vec{v} = \nabla\wedge\vec{A}$, $\nabla\cdot\vec{A} = 0$. Therefore

$$\vec{\omega} = -\nabla^2\vec{A}$$

which is integrated by a convolution of the vorticity with the Green's function appropriate to the relevant boundary conditions:

$$\vec{A} = G * \vec{\omega}.$$

For an infinitely extended fluid at rest at infinity, this gives

$$\vec{v}(\vec{x}) = \frac{-1}{4\pi}\int d^3x' \frac{(\vec{x}-\vec{x}')\wedge\vec{\omega}(\vec{x}')}{|\vec{x}-\vec{x}'|^3}. \tag{1.6}$$

A particularly simple situation occurs when vorticity is confined to tubes whose radius is small compared to any relevant lenght scale. The extreme case of such a thin tube is a vortex filament, a curve which is mathematically described as a three-vector $\vec{X}(\sigma,t)$ depending on time t and a Lagrangean parameter σ labeling the points of the filament. The vorticity associated with one such filament is

$$\vec{\omega} = \Gamma \oint_C d\sigma \frac{\partial \vec{X}}{\partial \sigma} \delta[\vec{x} - \vec{X}(\sigma,t)] \tag{1.7}$$

where the circulation Γ is a fixed parameter characterizing the vortex. The problem of finding the evolution of the vorticity field becomes then that of finding the evolution of the curve(s) C. The mathematics here is the same as that of the magnetic field (here the velocity) generated by a current (here the vorticity) carring wire (here the loop C). Thus, the velocity generated by a vortex filament of circulation Γ, obtained by inserting (1.7) into (1.6), is nothing more than a transcription of the Biot-Savart Law:

$$\vec{v}(x,t) = \frac{\Gamma}{4\pi} \int_C d\sigma \frac{\partial \vec{X}}{\partial \sigma} \wedge \frac{\{\vec{x} - \vec{X}(\sigma,t)\}}{|\vec{x} - \vec{X}(\sigma,t)|} . \tag{1.8}$$

In regions of vanishing vorticity it is possible to introduce a scalar potential, $\vec{v} = \nabla\phi$ and so (1.8) must be such a gradient. The proviso that one must keep in mind, of course, is that ϕ will not be a single valued function since the region of vanishing vorticity is not simply connected. To find ϕ for (1.8) write it as $(\vec{X}' \equiv \partial\vec{X}/\partial\sigma)$

$$v_i = \frac{\Gamma}{4\pi} \oint_C d\sigma \varepsilon_{ijk} \partial_j \left(\frac{1}{|\vec{x} - \vec{X}|}\right) \vec{X}'_k$$

$$= \frac{\Gamma}{4\pi} \int_S dS^k \varepsilon_{k\ell m} \varepsilon_{ijm} \partial_\ell \partial_j \left(\frac{1}{|\vec{x} - \vec{X}|}\right)$$

where S is any surface bounded by C. Using $\varepsilon_{k\ell m}\varepsilon_{ijm} = \delta_{ki}\delta_{\ell j} - \delta_{kj}\delta_{i\ell}$ we get

$$v_i = - \partial_i \left[\frac{\Gamma}{4\pi} \int_S dS^j \partial_j \left(\frac{1}{|\vec{x} - \vec{X}|}\right)\right] , \quad \vec{x} \notin S$$

and the scalar potential is $(\Gamma/4\pi)$ times the solid angle subtended by C at the point of observation, which is indeed a multivalued function.

Even more algebraic simplicity is obtained when vortices are straight, infinitely long and parallel so that everything is symmetric under translations along one direction. The problem becomes then effectively two dimensional, N vortex filaments in this case being parametrized as

$$\vec{X}_A(\sigma,t) = \{X_A(t), Y_A(t), \sigma\} \qquad A = 1, \ldots, N$$

and the vorticity becomes localized at points:

$$\vec{\omega} = \hat{x}_3 \sum_A \Gamma_A \delta\{x - X_A(t)\}\delta\{y - Y_A(t)\}. \tag{1.9}$$

The equation $D\vec{\omega}/Dt = 0$ then turns into

$$\sum_A \Gamma_A \left[- \dot{\vec{X}}_A \cdot \nabla\delta\{\vec{x} - \vec{X}_A(t)\} + v \cdot \nabla\delta\{\vec{x} - \vec{X}_A(t)\}\right] = 0,$$

and letting $\vec{x} = \vec{X}_B(t) + \vec{\varepsilon}$ with $\vec{\varepsilon}$ small, multiplying by $\vec{\varepsilon}$ and integrating on a circular disk of radius ε one has

$$\dot{\vec{X}}_A(t) = \vec{v}\{X_A(t)\}. \tag{1.10}$$

Namely, point vortices move with the local flow velocity. From (1.8) one has moreover the important consequence that for an incompressible fluid, a point vortex, say X_A, does not contribute to the right hand side of (1.10) or, in other words, it does not act on itself. Indeed, the velocity generated by a straight vortex is, from (1.8)

$$\vec{v}(\vec{x},t) = \frac{\Gamma}{2\pi} \frac{\hat{k} \wedge \{\vec{x} - \vec{X}(t)\}}{|\vec{x} - \vec{X}(t)|^2}$$

where $\hat{k} = (0,0,1)$ and near the vortex, at $\vec{x} = \vec{X}(t) + \vec{\varepsilon}$ this is

$$\vec{v}\{\vec{X}(t) + \vec{\varepsilon}, t\} \approx \frac{\Gamma}{2\pi} \frac{\hat{k} \wedge \vec{\varepsilon}}{\varepsilon^2}$$

which, although large for small ε, averages out to zero. The important point is that, in the incompressible approximation, the flow $\vec{v}(\vec{x},t)$ generated at time t by the vortex everywhere, including its own neighborhood, depends on its instantaneous position at the same instant t. For a set of isolated vortices the velocity is obtained from (1.9) and substituted into (1.10) to obtain a set of equations for their dynamics:

$$\dot{\vec{X}}_A(t) = \sum_{B \neq A} \frac{\Gamma_B}{2\pi} \frac{\hat{k} \wedge \{\vec{X}_A(t) - \vec{X}_B(t)\}}{|\vec{X}_A(t) - \vec{X}_B(t)|^2}$$

which is well known [4] to be Hamiltonian with

$$H = \frac{1}{4\pi} \sum_{A \neq B} \Gamma_A \Gamma_B \ln|\vec{X}_A - \vec{X}_B|$$

and Poisson brackets

$$\{Y_B, X_A\} = \Gamma_A^{-1} \delta_{AB},$$

with the feature that the two coordinates of the "point" vortex are canonically conjugate variables, which is due to the dynamical equations being first (as opposed to second) order in time in configuration space.

Filament dynamics in three dimensions is similarly obtained by substitution of (1.7) into $D\vec{\omega}/Dt = (\vec{\omega} \cdot \nabla)\vec{v}$ which gives

$$\int d\sigma(\dot{X}_i X_j' - X_i' \dot{X}_j)\nabla_j \delta(x - X) = \int d\sigma\{X_j'(\nabla_j v_i)\delta(x - X) - X_i' v_j \nabla_j \delta(x - X)\}.$$

Evaluating this at $\vec{x} = \vec{X}(s,t) + \vec{\varepsilon}$ and integrating over a small disk of radius ε yields

$$\dot{X}_i X_k' - X_i' \dot{X}_k = v_i(X) X_k' - v_k(X) X_i' \tag{1.11}$$

which says again that each point of the vortex filament moves with the local fluid velocity. More precisely, this is true of the component of velocity perpendicular to the filament, as the component along the filament has no physical meaning. It is worthwile to point out here that inclusion of the compressibility term $(\nabla \cdot \vec{v})\vec{\omega}$ in the vorticity equation leaves the fact that a vortex filament moves with the local fluid velocity absolutely unchanged, both in two and three dimensions. What is changed, of course, is the fluid velocity generated by the vortex. There will be now retardation effects due to the finite sound speed and in particular, a point vortex will have a non-vanishing action on itself.

Still in the incompressible approximation, the motion of a filament under its own induction is obtained substituting (1.8) in (1.11):

$$\dot{\vec{X}}(\sigma,t) \wedge \vec{X}'(\sigma,t) = \frac{\Gamma}{4\pi} \left[\int d\sigma' \frac{\vec{X}'(\sigma',t) \wedge [\vec{X}(\sigma',t) - \vec{X}(\sigma,t)]}{|\vec{X}(\sigma',t) - \vec{X}(\sigma,t)|^3}\right] \wedge \vec{X}'(\sigma,t). \tag{1.12}$$

This equation can be obtained from a variational principle: it is the

Euler-Lagrange equation obtained by extremizing the action

$$S = \frac{1}{3}\int dt d\sigma \vec{X}\cdot(\dot{\vec{X}}\wedge\vec{X}') + \frac{\Gamma}{8\pi}\int dt d\sigma d\sigma' \frac{\vec{X}'(\sigma',t)\cdot\vec{X}'(\sigma,t)}{|\vec{X}(\sigma',t)-\vec{X}(\sigma,t)|}$$

which has been used to construct a canonical formalism by Rasetti and Regge [5]. Such a formalism has to deal with the fact that the kinetic term in the action is linear in the velocities and that there is a gauge freedom due to the reparametrization invariance of the system: the dynamics of the filament cannot depend on the parameter σ chosen to describe it.

Equation (1.12) for the dynamics of a vortex filament points to an essential difficulty inherent to the approximation of treating a vortex as infinitely thin: the velocity (1.8) generated by a filamentary vortex is singular at the filament location and this gives a divergent contribution to (1.12) except, as we have already seen, in the very special case of a point vortex in the incompressible limit. This divergence of course does not make physical sense since any real vortex has a core of finite thickness and the filament approximation holds as long as this core thickness can be neglected in comparison with any other relevant length scale. The fact that, as long as one doesn't look at small length scales, the physics should be independent of the details of the core is implemented introducing a cut-off parameter h with no independent dynamics so that the short distance behaviour of the Green's function is modified. For example the Green's function for $-\nabla^2$, instead of $(4\pi)^{-1}|\vec{x}-\vec{x}'|^{-1}$ becomes $(4\pi)^{-1}\{(\vec{x}-\vec{x}')^2+h^2\}^{-1/2}$. The extent to which this procedure is justified has been discussed in detail by Moore and Saffman [6]. There is a large degree of arbitrariness in the choice of cut-off: it can be a constant, as in a cylinder of uniform cross-section that doesn't change in time, or it can be a variable, both along the filament and time. The latter possibility has been used to great effect by Siggia [7] and Siggia and Pumir [8] to study the pairing of vortex filaments and their subsequent evolution.

Whenever the curvature of a filament is not large and the filament does not bend back on itself, the integral on the right hand side of (1.12) will be dominated by the large contribution from those points that are near the point whose velocity is on the left hand side. The localized induction approximation consists in keeping only that large contribution and neglecting the rest. More precisely, if $\vec{X}(\sigma)$ feels the presence of those points $\vec{X}(\sigma')$ with $|\sigma'-\sigma|<\delta$, Eqn. (1.12) becomes, with the cut-off procedure already mentioned

$$\dot{\vec{X}}\wedge\vec{X}' \approx \frac{\Gamma}{8\pi}\frac{(\vec{X}'\wedge\vec{X}'')\wedge\vec{X}'}{|\vec{X}'|^3}\int_{-\delta}^{\delta}\frac{\Delta^2 d\Delta}{(\Delta^2+h^2|X'|^{-2})^{3/2}}$$

$$\dot{\vec{X}}\wedge\vec{X}' = \frac{\Gamma}{4\pi|\vec{X}'|^3}\{(\vec{X}'\wedge\vec{X}'')\wedge\vec{X}'\}\ln\frac{\delta}{h} + O(1), \qquad (1.13)$$

in which it is supposed that the local induction cut-off δ is large compared to the core cut-off h. It thus appears that, to leading order, the velocity of a filament is proportional to its local curvature.

1.3. Generation of Sound by Vorticity

Sound, as usually understood, is a propagating pressure disturbance in a fluid which, from the point of view of velocity, is longitudinal. Vorticity, on the other hand, is transverse. In other words, if we write, as is always possible, the velocity as a sum of a longitudinal and transverse part $\vec{v} = \vec{v}_L + \vec{v}_T$ with $\nabla \cdot \vec{v}_T = 0 = \nabla \wedge \vec{v}_L$, sound is associated with $\vec{v}_L$ and vorticity with $\vec{v}_T$. If $p_1 = c^2 \rho_1$ with c a constant speed of sound as is the case for small density variations ($p = p_0 + p_1$, $\rho = \rho_0 + \rho_1$, $\rho_1 << \rho_0$, $p_1 << p_0$, ρ_0 and p_0 are constants) the linearized version of Eqs. (1.1) - (1.2) is

$$\frac{\partial^2 \rho_1}{\partial t^2} - c^2 \nabla^2 \rho_1 = 0$$

$$\frac{\partial \rho_1}{\partial t} + \rho_0 \nabla \cdot \vec{v}_L = 0 \qquad (1.14)$$

$$\frac{\partial \vec{v}_T}{\partial t} = 0$$

where the last equation comes form the momentum equation noting there is a scalar potential for the longitudinal velocity $\vec{v}_L = \nabla\phi$ taking a time derivative of the second equation and substituting the first into the result, which leads to

$$\nabla^2 (c^2 \rho_1 + \rho_0 \frac{\partial \phi}{\partial t}) = 0$$

which means $\rho_1 = - c^{-2} \rho_0 \partial\phi/\partial t$ whenever both are single valued. The point of the linear equations (1.14) is that $\vec{v}_L$ and $\vec{v}_T$ are decoupled and the two modes evolve independently. This independence is of course lost when the quadratically nonlinear terms in the equations are retained: vorticity will generate sound, and sound will affect vorticity with the constraint of conservation of circulation (Kelvin's theorem) which, as was shown at the beginning of this section, holds for compressible, adiabatic, fluids.

It seems to be Lighthill [9], [10] who, prompted by the development of jet aircraft, first made considerable quantitative progress in the study of sound generated aerodynamically, that is, by fluid flow rather than by the vibration of solid bodies. He noted that Equations (1.1) - (1.2) imply

$$\frac{\partial^2 \rho}{\partial t^2} - c^2 \nabla^2 \rho = \partial_i \partial_j T_{ij} \tag{1.15}$$

where c is a constant and

$$T_{ij} = \rho v_i v_j + (p - c^2 \rho)\delta_{ij}. \tag{1.16}$$

This looks then like a wave equation for the fluid density with a source term. If one knew the behaviour of the source, Eqn. (1.14) would describe sound waves through density fluctuations. The catch is, of course, that the source term is inextricably mixed with the density and in general (1.14) cannot be solved any more than (1.1) - (1.2) can. However, a new way of looking at the aerodynamic sound generation mechanism emerges: perhaps a region of the fluid evolves to a good approximation according to the laws of incompressible flow, and perhaps the source (1.16) is significantly different from zero within that region only; in this case, (1.15) can be used to study the propagation of sound away from this localized source. Particularly lucid expositions of this train of thought have been given by Crow [11] and Howe [12].

If incompressibility is a good approximation in a source region (i.e. its flow is of low Mach number), then

$$T_{ij} \approx \rho_0 v_i v_j$$

and

$$\partial_i \partial_j T_{ij} = \rho_0 \{\nabla \cdot (\vec{\omega} \wedge \vec{v}) + \frac{1}{2} \nabla^2 v^2\}.$$

Moreover, the vorticity equation can be written

$$\frac{\partial \vec{\omega}}{\partial t} = \nabla \wedge (\vec{v} \wedge \vec{\omega}).$$

Now, the (formal) solution of (1.15) is

$$\rho_1 = \int G \partial_i \partial_j T_{ij}$$

where the integral on the right hand side is carried over the source region and G is the Green's function for the scalar wave equation. If the last few considerations apply,

$$\rho_1 = \int \nabla G \cdot (\vec{\omega} \wedge \vec{v} + \frac{1}{2} \nabla v^2).$$

Next, if it were possible to find a vector Green's function $\vec{G}$ with the property

$$\nabla \wedge \vec{G} = \nabla G$$

one would end up with

$$\rho_1 = \int \vec{G} \cdot \frac{\partial \vec{\omega}}{\partial t} ,$$

which locates the sound source at the regions of time-dependent vorticity. This procedure was employed by Mohring [13] who was able to give a general expression for the far field pressure generated by a localized distribution of vorticity:

$$p_1(\vec{x},t) \underset{|x|\to\infty}{\sim} \frac{\rho_0}{12\pi c^2} \frac{x_i x_j}{|x|^3} \frac{\partial^3}{\partial t^3} \int d^3y \varepsilon_{k\ell j} y_i y_k \omega_\ell (\vec{y}, t - \frac{|x|}{c}). \quad (1.17)$$

This relation is valid at distances that are large compared to typical source dimensions, which are also supposed to be small compared to typical wavelengths so that a single retarded time can be used.

The idea that the sound generated by vorticity in a finite region at low Mach number should be computable in a succesive approximation scheme was formulated in characteristically precise fashion by Crow [11] in terms of matched asymptotic expansions with the Mach number M as a small parameter. In a slightly compressible fluid, the flow near the vorticity bearing region should differ form the incompressible solution by succesively higher powers of M. Far away from that region one should have outgoing waves. Matching in an intermediate region gives the final answer. This matching, as carried out by Crow [11], depends on a multipole expansion and the result is thus limited to source dimensions small compared to typical wave lengths. Various reformulations of this singular perturbation scheme have been given by Obermeier [14] and Kambe and Minota [15] who have also carried out experiments to measure the sound radiated by vortex rings colliding head-on [16].

2. RECENT DEVELOPMENTS IN SOUND-VORTEX INTERACTION DYNAMICS

We now adress the problem of finding the flow generated by a vortex filament whose typical dimensions need not be small compared to the relevant wavelengths, as well as the response of such a filament to an outside flow [17]. The method of attack used on the second question will be motivated with an example from electromagnetism.

2.1. Sound from a Vortex Filament

Consider a vortex filament $\vec{X}(\sigma,t)$, as introduced in section 1, moving in some prescribed fashion, for instance according to the laws of incompressible fluid dynamics. What is the velocity of the fluid everywhere outside the filament? In the incompressible approximation, the answer is given by the Biot-Savart law (1.8). When compressibility is considered, this is no longer true due to retardation effects: a change in velocity of the filament, for instance, will be felt at a retarded time given by the distance to the point of observation divided by the speed of sound c. The incompressible limit is obtained when $c \to \infty$ and retardation effects disappear. There is however one very important statement that holds for a compressible fluid: Kelvin's theorem on the conservation of circulation which guarantees that vortices retain their identity as long as viscosity effects can be neglected.

The technical advantage of working with a filamentary vortex is that almost everywhere (i.e. everywhere except along the vortex itself) there exists a velocity potential, $\vec{v} = \nabla\phi$, which need not be single valued as it is defined in a domain that is not simply connected. Indeed, from (1.7) one has that ϕ has a discontinuity equal to Γ across some (arbitrary) surface $S(t)$ bounded by the filament:

$$\phi(x,t)\Big|_{S^+} - \phi(x,t)\Big|_{S^-} = \Gamma. \tag{2.1}$$

This is a purely kinematic statement resting on Kelvin's theorem without any additional asumptions. To make further progress, we assume that the nonlinear terms of the fluid dynamics equations can be neglected everywhere outside the filament, where velocities as well as density and pressure variations are supposed to be small. From (1.1) - (1.2) one has then

$$\frac{\partial\rho_1}{\partial t} + \rho_0\nabla^2\phi = 0$$

$$\nabla\left(\frac{\partial\phi}{\partial t}\right) = -\frac{c^2}{\rho_0}\nabla\rho_1 \tag{2.2}$$

everywhere outside the vortex, and ϕ is a solution of the wave equation

$$c^{-2}\ddot{\phi} - \nabla^2\phi = 0. \tag{2.3}$$

We are then in the position of solving (2.3) with the boundary condition (2.1) and whatever additional conditions are needed at any solid boundaries that might be present. Here we shall take the simplest possible situation: an infinitely extended fluid with no solid boundaries, at rest at infinity. Under these conditions the Green's function for (2.3) is well known to be

$$G(\vec{x}-\vec{x}',t-t') = \frac{1}{|\vec{x}-\vec{x}'|}\,\delta(t-t'-c^{-1}|\vec{x}-\vec{x}'|).$$

Replacing the vortex by a boundary condition on the velocity potential makes possible the use of standard Green's function techniques. Consider the integral

$$\int d^3x'dt' \frac{\partial\phi}{\partial x'_i}\frac{\partial G}{\partial x'_i}$$

where $\infty < t' < \infty$ and the volume of integration is all space with the exclusion of a thin slab around S(t), the surface where ϕ is discontinuous (with continuous derivatives, velocity and pressure, however). Using Gauss' theorem for the derivatives of the Green's function one has

$$\int d^3x'dt' \frac{\partial\phi}{\partial x'_i}\frac{\partial G}{\partial x'_i} = \int dt' \int dS'_i \frac{\partial\phi}{\partial x'_i} G - c^{-2}\int d^3x'dt' G\ddot{\phi}$$

$$= - c^{-2}\int d^3x'dt' G\ddot{\phi} \qquad (2.4)$$

since the surface integral does not contribute: it is carried over the upper and lower faces of the thin slab sorrounding S(t) and both $\nabla\phi$ and G are continuous across it. It is expected that ϕ decays at least as $|x|^{-1}$ at infinity so there are no contributions form that region either. Integrating the derivatives of ϕ yields

$$\int d^3x'dt' \frac{\partial\phi}{\partial x'_i}\frac{\partial \phi}{\partial x'_i} = \Gamma\int dt' \int dS'_i \frac{\partial G}{\partial x'_i} -$$

$$- \int d^3x'dt'\phi\{c^{-2}\ddot{G} - 4\pi\delta(\vec{x}-\vec{x}')\delta(t-t')\} \qquad (2.5)$$

where (2.1) has been used. Equating (2.4) and (2.5) then gives

$$4\pi\phi(x,t) = -\Gamma\int dt' \int_{S(t)} dS'_i \frac{\partial G}{\partial x'_i} + c^{-2}\int d^3x'dt'(G\ddot{\phi} - \ddot{G}\phi).$$

Noting that, because of the asymptotic behaviour of G

$$0 = \int dt' \frac{d}{dt'}\int d^3x'(G\dot{\phi} - \dot{G}\phi)$$

$$= \int dt'd^3x'(G\ddot{\phi} - \ddot{G}\phi) + \int dt' \int dS'_i V^i(G\dot{\phi} - \dot{G}\phi)$$

where V^i is the velocity of the points on the surface bounding the slab that surrounds S(t). Again using (2.1) and the singlevaluedness of $\dot{\phi}$

across S(t) one finally has

$$4\pi\phi(x,t) = \Gamma \int dt' \int_{S(t')} dS_i' \frac{\partial G}{\partial x_i} + \frac{\Gamma}{c^2} \int dt' \int_{S(t')} dS_i' V^i \frac{\partial G}{\partial t}\,, \qquad (2.6)$$

which represents the scalar potential as a convolution of a Green's function with a source localized on the surface S(t) whose boundary is the vortex filament $\vec{X}(\sigma,t)$. Indeed, (2.6) can be written

$$4\pi\phi = G * f \equiv \int dt'' d^3x'' G(\vec{x} - \vec{x}'', t - t'') f(\vec{x}'', t'')$$

with

$$f(x'',t'') = \int dt' \int_{\vec{x}'\varepsilon S(t')} dS_i' \nabla_i \delta(\vec{x}'' - \vec{x}')\delta(t' - t'') +$$

$$+ \frac{1}{c^2} \int dt' \int_{\vec{x}'\varepsilon S(t)} dS_i' V^i \delta(\vec{x}' - \vec{x}'')\delta'(t'' - t')$$

where δ' means the derivative of δ with respect to its argument, explicitely displaying the source localization at the surface S(t). This source is the sum of two terms, the first of which is what one would expect from a dipole layer at S(t) but this is not enough, a time component has to be added. The surface S(t), although artificial, is unavoidable in a description in terms of the velocity potential ϕ due to its multivaluedness. It should be possible to do away with it when looking at the derivatives of ϕ such as velocity and pressure. This expectation is fulfilled, as is readily checked by straightforward derivation of (2.6) and using Stokes' theorem to turn surface integrals over S(t) into line integrals over $\vec{X}(\sigma,t)$. The result is [17]

$$\nabla\phi = \frac{\Gamma}{4\pi} \int dt' d\sigma (\nabla G \wedge \frac{\partial \vec{X}}{\partial \sigma}) + \frac{\Gamma}{4\pi c^2} \int dt' d\sigma \frac{\partial G}{\partial t} (\frac{\partial \vec{X}}{\partial t'} \wedge \frac{\partial \vec{X}}{\partial \sigma}) \qquad (2.7a)$$

$$\frac{\partial \phi}{\partial t} = \frac{\Gamma}{4\pi} \int dt' d\sigma (\frac{\partial \vec{X}}{\partial t'} \wedge \frac{\partial \vec{X}}{\partial \sigma}) \cdot \nabla G, \qquad (2.7b)$$

which give velocity and pressure as line integrals along the vortex filament. These expressions are valid everywhere outside the vortex core, not just in a far field region, and although the core is supposed to be very thin, i.e., very small compared to relevant wavelengths as well as to typical filament dimensions, these typical dimensions (like the diameter of a vortex ring) need not be small compared to typical wavelengths. The derivation has been carried out without recourse to a vector Green's function and without matched asymptotic expansions. The only ingredients are Kelvin's circulation theorem hidden in the assumption that it makes sense to speak of a vortex filament with constant circulation, and the linear wave equation for the velocity potential ϕ.

Results for point vortices are obtained with the parametrization $\vec{X}(\sigma,t) = \{X(t),Y(t),\sigma\}$. The Biot-Savart law (1.8) is obtained from (2.7a) when $c \to \infty$, as in this case $G \to |\vec{x} - \vec{x}'|^{-1}\delta(t - t')$. Thus, (2.7a) is a correction to the Biot-Savart law for the flow generated by a vortex filament when compressibility effects are small but non-vanishing. Results for a vorticity distribution, discrete (several filaments) or continuous, are obtained with the substitution

$$\Gamma \int d\sigma \, \frac{\partial \vec{X}(\sigma,t')}{\partial \sigma} \to \sum_{\alpha} \Gamma_{\alpha} \int d\sigma \, \frac{\partial \vec{X}^{\alpha}(\sigma,t')}{\partial \sigma} \to \int d^3y \vec{\omega}(\vec{y},t').$$

The far field results described in section 1 in the case of compact vorticity distributions are recovered in the appropriate limits (Lund and Zabusky, in preparation).

It is of interest to note that (2.7) can be obtained in a more compact form thinking of the trajectory of the vortex as a surface in four dimensional space labeled by time and three space coordinates. For a homogeneous unbounded fluid the speed of sound is a constant, the wave equation is Lorentz invariant, and the machinery of Minkowsky geometry may be used [18]. The wave equation may be written

$$\partial^{\mu}\partial_{\mu}\phi = 0 \qquad (\mu = 0,1,2,3)$$

and we look for a solution with discontinuity Γ (but with continuous gradient) across a three dimensional manifold Ω which is the volume enclosed by $\partial\Omega$, the world history of the vortex loop. The result is

$$\phi(y) = \frac{\Gamma}{4\pi} \int \varepsilon_{\mu\nu\lambda\rho} dx^{\nu} \wedge dx^{\lambda} \wedge dx^{\rho} \partial^{\mu} G(y - x)$$

which is (2.6) when $x^0 = ct$ and time and space components are separated. The gradients can be quickly computed in the four dimensional formulation using Stokes' theorem to find

$$\partial_{\mu}\phi(y) = \Gamma \int_{\partial\Omega} dx^{\nu} \wedge dx^{\lambda} \varepsilon_{\mu\alpha\nu\lambda} \partial^{\alpha} G(y - x) \qquad (2.8)$$

away from Ω which, of course, is nothing but (2.7) in four dimensional language. This geometrical way of looking at the flow generated by a vortex loop will prove useful later when studying the response of a vortex filament to sound.

2.2. An Aside: Response of a Singularity to an External Disturbance in Classical Electrodynamics

This section is based on work of Dirac [19], Feynman [20], Teitelboim et. al. [21] and Tabensky [22].

Consider the electromagnetic field as described by the four-vector potential A_μ. In the Lorentz gauge $A^\mu{}_{,\mu} = 0$ it satisfies the wave equation

$$A_{\nu,\mu}{}^{\mu} = 0 \tag{2.9}$$

when no sources are present. The fields generated by a point charge q undergoing a prescribed motion described by a world-line $x^\mu(\tau)$ are given by the Liénard-Wiechert solution

$$a_\mu(y) = \int d^4y' G(y - y') j_\mu(y') \tag{2.10}$$

where

$$j_\mu(y') = q \int d\tau \delta\{y' - x(\tau)\} \frac{dx^\mu}{d\tau}$$

and

$$G(y) = 2\theta(y^0)\delta(y^2)$$

is the retarded Green's function for the wave equation $\partial_\mu \partial^\mu G = 4\pi\delta(y)$ written in Lorentz covariant form. The Liénard-Wiechert solution (2.9) is singular along the world-line of the point charge (the well-known ~ $(\text{distance})^{-2}$ behaviour of Coulomb's law) and thus, strictly speaking, it is a solution of the wave equation (2.9) on a four dimensional manifold which is Minkowsky space less the world line of the particle. The following question arises: Is it possible to determine the response of this singularity to prescribed external field? The answer is yes, with qualifications that will be spelled out below. The key ingredient to understand how this comes about is a variational principle: Eqn's (2.9) are Euler-Lagrange equations for the action

$$S = \frac{1}{4} \int d^4y F_{\mu\nu} F^{\mu\nu} \tag{2.11}$$

where $F_{\mu\nu} \equiv A_{\mu,\nu} - A_{\nu,\mu}$ is the electromagnetic field tensor. Suppose that

$F_{\mu\nu}$ has a singularity that makes $F^2 \equiv F_{\mu\nu}F^{\mu\nu}$ non-integrable. In that case the volume of integration in (2.11) becomes a functional of the singularity, and extrema of S with respect to world histories of the latter will provide equations to determine its evolution. This program can be carried out as follows: suppose we have a test charge and an external field $\tilde{F}_{\mu\nu} = \tilde{A}_{\mu,\nu} - \tilde{A}_{\nu,\mu}$ (it makes sense to keep them separate as the former is not supposed to influence the latter) and the total field at any point off the singularity will be the sum of the two. More precisely, this will hold in all regions where the superposition principle is valid, which means outside some small cut-off region around the singularity. Experiment of course shows that quantum effects becomes important before anything like this classical cut-off region is relevant. Nothing prevents one, however, from asking what happens if one pushes the classical theory to the limit. The answer is that near a singularity the linear theory must break down. How, one doesn't know, and the task at hand is to give a quantitative description of the dynamics of this region of breakdown of the linear theory without knowing what goes on inside it. This is possible only when looking at wavelengths that are large compared to the dimensions of this region, and the question, although it appears today as academic to a certain extent, is not so when the field theory at hand is fluid mechanics, as will be discussed below, or elasticity [23].

So, we have a field $A_\mu = a_\mu + \tilde{A}_\mu$ which is the sum of that generated by a test particle, given by the Liénard-Wiechert expression (2.10), and an external field $\tilde{A}_\mu$ assumed to be well-behaved at the singularity of a_μ. Substitution of this sum into the action (2.11) splits the latter into three terms

$$S = S_s + S_m + S_e,$$

where the self-action S_s involves only a_μ, the mixed action S_m involves both a_μ and $\tilde{A}_\mu$, and the external action S_e involves only $\tilde{A}_\mu$. The integrals in both S_s and S_m must therefore exclude the singularity. The volume of integration will be Minkowski space less a tube around the singularity:

$$x^\mu(\tau) + \varepsilon^\mu$$

where ε^μ is in general a function of proper time τ and two angles, and we shall study what happens when this tube is very thin.

First, take the mixed action:

$$S_m = \frac{1}{2}\int d^4y(\tilde{A}_{\mu,\nu}a^{\mu,\nu} - \tilde{A}_{\mu,\nu}a^{\nu,\mu}).$$

It can be integrated by parts to get

$$S_m = \int dS^\nu \tilde{A}_\mu a^\mu{}_{,\nu}$$

involving an integral over the surface of the tube described above. It is obtained noting that both $\tilde{A}_\mu$ and a_μ are solutions of the wave equation (2.8) in the Lorentz gauge throughout the volume of integration and adjusting $\tilde{A}_\mu$ so the surface at infinity does not contribute. Space like surfaces bounding the action do not contribute either. Substitution of (2.10) into S_m yields

$$S_m = q\int dS^\nu \tilde{A}_\mu (\dot{x}^2)^{-1/2}(\varepsilon_\perp^2)^{-3/2}(\varepsilon_\perp^\nu \frac{dx^\mu}{d\tau} - \varepsilon_\perp^\mu \frac{dx^\nu}{d\tau})$$

where $\dot{x}^\mu \equiv dx^\mu/d\tau$ and $\varepsilon_\perp^\mu \equiv \varepsilon^\mu - (\dot{x}^2)^{-1}(\varepsilon \cdot \dot{x})\dot{x}^\mu$ is the projection of ε^μ on a (hyper) plane perpendicular to the world-line of the particle. The external field is evaluated on the world-line of the particle, $\tilde{A}_\mu = \tilde{A}_\mu\{x^\nu(\tau)\}$. The resulting surface integral along the thin tube can be evaluated and it is found to be independent of the shape of the tube:

$$S_m = 4\pi q \int d\tau \tilde{A}_\mu \frac{dx^\mu}{d\tau}\,. \tag{2.12}$$

Similarly, the self-action can be turned into a surface integral. The contribution from the tube surface will diverge like ε^{-1} where ε is a typical thickness of the tube while the contribution from the surface at infinity will be finite. We now take the approach of studying only those motions in which the energy radiated by the charge is negligible compared to the work done on it by the external field (which means that the accelerations involved are not too large) and the response of the test charge is thus governed by the most singular term in the action. One has then

$$S_s = \frac{1}{2}\int dS^\nu a_\mu a^\mu{}_{,\nu}$$

and substitution of (2.10) gives different terms that vanish, are finite, or grow like ε^{-1} when $\varepsilon \to 0$. In keeping with the approximation mentioned above, only the latter are retained and

$$S_s = \frac{q^2}{2} \int dS^\nu (\dot{x}^2)^{-1} (\varepsilon_\perp^2)^{-2} \frac{dx^\mu}{d\tau} (\varepsilon_\perp^\nu \frac{dx_\mu}{d\tau} - \varepsilon_\perp^\mu \frac{dx^\nu}{d\tau})$$

+ (finite terms).

This is as far as a theory based on Maxwell's equations alone allows one to go. The response of a singularity to an external disturbance depends on an unspecified cut-off parameter, the shape of the world tube we have used to isolate the singularity, which is a measure of our ignorance of what goes on inside that region. To make further progress a choice of tube must be introduced by hand, a convenient one being $\varepsilon^2 = \text{constant}$, a cylindrical tube whose cross section perpendicular to the world line of the particle is uniform and spherical. In this case

$$S_s = \frac{2\pi q^2}{|\varepsilon_\perp|} \int d\tau \, (\frac{dx^\mu}{d\tau} \frac{dx_\mu}{d\tau})^{1/2}$$

which is the action for a point particle of mass $m = 2\pi q^2/|\varepsilon\ |$. This would give a satisfactory explanation to the origin of mass of charged particles as being entirely electromagnetic if they all had the same mass. Experiment shows this is not so and this electromagnetic mass is added to an unknown mechanical mass ("classical mass renormalization") to give the observed inertial mass. Thus, observation privileges one cut-off procedure over others. The resulting action, a functional of both the external field $\tilde{A}_\mu$ and point charge world line $x^\mu(\tau)$ is

$$S = 4\pi m \int d\tau (\dot{x}_\mu \dot{x}^\mu)^{1/2} + 4\pi q \int d\tau \tilde{A}_\mu \dot{x}^\mu + \frac{1}{4} \int d^4 y F^2$$

whose extrema with respect to world line variations are given by the equation

$$m\ddot{x}^\mu = qF^{\mu\nu}\dot{x}_\nu ,$$

the well known relation between acceleration of a point charge and the Lorentz force which, to the extent specified above, has been here derived form the Maxwell's equations.

We now wish to apply these ideas on finding the response of a singularity to an external field to the case of a vortex filament.

2.3. Response of a Vortex Filament to Sound

It was mentioned in section 1 that a vortex filament in a compressible fluid satisfied the equation

$$\dot{\vec{X}}(\sigma,t) \wedge \vec{X}'(\sigma,t) = \vec{v}\{\vec{X}(\sigma,t)\} \wedge \vec{X}',$$

namely, the filament moves with the local fluid velocity and at first sight there is no distinction with the incompressible case. There is one essential difference however: it is due to the contribution of the filament itself to the fluid velocity. In the incompressible approximation the velocity generated by the vortex at one instant of time depends on its position at that same instant and we can speak of an instantaneous effect . This is no longer so when compressibility is considered because the finite speed of sound introduces retardation effects and the motion of one portion of the vortex filament will be affected not by what other portions are doing at the same instant of time but by what they were doing at various previous times. In general, then, we expect the evolution to be governed not by a differential (in time) equation such as (1.12) but by an integrodifferential equation. This is true even in two dimensions, where so called vortex "points" are really cuts of infinite straight lines by a plane, something that finds its mathematical expression in the fact that the Green's function for the wave equation in two dimensions has an infinitely long tail in time (which does not happen in three dimensions) and the motion of a point is indeed affected by its whole previous history.

In order to be quantitative, as well as to gain physical insight into the processes involved in the reaction of vorticity to sound, we shall consider the following idealized situation: a test vortex filament is acted upon by an external flow in a slightly compressible fluid. The point of the filament being a "test" one is that it is possible to keep both components separate. As a result of the external flow, as well as because of its own self-induction, the filament will in general radiate. This radiation will be neglected. Moreover, we shall take a local induction approximation, in which a point on the filament is acted upon only by those other points of the filament that are a small distance away. Thus, any effect of radiation from remote regions of the vortex on itself is not considered. All this is possible as long as the filament has small accelerations and curvature, and it does not bend back on itself. The neglect of radiation means that we consider the vortex sufficiently thin that its response to an outside flow is dominated by the most divergent (it turns out to be of order of the log of the cut-off radius as could be guessed from the incompressible approximation) term in the energy and momentum balance. Within this scheme, it is possible to look at the interaction of sound and vorticity as a theory of fields (fluid flow) and sources (vortex filaments) much as was done in electromagnetism.

To find the equation of motion of a singularity in fluid dynamics with the same strategy that was shown above to be succesful in electrodynamics we need a variational principle. The equations (1.1)-(1.2) are Euler-Lagrange equation for the action

$$S = \int dtd^3x\{\rho \frac{\partial\Phi}{\partial t} + \frac{1}{2}\rho(\nabla\Phi)^2 + \rho E\} \tag{2.13}$$

where $E = H - (p/\rho)$ is the internal energy, for independent variation of ρ and Φ. The volume of integration excludes thin regions around any singularities, which are supposed to enclose all regions of nonlinear behaviour. We may thus approximate

$$S \approx S_A = \int dtd^3x\{(\rho_0 + \rho_1)\frac{\partial\Phi}{\partial t} + \frac{1}{2}\rho_0(\nabla\Phi)^2 + \frac{c^2}{2\rho_0}\rho_1{}^2\}$$

which can be written in terms of Φ only.

$$S_A = \rho_0 \int dtd^3x\{\frac{\partial\Phi}{\partial t} + \frac{1}{2}(\nabla\Phi)^2 - \frac{1}{2c^2}(\frac{\partial\Phi}{\partial t})^2\}.$$

From now on ρ_0, a constant, will be omitted. We now write $\Phi = \phi + \tilde{\Phi}$ where ϕ is the velocity potential of a test vortex filament, given by (2.6), and $\tilde{\Phi}$ is some prescribed "external" velocity potential which is of course non-singular at the location of the test filament. The action splits in three terms: self S_s, mixed S_m and external S_e. Substitution of (2.7) for $\nabla\phi$ and $\partial\phi/\partial t$ lead to an action which is a functional of the filament history $\vec{X}(\sigma,t)$ and the external flow $\tilde{\Phi}(\vec{x},t)$. Computations involving the quadratic terms of both S_s and S_m are greatly simplified by introducing a fourth coordinate $x^0 = ct$ and working in Minkowsky space. In this language one has

$$S_m = \int d^4x\partial_\mu\phi\partial^\mu\tilde{\Phi}, \tag{2.14}$$

$$S_s = \int dtd^3x \frac{\partial\phi}{\partial t} + \frac{1}{2}\int d^4x\partial_\mu\phi\partial^\mu\phi. \tag{2.15}$$

Substitution of (2.8) into S_m gives

$$S_m = \frac{\Gamma}{6}\int_\Omega \partial^\lambda\tilde{\Phi}\varepsilon_{\lambda\mu\nu\rho}dx^\mu \wedge dx^\nu \wedge dx^\rho \tag{2.16}$$

where Ω, as in section 2.1, is the interior of the world history of the vortex filament. This is most directly seen turning (2.14) into a surface integral becuase $\partial_\mu\partial^\mu\tilde{\Phi} = 0$ throughout the volume of integration, adjusting the behaviour of $\tilde{\Phi}$ so that surface integrals at spacelike infinity do not contribute, checking that neither do the spacelike boundaries in the action and the only term left is a surface integral around a thin slab surrounding Ω, across which ϕ has a discontinuity Γ, resulting in (2.16). Although S_m involves Ω, its variations with respect to x have support only along $\partial\Omega$, the boundary of Ω, the filament history [18]:

$$\delta S_m = \Gamma\varepsilon_{\nu\lambda\rho\alpha}\int_{\partial\Omega} \partial^{\alpha}\tilde{\Phi} dx^{\nu} \wedge dx^{\lambda}\delta x^{\rho} \tag{2.17}$$

which, with time separated out is

$$\frac{\partial S_m}{\partial\vec{X}(\sigma,t)} = \Gamma\left(\nabla\tilde{\Phi}\wedge\frac{\partial\vec{X}}{\partial\sigma}\right) + \frac{\Gamma}{c^2}\frac{\partial\tilde{\Phi}}{\partial t}\left(\frac{\partial\vec{X}}{\partial t}\wedge\frac{\partial\vec{X}}{\partial\sigma}\right) \tag{2.18}$$

Next we take the second term on the right of the self action S_s (2.15) and substitute (2.8). In keeping with the approximation scheme discussed above, we retain only the leading term as the cut-off tube becomes very thin, in a local induction approximation. The result is [18]; see also [24]

$$\Gamma^2 \ln\left(\frac{\delta}{\varepsilon}\right)\int d\sigma d\tau\sqrt{-g} + \underset{\text{when} \quad \varepsilon\to 0}{\text{finite term}} \tag{2.19}$$

where $\sqrt{-g}$ is the determinant of the metric of the surface (embedded in Minkowsky space) described by the world-history of the vortex filament. An additional cut-off $\delta \gg \varepsilon$ has been introduced by the local induction approximation: it is the distance (in parameter σ) beyond which points of the filamentary vortex do not influence each other. As it was the case in electrodynamics, the self-action (2.19) is cut-off dependent not only through the explicit ε dependence exhibited but also through the way the limit $\varepsilon \to 0$ is taken. As opposed to electrodynamics, there is no additional information to privilege one choice of limiting procedure over another and we have taken what appears to be the simplest one. A precise discussion of the delicacies involved is deferred to a future publication (Lund and Zabusky, in preparation). Separating the time out (2.19) becomes

$$\Gamma^2 \ln\left(\frac{\delta}{\varepsilon}\right)\int d\sigma dt\{(1-\frac{1}{c^2}\dot{\vec{X}}^2)(\vec{X}')^2+\frac{1}{c^2}(\dot{\vec{X}}\cdot\vec{X}')^2\}^{1/2}. \tag{2.20}$$

Finally, the first term on the right of the self-action S_s (2.15) is

$$\int dtd^3x \frac{\partial\phi}{\partial t} = -\Gamma \int dt \int dS_i V^i \tag{2.21}$$

noting that total time derivatives do not contribute to the equations of motion. It might seem odd to keep a finite term in the self-action since only the leading terms when $\varepsilon \to 0$ were supposed to be retained. To understand this it is worthwile to pause and take a look at the physics of what we are doing. The idea is that to a first approximation filamentary vortex dynamics is governed by the laws of incompressible flow as embodied by the Biot-Savart law in the local induction approximation (1.13). What we are after is an equation that will describe deviations from this behaviour in a slightly compressible fluid, namely, one having a very large speed of sound and we thus expect an equation with additional terms that will vanish in the limit $c \to \infty$. It is these additional terms that will have a varied behaviour in the limit of a very thin vortex, and, for the vortex filament subject to outside radiation that we are considering, we keep only the largest of them. To repeat, it is the corrections to the incompressible behaviour in the response of a test vortex filament to an external flow that we are taking as being dominated by the most divergent (as the vortex thickness vanishes) term in the action. Going back to (2.21), integrating by parts and discarding total time derivatives it is easily found that

$$\int dtd^3x \frac{\partial\phi}{\partial t} = \frac{\Gamma}{3} \int dtd\sigma \vec{X} \cdot \left(\frac{\partial\vec{X}}{\partial t} \wedge \frac{\partial\vec{X}}{\partial\sigma}\right). \tag{2.22}$$

Writing the mixed action (2.17) with time explicitely separated and collecting with the self-action given by (2.20) and (2.22) we get [17] the following action integral, a functional of the external field $\tilde{\Phi}$ and vortex filament history $\vec{X}(\sigma,t)$:

$$S_A = \frac{\Gamma}{3}\int dtd\sigma\vec{X} \cdot \left(\frac{\partial\vec{X}}{\partial t} \wedge \frac{\partial\vec{X}}{\partial\sigma}\right) + \int dtd\sigma L + \Gamma \int dt \int_{S(t)} d\vec{S} \cdot \nabla\tilde{\Phi} +$$

$$+ \frac{\Gamma}{c^2}\int dt \int_{S(t)} d\vec{S} \cdot \vec{V} \frac{\partial\tilde{\Phi}}{\partial t} + \frac{1}{2}\int dtd^3x\left\{(\nabla\tilde{\Phi})^2 - \frac{1}{c^2}\left(\frac{\partial\tilde{\Phi}}{\partial t}\right)^2\right\}, \tag{2.23}$$

where

$$L = \Gamma^2 \ln\frac{\delta}{\varepsilon}\left[\left(1 - \frac{1}{c^2}\dot{\vec{X}}^2\right)(\vec{X}')^2 + \frac{1}{c^2}(\dot{\vec{X}} \cdot \vec{X}')^2\right]^{1/2}.$$

Extrema of this action with respect to variations of filament trajectories give, noting (2.18), a differential equation governing its evolution

$$\frac{\partial}{\partial t}\left(\frac{\partial L}{\partial \dot{\vec{X}}}\right) + \frac{\partial}{\partial \sigma}\left(\frac{\partial L}{\partial \vec{X}'}\right) = \Gamma(\nabla\tilde{\Phi} \wedge \vec{X}') + \Gamma\left(\frac{1}{c^2}\frac{\partial\tilde{\Phi}}{\partial t} - 1\right)\dot{\vec{X}} \wedge \vec{X}'. \tag{2.24}$$

It is valid in a localized induction approximation, for a test vortex, and with the neglect of radiation as has been spelled out in detail above. The local induction approximation for the incompressible case (1.13) is recovered when $c \to \infty$. Incompressible point vortex dynamics is also a special case, for which the left-hand side of (2.24) vanishes. Examples of solutions to (2.24) will be given in a future publication.

ACKNOWLEDGMENTS

I am grateful to N. J. Zabusky for a most profitable collaboration. This work has been supported by the Fondo Nacional de Ciencias and The Departamento de Investigación y Bibliotecas de la Universidad de Chile.

REFERENCES

[1] G. K. Batchelor. An Introduction to fluid dynamics, (Cambridge U. P. 1967).
[2] A. L. Fetter and J. D. Walecka. Theoretical Mechanics of particles and continua, (McGraw Hill 1980).
[3] B. T. Chu and L. S. Kovasznay. Non-linear interactions in a viscous heat-conducting compressible gas. J. Fluid Mech. 3, 494-514 (1958).
[4] H. Aref. Integrable, chaotic and turbulent vortex motion in two-dimensional flows. Ann. Rev. Fluid Mech, 15, 345-389 (1985).
[5] M. Rasetti and T. Regge. Vortices in He II, current algebras and quantum knots. Physica 80A, 217-233 (1975).
[6] D. W. Moore and P. G. Saffman. The motion of a vortex filament with axial flow. Philos. Trans. Roy. Soc. London A272, 403-429 (1972).
[7] E. D. Siggia. Collapse and amplification of a vortex filament. Phys. Fluids 28, 794-805 (1985)
[8] E. D. Siggia and A. Pumir. Incipient singularities in the Navier-Stokes equations. Phys. Rev. Lett. 55, 1749-1752 (1985).
[9] M. J. Lighthill. On sound generated aerodynamically I. General theory. Proc. Roy. Soc. London A211, 564-587 (1952).
[10] M. J. Lighthill. On sound generated aerodynamically II. Turbulence as a source of sound. Proc. Roy. Soc. London A222, 1-32 (1954).
[11] S. C. Crow. Aerodynamic sound emission as a singular perturbation problem. Stud. App. Math. 49, 21-44 (1970).
[12] M. S. Howe, Contributions to the theory of aerodynamic sound, with application to excess jet noise and the theory of the flute. J. Fluid Mech. 71, 625-673 (1975).
[13] W. Mohring. On vortex sound at low Mach number. J. Fluid Mech.

85, 685-691 (1978).
[14] F. Obermeier. On a new representation of aeroacoustic source distribution I. General Theory. Acustica 42, 56-61 (1979).
[15] T. Kambe and T. Minota. Sound radiation from vortex systems. J. Sound Vib. 74, 61-72 (1981).
[16] T. Kambe and T. Minota. Acoustic wave radiated by head-on collision of two vortex rings. Proc. Roy Soc. London A386, 277-308 (1983).
[17] F. Lund and N. J. Zabusky. Compressibility effects in vortex dynamics at constant sound speed. Submitted to Phys. Rev. Lett.
[18] F. Lund. Relativistic string coupled to massless scalar field. Phys. Rev. D33, 3124-3126 (1986).
[19] P. A. M. Dirac. Classical Theory of radiating electrons. Proc. Roy. Soc. London A167, 148-169 (1938).
[20] R. P. Feynmann. A relativistic cut-off for classical electrodynamics. Phys. Rev. 74, 939-946 (1948).
[21] C. Teitelboim, D. Villarroel and Ch. G. Van Weert. Classical electrodynamics of retarded fields and point particles. Rivista Nuovo Cimento, Vol. 3, 1 (1980).
[22] R. Tabensky. Electrodynamics and the electron equation of motion Phys. Rev. D13, 267-273 (1976).
[23] F. Lund. Equation of motion of stringlike dislocation. Phys. Rev. Lett. 54, 14-17 (1985).
[24] F. Lund and T. Regge. Unified approach to string and vortices with soliton solutions. Phys. Rev. D14, 1524-1535 (1976).

SPATIO-TEMPORAL INSTABILITIES IN CLOSED AND OPEN FLOWS

P. Huerre
Department of Aerospace Engineering
University of Southern California
Los Angeles, California 90089-0192
U.S.A.

ABSTRACT. A review is given of the general theory describing the linear evolution of spatio-temporal instability waves in fluid media. According to the character of the impulse response, one can distinguish between absolutely unstable (closed) flows and convectively unstable (open) flows. These notions are then applied to several evolution models of interest in weakly nonlinear stability theory. It is argued that absolutely unstable flows, convectively unstable flows and mixed flows exhibit a very different sensitivity to external perturbations. Implications of these concepts to frequency selection mechanisms in mixed flows and the onset of chaos in convectively unstable flows are also discussed.

1. INTRODUCTION

In hydrodynamics, one often distinguishes between the behavior of closed systems and open systems without being too specific about the meaning of these concepts. From a purely kinematic point of view, one is tempted to say that a flow is closed when fluid particles are recycled within the physical domain of interest. If all fluid particles ultimately leave the domain, the flow is said to be open. On such grounds, Bénard convection within a horizontal fluid layer is a closed system whereas plane Poiseuille flow, wakes, jets, etc... are open systems.

In the present review, closed and open flows will be defined instead with respect to the character of the hydrodynamic instabilities which they can support. Thus, the motion of waves is claimed to be more relevant than the motion of fluid *particles*. Following Briggs [1] and Bers [2], a flow will be said to be absolutely unstable (closed) if its impulse response is unbounded everywhere for large time (Figure 2b). If the impulse response decays to zero at all spatial locations, the flow will be said to be convectively unstable (open) (Figure 2c). These notions will be defined and reviewed in detail in section 2. In other words, from a hydrodynamic stability point of view, closed and open systems refer to absolutely unstable and

E. Tirapegui and D. Villarroel (eds.), Instabilities and Nonequilibrium Structures, 141–177.

convectively unstable flows, respectively. In the core of this paper, we shall use this latter terminology only.

It should be made clear that a given flow can be open with respect to fluid particle trajectories but closed (absolutely unstable) with respect to instability waves. For instance, a parallel wake flow with a $\mathrm{sech}^2 y$ velocity profile is absolutely unstable for large values of the velocity deficit [3], see discussion of section 4.3. Yet, fluid particles clearly move out of the physical domain and no recycling of particles takes place. Similarly, parallel axisymmetric hot jets are absolutely unstable if the core temperature is sufficiently higher than the ambient temperature [4,5]. Here again, fluid particles move downstream but hydrodynamic instability waves contaminate the entire medium. Thus, in many cases, one must rely on *mathematical* criteria to determine whether a given flow is absolutely unstable or convectively unstable. These criteria which were first proposed by Briggs [1] in the context of plasma instabilities are discussed in section 2.

In section 3, the general method is illustrated on three amplitude evolution models commonly arising in the context of hydrodynamic instabilities: the Ginzburg-Landau equation, the Klein-Gordon equation and the long-wavelength integro-differential equation obtained in reference [6]. The implications of these notions to receptivity issues, frequency selection mechanisms and chaos in open flows are critically reviewed in section 4.

For a recent review of similar topics within the context of shear flows, the reader is referred to [7].

2. GENERAL FORMALISM

The main concepts will first be developed in the general context of spatio-temporal waves in one-dimensional space. The development broadly follows the review of Bers [2]. Let $A(x,t)$ be a scalar function of space coordinate x and time t, which may represent for instance the fluctuating velocity or temperature field in a fluid medium. More generally, $A(x,t)$ could be a column vector with an arbitrary number of components. For simplicity, $A(x,t)$ is assumed here to have a single component only and to satisfy a differential equation of the form

$$D\left[-i\frac{\partial}{\partial x}, i\frac{\partial}{\partial t}; R\right] A(x,t) = S(x,t), \tag{1}$$

where D is a *linear* differential operator arising from a perturbation analysis around some basic state of the medium. The function $S(x,t)$ specifies the excitation imposed on the system in some region of space for a given length of time. We assume that $S(x,t) = 0$ everywhere when $t < 0$. The flow may depend on an external control parameter R, such as the Reynolds number, Rayleigh number, etc...

Fourier-transform pairs in space x and time t are then introduced, according to the definition

$$A(x,t) = \frac{1}{(2\pi)^2} \int_F \int_L A(k,\omega) e^{i(kx-\omega t)} \, d\omega dk. \tag{2}$$

In equation (2), the path F in the complex plane of wavenumbers k is initially taken to be the real axis. The contour L in the complex frequency plane is chosen so that causality is satisfied, namely $A(x,t) = 0$ everywhere when $t < 0$: it is a straight line lying above all the singularities of the complex ω plane. When $t < 0$, the L contour can then be closed by a semicircle above L and the response $A(x,t)$ is identically zero. A sketch of the paths of integration is included in Figure 1.

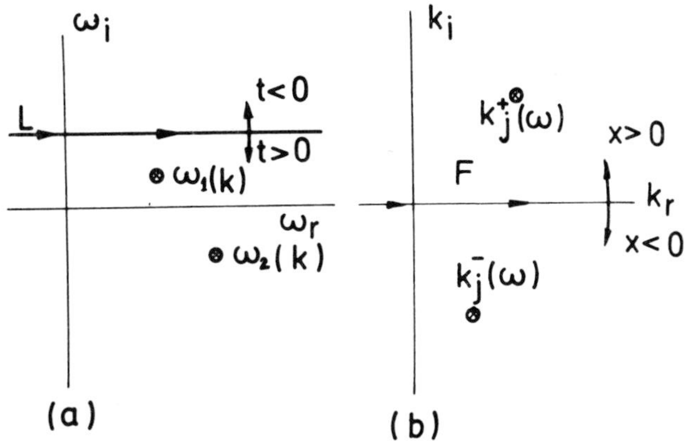

Figure 1. Paths of integration of the inverse Fourier transforms (2): (a) complex frequency plane; (b) complex wavenumber plane.

The Fourier transform of equation (1) readily provides a statement of the problem in the spectral domain:

$$D[k,\omega;R]A(k,\omega) = S(k,\omega). \tag{3}$$

In the absence of forcing, $S(x,t) = S(k,\omega) \equiv 0$, one obtains the *normal modes* of the system in the form of a complex dispersion relation

$$D[k,\omega;R] = 0 \tag{4}$$

between wavenumber k and frequency ω. When k is taken to be real, the zeroes of the dispersion relation yield a collection of *temporal* normal modes with complex eigenvalues $\omega_j(k)$, $j = 1,2,3,...$ Conversely, when ω is given real, one obtains *spatial* branches $k_j^+(\omega), k_j^-(\omega)$, etc. Both sets play a crucial role in the evaluation of the Green's function as discussed in the next section.

In the forced problem, one can immediately solve for $A(k,\omega)$ in (3) to arrive at a purely formal expression of the solution in terms of the inverse Fourier transform (2):

$$A(x,t) = \frac{1}{(2\pi)^2} \int_F \int_L \frac{S(k,\omega)}{D[k,\omega;R]} e^{i(kx-\omega t)} d\omega dk, \tag{5}$$

the essential problem being the evaluation of the integrals along L and F!

2.1. Spatio-temporal evolution: The Green's function or Impulse response

Let $G(x,t)$ be the response to a unit impulse $\delta(x)\delta(t)$, i.e. the Green's function of the operator D. We recall that the response of a linear system to an arbitrary forcing input $S(x,t)$ can be determined by convolution of G with S over space and time. From equation (1), the Green's function $G(x,t)$ satisfies

$$D\left[-i\frac{\partial}{\partial x}, i\frac{\partial}{\partial t}; R\right]G(x,t) = \delta(x)\delta(t). \tag{6}$$

Equivalently, one obtains in the spectral domain (see (3)):

$$D[k,\omega;R]G(k,\omega) = 1 \tag{7}$$

At this stage, one may follow one's personal taste to choose the order of the inverse Fourier transforms in (5). We shall first perform the integration in ω-space

$$G(k,t) = \frac{1}{2\pi} \int_L \frac{e^{-i\omega t}}{D[k,\omega;R]} d\omega. \tag{8}$$

When $t < 0$, the contour L is closed from above and no residues contribute to $G(k,t)$ so that $G(k,t) = 0$. When $t > 0$, the contour L is closed by a semi-circle at infinity in the lower half ω-plane bounded by L (see Figure 1a). We shall assume, for simplicity, that the only singularities in the integrand are poles arising from the zeroes of the dispersion relation $D[k,\omega;R]$. Since k is given on the real axis F in the complex k-plane, these poles are located at the temporal (k real, ω complex) eigenmodes $\omega_j(k)$, $j = 1,2,3,\ldots$ Thus the normal modes of the unforced problem naturally arise in the calculation of G to give the residue contributions

$$G(k,t) = -i \sum_j \frac{e^{-i\omega_j(k)t}}{(\partial D/\partial\omega)[k,\omega_j(k);R]} . \tag{9}$$

Inversion of the Fourier transform with respect to k then leads to the wave packet integrals

$$G(x,t) = -\frac{i}{2\pi}\sum_j \int_{-\infty}^{\infty} \frac{e^{i[kx-\omega_j(k)t]}}{(\partial D/\partial\omega)[k,\omega_j(k);R]}\,dk \tag{10}$$

Each integral can be evaluated for large time, x/t fixed, by applying the method of steepest descent [8,9]. Details very much depend on the nature of $\omega_j(k)$ (see specific applications in section 3). To pursue the formal development, it is assumed that each branch $\omega_j(k;R)$ gives rise to a single stationary point k_j^* satisfying

$$\frac{d\omega_j}{dk}(k_j^*) = \frac{x}{t}\,. \tag{11}$$

The surface Σ defined by $\phi(k_r,k_i) = \mathrm{Re}\{i(kx/t - \omega_j)\}$ exhibits in (k_r,k_i,ϕ) space a saddle point at $k_r = k_{jr}^*$, $k_i = k_{ji}^*$. The original contour F can be deformed into a steepest descent path issuing from each saddle point k_j^*, provided no hills of the surface Σ are crossed at infinity. Using standard arguments one obtains

$$G(x,t) \sim -(2\pi)^{-1/2}e^{i\pi/4}\sum_j \frac{e^{i[k_j^*x - \omega_j(k_j^*)t]}}{(\partial D/\partial\omega)[k_j^*,\omega_j(k_j^*)][(d^2\omega_j/dk^2)(k_j^*)t]^{1/2}}\,. \tag{12}$$

The Green's function takes the form of a train of wavepackets in the (x,t)-plane. For a given packet, the flow selects, along each ray x/t = const., a particular wavenumber k_j^* given by (11). The temporal amplification rate of k_j^* along the ray reduces to $\sigma = \mathrm{Im}\{\omega_j(k_j^*) - k_j^*(d\omega_j/dk)(k_j^*)\}$ and may in general be positive or negative. Different situations are now examined.

2.2. A criterion for convective or absolute instability

Consider, for a moment, a dispersion relation with a single normal mode $\omega(k)$. Definitions and results can readily be extended to systems with multiple modes. According to the character of the impulse response G(x,t), one may first distinguish between stable and unstable flows.

A flow is defined as *stable* if

$$\lim_{t\to\infty} G(x,t) = 0, \text{ along all rays } x/t = \text{const.} \tag{13}$$

In other words, the temporal growth rate $\omega_i(k) \equiv \mathrm{Im}\,\omega(k)$ of the normal mode is negative for all real wavenumbers, and the response of the system takes the form of a decaying wavepacket, as sketched in Figure 2a.

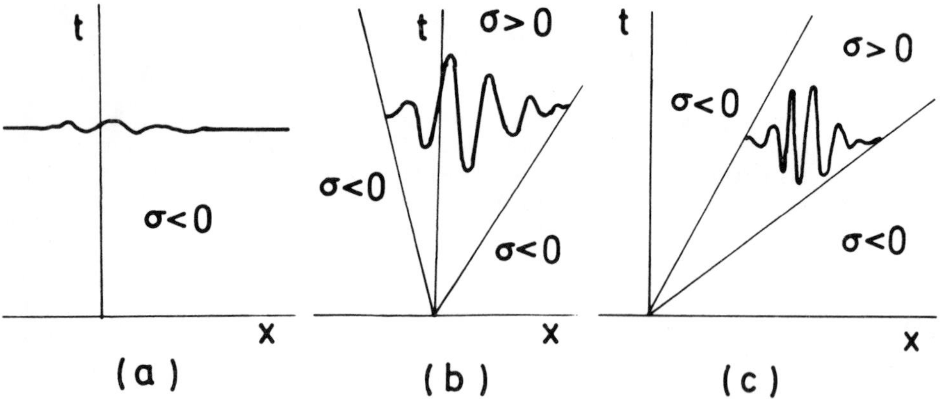

Figure 2. Sketch of impulse responses: (a) stable flow; (b) absolutely unstable flow; (c) convectively unstable flow.

A flow is *unstable* if *there exists rays* x/t = *const., along which*

$$\lim_{t\to\infty} G(x,t) = \infty \;. \tag{14}$$

In such a situation, the temporal growth rate $\omega_i(k)$ is positive in some range of real wavenumbers, and the impulse excitation gives rise to an unstable wavepacket confined within a wedge bounded by the two rays of zero amplification rate (see Figures 2b,c). Disturbances grow exponentially in the wedge as given by equation (12). Let k_{max} denote the wavenumber of maximum temporal growth rate such that $(d\omega_i/dk)(k_{max}) = 0$. As seen from (11), k_{max} travels along the particular ray $x/t = (d\omega_r/dk)(k_{max})$. The peak of the wavepacket, which grows at the maximum allowable rate $\omega_i(k_{max})$, therefore moves in that same direction.

But, within the unstable flow category, one must further distinguish between absolute instability and convective instability, as illustrated in Figures 2b and 2c.

An unstable flow is said to be *absolutely unstable* if *for all fixed* x,

$$\lim_{t\to\infty} G(x,t) = \infty \;. \tag{15}$$

An unstable flow is said to be *convectively unstable* if, *for all fixed* x,

$$\lim_{t\to\infty} G(x,t) = 0 \;. \tag{16}$$

The nature of the instability, is determined by the *long-time* behavior of the wavenumber k_0 staying at a *fixed* spatial location along the ray $x/t = 0$, i.e. the wavenumber of zero group velocity $(d\omega/dk)(k_0) = 0$. Note that, in general, the corresponding complex frecuency $\omega_0 = \omega(k_0)$ will be an algebraic branch point of the function $k(\omega)$. In absolutely unstable flows (Figure 2b) the edges of the wavepacket travel in opposite directions and the ray $x/t = 0$ remains in the unstable wedge. Thus, k_0 must have a positive growth rate $\omega_i(k_0) > 0$. Conversely, in convectively unstable flows (Figure 2c) the edges of the packet travel in the same direction, leaving the ray $x/t = 0$ outside the wedge so that $\omega_i(k_0) < 0$. Following Pierrehumbert [10], the quantity $\omega_i(k_0)$ can be called the *absolute growth rate*: $\omega_i(k_0)$ denotes the temporal growth rate of the wavenumber k_0 staying at a fixed x location, whereas $\omega_i(k_{max})$, defined previously, is the temporal growth rate following the peak of the wavepacket. We are led to conclude from the above argument that an unstable flow is convectively unstable when its absolute growth rate is negative: the branch-point singularities of $k(\omega)$ lie in the lower half ω-plane. When the absolute growth rate is positive, the branch points of $k(\omega)$ lie in the upper half ω-plane and the flow is absolutely unstable.

This criterion, however, is not explicit enough as it stands, and one needs to carefully monitor the positions of the zeroes $\omega_i(k)$ and $k_j^{\pm}(\omega)$ of $D(k,\omega)$ located in the ω-plane and k-plane, respectively. To satisfy causality, the contour L in the ω-plane (Figure 1a) can always be placed high enough so that the zeroes $\omega_j(k)$ lie below L when k is real on the initial contour F of Figure 1b. Conversely [2], when ω is on L, none of the zeroes $k_j^+(\omega)$, $k_j^-(\omega)$, etc... of $D(k,\omega)$ in the k-plane can then cross the original contour F. If they did, L itself would intersect one of the curves $\omega_j(k)$ in the ω-plane, which leads to a contradiction. Thus, provided L is high enough, the original F contour neatly separates the spatial branches $k_j^+(\omega)$ and $k_j^-(\omega)$ located in the upper and lower half k-planes. When $x > 0$ $(x < 0)$, the contour F is closed in the upper (lower) half k-plane and the residues of the spatial branches $k_j^+(\omega)$ $(k_j^-(\omega))$ contribute. Assume as in the rest of our discussion that a single second-order branch point $\omega_{0,j}$ is associated with each mode $\omega_j(k)$. Two radically distinct situations may then take place.

First, the two Riemann sheets of the branch point $\omega_{0,j}$ may correspond to *spatial branches* $k_j^+(\omega)$ *and* $k_j^-(\omega)$ *located, when* L *is high enough*, on opposite sides of F, i.e., *in the upper and lower half k-planes respectively.* Then, as L is displaced downward, the curves $k_j^+(\omega)$ and $k_j^-(\omega)$ move towards each other in the k-plane (Figure 3a,b). In this process, one must correspondingly deform the original contour F so as to retain the same number of spatial branches for $x > 0$ and $x < 0$, while at the same time lowering the curve of zeroes $\omega_j(k)$ in the ω-plane. Of course, the simultaneous deformation of L and F must stop when L touches $\omega_j(k)$ and F becomes "pinched" between the branches $k_j^+(\omega)$ and $k_j^-(\omega)$ [2], as sketched in Figure 3c. This is precisely the point (k_{0j},ω_{0j}) identified previously, where the group velocity $d\omega/dk$ is zero. When pinching takes place with ω_0 still located in the upper ω-plane, the instability is absolute. Otherwise, it is convective.

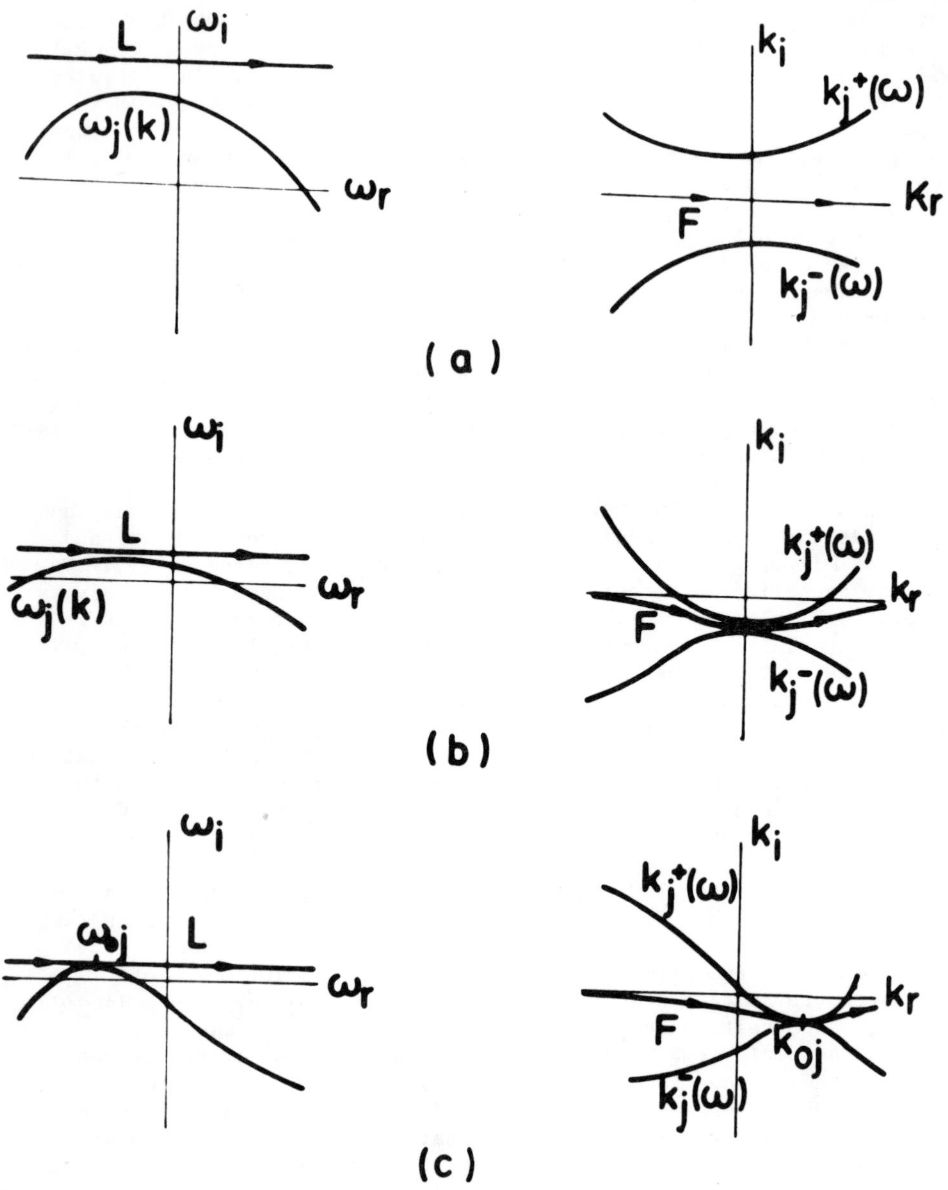

Figure 3. Locus of spatial branches $k_j^{+,-}(\omega)$ as L contour is displaced downward in complex ω-plane. (a), (b), and (c) describe different stages of pinching process.

As emphasized in a recent paper by Pierrehumbert [11], it may also happen that the two Riemann sheets of the branch point ω_{0j} pertain to *spatial branches, say* $k_j^+(\omega)$ *and* $k_j^{+'}(\omega)$ *located, for high enough* L, *in the same half* k-*plane*. When the contour L is lowered, no pinching of the F contour can therefore occur. The corresponding branch point (ω_{0j}, k_{0j}) is not associated with an absolute growth rate. Presumably, the contour F cannot be deformed continuously to go through the saddle point along a path of steepest descent and the mode ω_{0j} does not appear in the response (12). An example of this type of behavior is given in section 3.2.

Thus a flow is convectively (absolutely) unstable when the branch points of $k(\omega)$, which pertain to spatial branches $k^+(\omega)$ and $k^-(\omega)$ originating from *distinct* halves of the k-plane, are located in the lower (upper) half ω-plane.

2.3. Spatial waves: the signalling problem in convectively unstable media.

Let us now consider the response of the flow to a simple monochromatic input of frequency ω_f located at $x = 0$ and switched on at $t = 0$. The initial state is assumed to be identically zero so that

$$D\left[-i\frac{\partial}{\partial x}, i\frac{\partial}{\partial t}; R\right]A(x,t) = \delta(x)H(t)e^{-i\omega_f t} \tag{17}$$

which, in the spectral domain, yields

$$A(k,\omega) = \frac{i}{D[k,\omega;R]\,(\omega - \omega_f)} \, . \tag{18}$$

The evaluation of the inverse Fourier transform in ω

$$A(k,t) = \frac{i}{2\pi}\int_L \frac{e^{-i\omega t}}{D[k,\omega;R]\,(\omega - \omega_f)}\, d\omega \tag{19}$$

leads us, when $t > 0$, to distinguish several contributions to the response arising from the poles of the integrand in (19). The transient signal, due to switch-on, is associated with the zeroes of the dispersion relation $D[k,\omega;R]$, whereas the steady-state response is related to the simple pole at $\omega = \omega_f$. A straightforward residue calculation gives

$$A(k,t) = 2\sum_j \frac{\omega_j(k)e^{-i\omega_j(k)t}}{[\omega_j^2(k) - \omega_f^2]\,(\partial D/\partial\omega)[k,\omega_j(k);R]} + \frac{e^{-i\omega_f t}}{D[k,\omega_f;R]} \, , \tag{20}$$

and the Fourier transform of (20) with respect to k reads

$$A(x,t) = \frac{1}{\pi} \sum_j \int_{-\infty}^{+\infty} \frac{\omega_j(k) e^{i[kx - \omega_j(k)t]}}{[\omega_j^2(k) - \omega_f^2](\partial D/\partial \omega)[k,\omega_j(k);R]} dk$$

$$+ \frac{e^{-i\omega_f t}}{2\pi} \int_{-\infty}^{+\infty} \frac{e^{ikx}}{D[k,\omega_f;R]} dk. \tag{21}$$

As $t \to \infty$, the first term's asymptotics can be obtained via the method of steepest descent, in exactly the same manner as in section 2.1. The second term is calculated by closing the contour F in the upper (lower) half k-plane for $x > 0$ ($x < 0$). Residue contributions arise from the zeroes $k_j^+(\omega_f)$ and $k_j^-(\omega_f)$ of the dispersion relation at a fixed real frequency ω_f. These are precisely the spatial branches encountered in many calculations. One arrives at the final result:

$$A(x,t) \sim \left(\frac{2}{\pi}\right)^{\frac{1}{2}} e^{-i\pi/4} \cdot \tag{22}$$

$$\cdot \sum_j \frac{\omega_j(k_j^*) e^{i[k_j^* x - \omega_j(k_j^*)t]}}{[\omega_j^2(k_j^*) - \omega_f^2](\partial D/\partial \omega)[k_j^*,\omega_j(k_j^*);R][d^2\omega_j)/dk^2)(k_j^*)t]^{\frac{1}{2}}}$$

$$+ i\sum_j \frac{e^{i[k_j^+(\omega_f)x - \omega_f t]}}{(\partial D/\partial k)[k_j^+(\omega_f),\omega_f;R]} H(x) - i\sum_j \frac{e^{i[k_j^-(\omega_f)x - \omega_f t]}}{(\partial D/\partial k)[k_j^-(\omega_f),\omega_f;R]} H(-x),$$

where the wavenumbers k_j^* are given, along the ray x/t, by equation (11) and H denotes the Heaviside unit step function. The solution is composed of a switch-on transient of the same qualitative form as the Green's function (12) and a "steady-state" response arising from forcing the flow at the frequency ω_f. The latter part takes the form of *spatially* growing and/or decaying waves located on either side of the source. The spatial branches $k_j^+(\omega_f)$ and $k_j^-(\omega_f)$ have unambiguously been assigned to the domains $x > 0$ and $x < 0$, respectively. This stems from the fact that they originate, for high enough L, from the upper and lower half k-plane respectively, as discussed in section 2.2.

If the flow is absolutely unstable, the transient contribution will progressively overwhelm the "steady-state" response at all spatial locations, thereby making the signalling problem meaningless. In a sense, spatially-growing waves are pathologically unstable to any kind of perturbations. However, if the flow is convectively unstable, transients will gradually move away from the source, leaving a genuinely observable steady-state signal. Spatially-growing waves are only relevant in convectively unstable physical systems.

3. APPLICATION TO AMPLITUDE EVOLUTION MODELS

In the following, the general concepts introduced in the previous sections are applied to three specific amplitude evolution models which commonly arise in the study of hydrodynamic instabilities.

3.1 The linearized Ginzburg-Landau equation

Perturbation analysis of fluid dynamical systems close to marginal stability at a finite wavenumber often lead to the Ginzburg-Landau

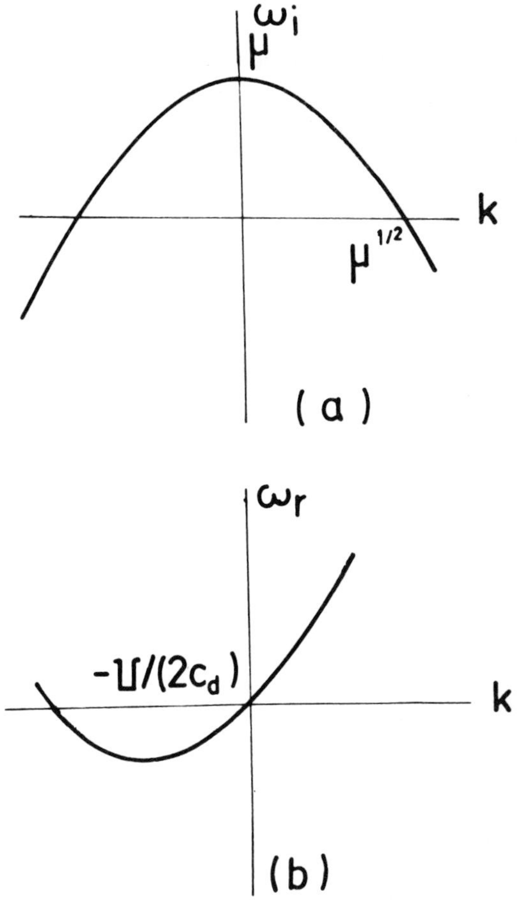

Figure 4. Temporal mode of the Ginzburg-Landau equation: (a) Temporal growth rate ω_i versus real wavenumber k; (b) ω_r versus real wavenumber k.

equation. In the particular context of Rayleigh-Bénard convection, this model bears the name of Newell-Whitehead [12] and Segel [13]. In plane Poiseuille flow, it has been studied and derived by Stewartson & Stuart [14]. If A(x,t) denotes the complex amplitude function of a wavepacket, the linear operator D takes the particular form

$$\frac{\partial A}{\partial t} + U\frac{\partial A}{\partial x} - \mu A - (1 + ic_d)\frac{\partial^2 A}{\partial x^2} = 0, \tag{23}$$

where U, μ and c_d are given real parameters. The parameter μ measures the "degree of supercriticality", i.e. how deep inside the unstable domain the system is. The constant c_d is calculated once and for all in a given flow situation, and U is a velocity which, when nonzero, breaks the reflectional symmetry $x \to -x$. The dispersion relation gives rise to a single temporal mode (Figures 4a,b)

$$\omega(k) = i\mu + Uk + (c_d - i)k^2, \tag{24}$$

of group velocity

$$\frac{d\omega}{dk} = U + 2(c_d - i)k. \tag{25}$$

When $\mu < 0$, the system is stable. When $\mu > 0$, it is unstable, the maximum growth rate $\omega_i(k_{max}) = \mu$ occurring at $k_{max} = 0$. The branch point of $k(\omega)$ is obtained by solving for $(d\omega/dk)(k_0) = 0$, which immediately yields

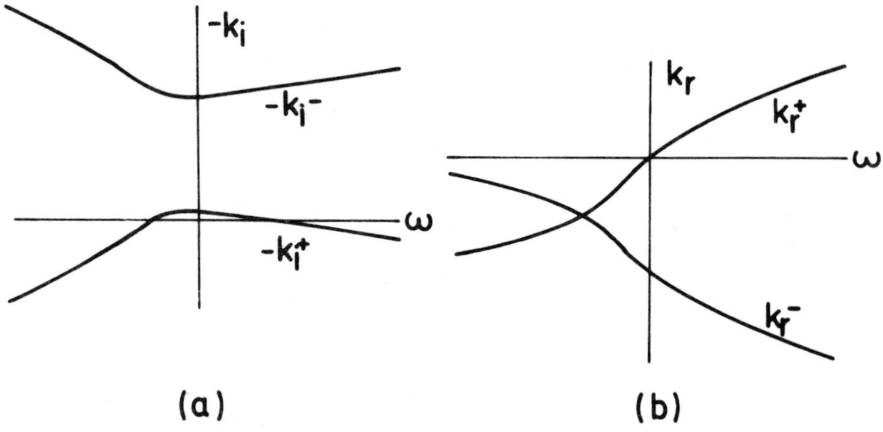

Figure 5. Spatial branches of the Ginzburg-Landau equation: (a) Spatial growth rate $(-k_i)$ versus real frequency ω; (b) k_r versus real frquency ω. The velocity U is such that $U > 2[\mu(1 + c_d^2)]^{1/2}$.

$$k_0 = -\frac{U}{2(c_d - i)} \ ; \quad \omega_0 = i\mu - \frac{U^2}{4(c_d - i)} \, . \tag{26a,b}$$

The temporal mode may then be rewritten in the convenient form

$$\omega(k) - \omega_0 = (c_d - i)(k - k_0)^2, \tag{27}$$

thereby clearly displaying the fact that the singularity is an algebraic branch point of order two. For a given value of ω, there exists two spatial branches (Figure 5a,b) given by

$$k^{+,-}(\omega) = k_0 \pm \left(\frac{\omega - \omega_0}{c_d - i}\right)^{\frac{1}{2}} \tag{28}$$

As the complex frequency $\omega = \omega_r + i\omega_{iL}$ varies along the straight line contour L, at the height ω_{iL}, the spatial branches are restricted to the hyperbola

$$(1 + c_d^2)(k_i - k_{0,i})^2 - (k_r - c_d k_i) = \omega_{iL} - \omega_{0,i} \, , \tag{29}$$

as sketched in Figure 6a,b. As ω_{iL} becomes sufficiently large, the two spatial branches $k^+(\omega)$ and $k^-(\omega)$ belong to distinct halves of the k-plane, and (k_0,ω_0) is a genuine singular point for the determination of absolute or convective instability (see section 2.2).

The absolute growth rate is equal to the imaginary part of (26b),

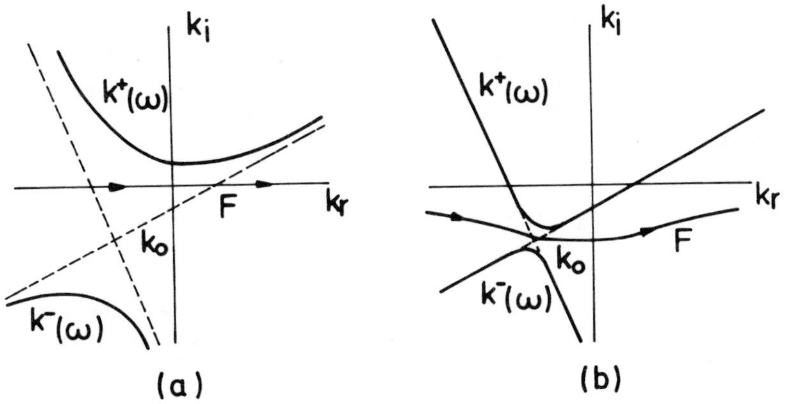

Figure 6. Locus of spatial branches as $\omega_{0,i}$ is decreased: (a) and (b) describe different stages of pinching process for the Ginzburg-Landau equation.

namely

$$\omega_i(k_0) = \omega_{0,i} = \mu - \frac{U^2}{4(1+c_d^2)} . \tag{30}$$

When $|U| < 2\mu^{1/2}(1+c_d^2)^{1/2}$, the absolute growth rate is positive and the system is absolutely unstable. When $|U| > 2\mu^{1/2}(1+c_d^2)^{1/2}$, it is negative and the system is convectively unstable.

The Green's function $G(x,t)$ is given by (10) with $\omega(k)$ specified by (27), so that

$$G(x,t) = \frac{1}{2\pi} e^{-i\omega_0 t} \int_{-\infty}^{+\infty} e^{i[kx-(c_d-i)(k-k_0)^2 t]} dk . \tag{31}$$

This integral can be evaluated exactly by a straightforward change of variable to give the closed form solution

$$G(x,t) = \frac{1}{2} (\pi t)^{-1/2} (1 + ic_d)^{-1/2} \exp[i(k_0 x - \omega_0 t) - \frac{x^2}{4(1+ic_d)t}] . \tag{32}$$

The method of steepest descent would yield in this case the exact result (32) by application of equation (12). The response to an impulse takes the form of a wavepacket which displays all of the main characteristics of the general case as discussed in section 2.1. The peak grows at the maximum growth rate μ along the ray $x/t = U$. The edges of the wavepacket, where the growth rate is equal to zero, move at the velocities $U \pm 2(1+c_d^2)^{\frac{1}{2}}\mu^{\frac{1}{2}}$. At a *fixed* x location, the asymptotic temporal growth rate is, as $t \to \infty$, ω_{0i}, the absolute growth rate at $k = k_0$. An alternate expression for (32) is

$$G(x,t) = \frac{1}{2} (\pi t)^{-\frac{1}{2}} (1+ic_d)^{-\frac{1}{2}} \exp[\mu t - \frac{(x-Ut)^2}{4(1+ic_d)t}] . \tag{33}$$

The reader may apply to this result all the definitions and concepts introduced in section 2, to recover the main conclusions regarding the stability or instability of the system. The parameter μ is seen to control the peak growth rate while the parameter U controls the absolute growth rate. Different responses are ilustrated in Figure 7.

In the convectively unstable case, the solution of the signalling problem satisfies (22), the spatially-evolving waves issuing from the monochromatic source having characteristics displayed on Figures 5a,b.

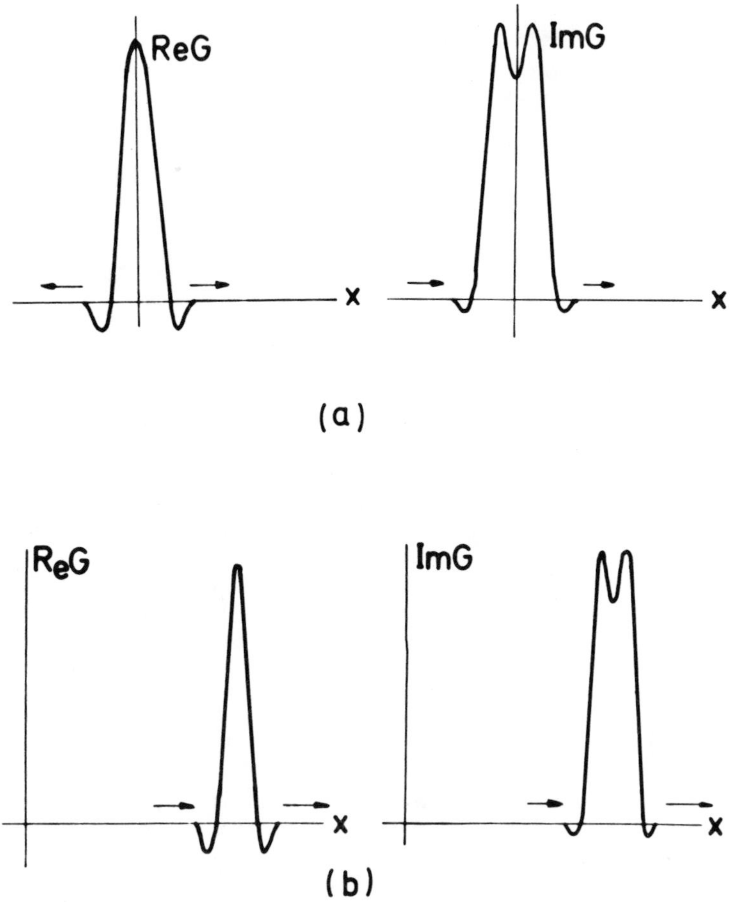

Figure 7. Green's function of the Ginzburg-Landau equation: (a) absolutely unstable case; (b) convectively unstable case.

3.2. The linear Klein-Gordon equation

The cubic nonlinear Klein-Gordon equation commonly arises in inviscid marginal stability analysis such as the baroclinic instability of a quasi-geostrophic two-layer model on the β plane [15], the Kelvin-Helmoltz instability of two layers of immiscible fluids [16], or the buckling of thin shells [17]. When linearized, the model takes the form

$$\left(\frac{\partial}{\partial t} + U\frac{\partial}{\partial x}\right)^2 A - \frac{\partial^2 A}{\partial x^2} - \mu A = 0 \ , \tag{34}$$

and the dispersion relation in Fourier space reduces to

$$D[k,\omega] \equiv -(\omega - Uk)^2 + k^2 - \mu = 0 . \tag{35}$$

There are two temporal modes (figures 8a,b)

$$\omega_{1,2}(k) = Uk \pm (k^2 - \mu)^{\frac{1}{2}} \ , \tag{36}$$

of group velocity

$$\frac{d\omega_{1,2}}{dk} = U \pm k(k^2 - \mu)^{-\frac{1}{2}} \ . \tag{37}$$

When $\mu < 0$, the system is neutrally stable, in the sense that $\omega_1(k)$ and $\omega_2(k)$ are purely real. When $\mu > 0$, the flow is unstable to wavenumbers in the range $|k| < \mu^{\frac{1}{2}}$. The value of U determines, as before, the character of the instability.

In the range $|U| < 1$, the two spatial branches are given by

$$k^{\pm}(\omega) = -\frac{\omega U}{1-U^2} \pm \frac{1}{1-U^2}[\omega^2 + \mu(1 - U^2)]^{\frac{1}{2}} \ , \tag{38}$$

and the branch points of $k(\omega)$ are located on the imaginary axis at

$$\omega_0 = \pm i[\mu(1 - U^2)]^{\frac{1}{2}} \ ; \ k_0 = \pm iU[\mu/(1 - U^2)]^{\frac{1}{2}} \ . \tag{39}$$

As in the previous example, one may check that, for ω_{iL} sufficiently large, the spatial branches are restricted to the upper and lower half k-plane respectively. Pinching first takes place at $\omega_0 = i[\mu(1 - U^2)]^{\frac{1}{2}}$ and $k_0 = -iU[\mu/(1 - U^2)]^{\frac{1}{2}}$, as ω_{iL} is gradually decreased (figures 9a,b,c). The absolute growth rate is positive and it must be concluded that the instability is absolute.

For parameter values such that $|U| > 1$, the spatial branches can be more conveniently written as

$$k^{+}_{1,2} = \frac{\omega U}{U^2-1} \mp \frac{1}{U^2-1}[\omega^2 - \mu(U^2 - 1)]^{\frac{1}{2}} \ , \tag{40}$$

and the branch points at

$$\omega_0 = \pm[\mu(U^2 - 1)]^{\frac{1}{2}} \ , \ k_0 = \pm U[\mu/(U^2 - 1)]^{\frac{1}{2}} \tag{41}$$

are confined to the real ω-axis. At large values of ω_{iL}, the spatial branches are located in the *same* half k-plane (upper half k-plane if $U > 1$, lower half k-plane if $U < -1$). As ω_{iL} is decreased to zero so that the L contour touches the branch points on the real ω-axis, the spatial roots "collide" *without* pinching the F contour (see Figures 10a, b,c). The branch points at $\omega_0 = \pm[\mu(U^2 - 1)]^{\frac{1}{2}}$ therefore do not contribute to an absolute growth rate which, at any rate, would be zero. The instability is convective.

The Green's function can be calculated exactly [18,16], by inverting $G(k,\omega) = 1/D(k,\omega)$, with $D(k,\omega)$ given by (35), For definitiveness, it is assumed that $|U| < 1$, but the same final result holds when $|U| > 1$. Instead of performing the ω- integration first (see section 2.1) we choose here, as an illustration, to first evaluate

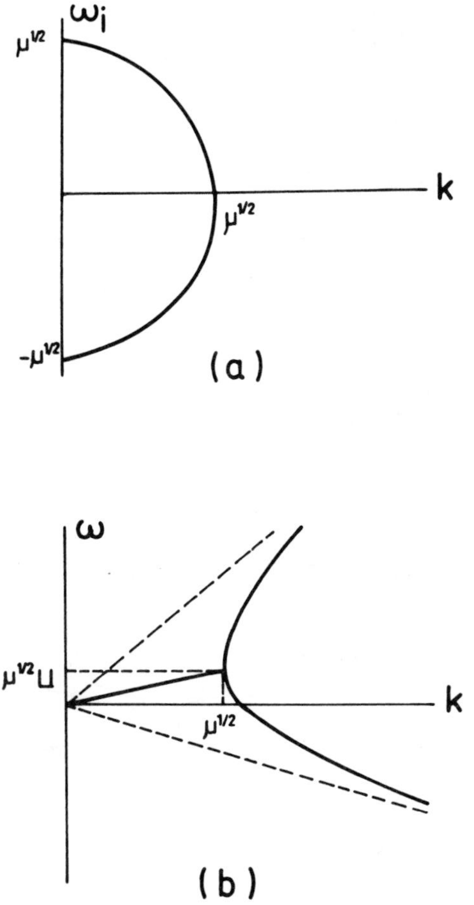

Figure 8. Temporal modes of the Klein-Gordon equation: (a) Temporal growth rate ω_i versus real wavenumber k; (b) ω_r versus real wavenumber k.

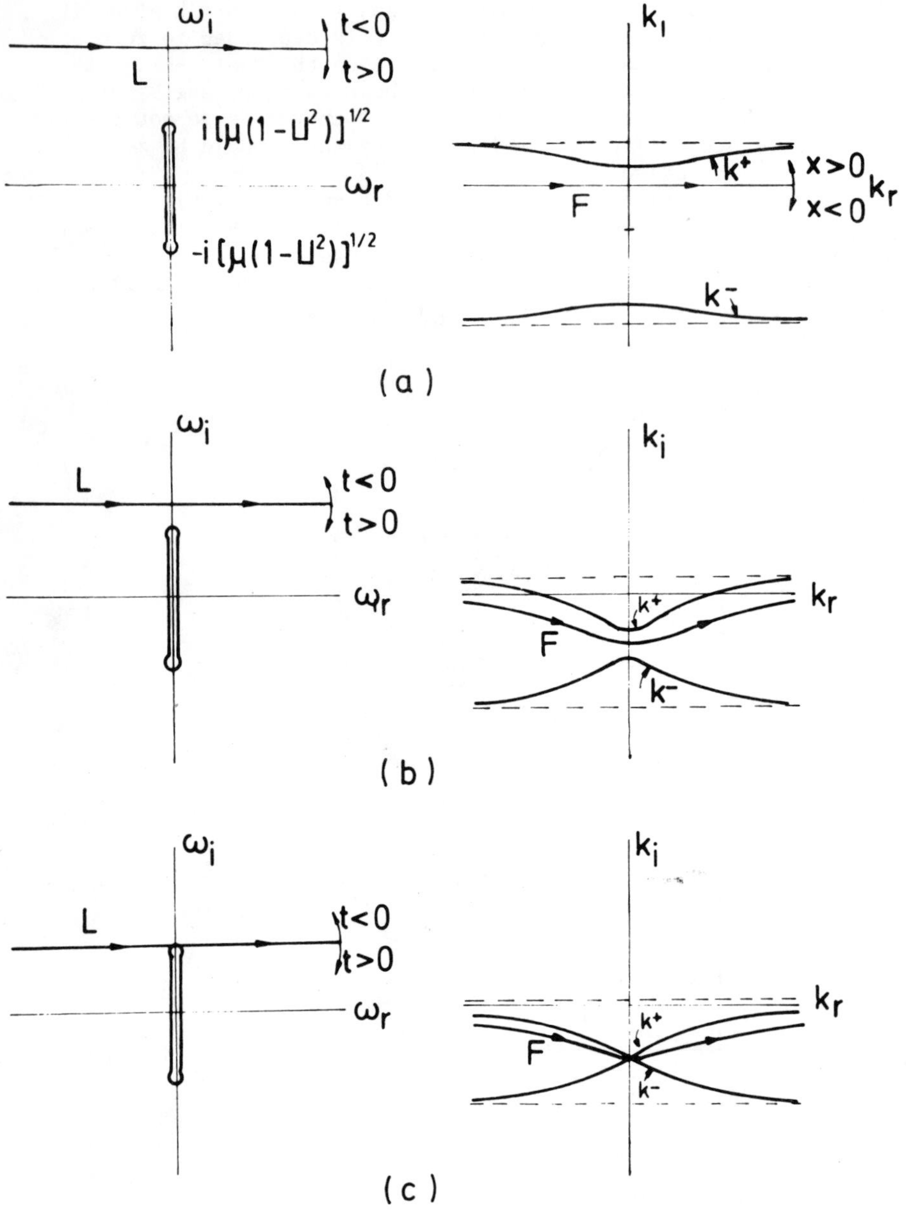

Figure 9. Locus of spatial branches in the k-plane, as ω_{iL} is decreased: (a), (b) and (c) describe different stages of pinching process in the Klein-Gordon equation. The velocity U is such that $|U| < 1$. Flow is absolutely unstable.

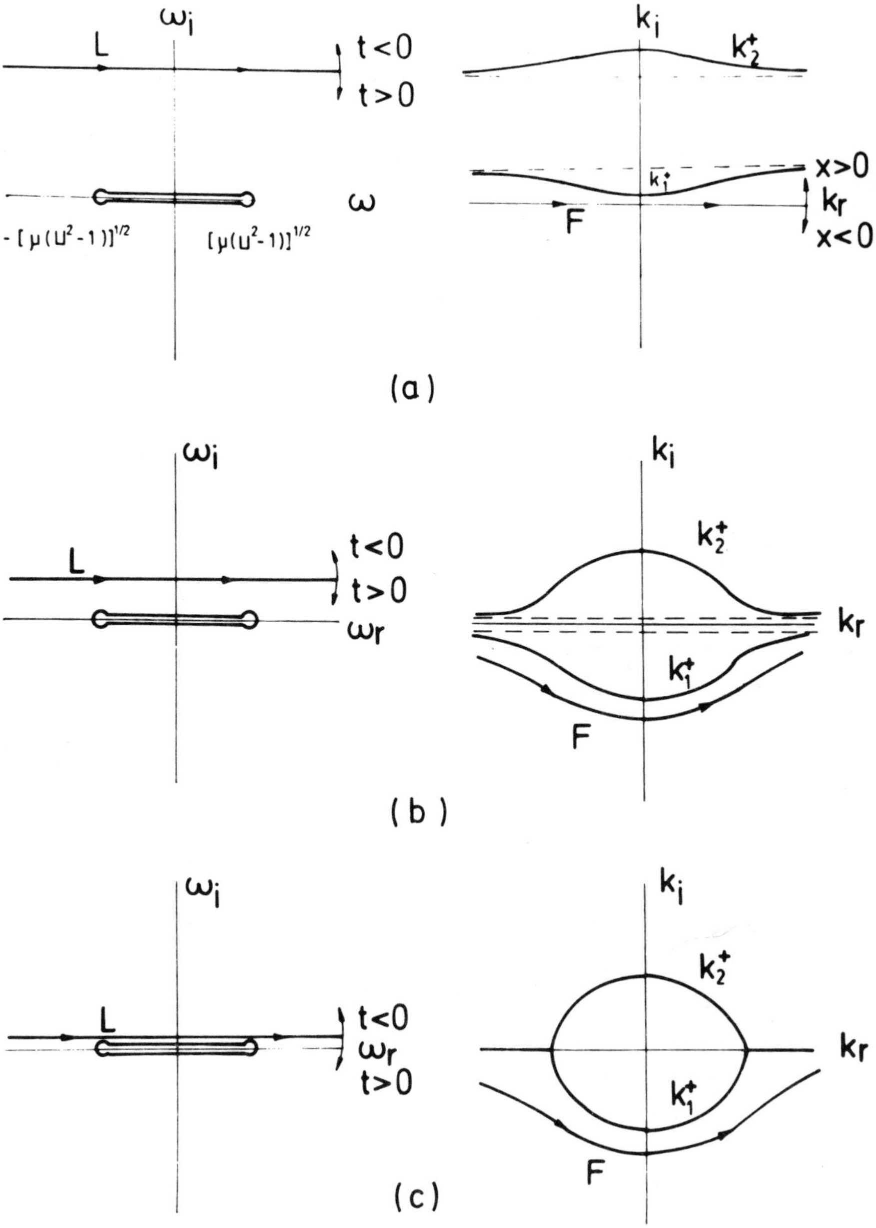

Figure 10. Same as figure 9, when $|U| > 1$. Flow is convectively unstable. Note absence of pinching of F contour by spatial branches.

the integral over all wavenumbers, namely

$$G(x,\omega) = \frac{1}{2\pi}\int_{-\infty}^{+\infty} \frac{e^{ikx}}{(1-U^2)[k-k^+(\omega)][k-k^-(\omega)]}\,dk\,. \tag{42}$$

There are two simple poles $k = k^{\pm}(\omega)$ located on either side of the real k axis when $|U| < 1$ (recall the previous discussion). Closing the contours as indicated in Figure 9a, one obtains the residue contributions

$$G(x,\omega) = \frac{i}{2(1-U^2)}\left[H(x)\,\frac{e^{ik^+(\omega)x}}{K(\omega)} + H(-x)\,\frac{e^{ik^-(\omega)x}}{K(\omega)}\right] \tag{43}$$

with

$$K(\omega) = \frac{1}{1-U^2}\,[\omega^2 + \mu(1-U^2)]^{1/2}, \tag{44}$$

and $k^{\pm}(\omega)$ given by (38).

There remains to evaluate the inverse transform with respect to ω

$$G(x,t) = \frac{i}{4\pi(1-U^2)}\,[H(x)I^+ + H(-x)I^-], \tag{45}$$

where

$$I^{\pm} \equiv \int_L \frac{\exp\{i[k^{\pm}(\omega)x-\omega t]\}}{K(\omega)}\,d\omega\,. \tag{46}$$

The contour L is chosen to be above the branch cut (Figure 9) linking the branch points at $\omega_0 = \pm i[\mu(1-U^2)]^{1/2}$. An elementary examination of the real part of each exponent indicates that the contour should be closed from above when $t^2 - (x-Ut)^2 < 0$. Cauchy's theorem then yields $G(x,t) = 0$. In the range $t^2 - (x-Ut)^2 > 0$, it is convenient to introduce the notation $U = \tanh\alpha$ and the parameters ξ and Θ such that

$$x - Ut = \xi\sinh\Theta, \quad t = \xi\cosh\Theta, \tag{47}$$

conversely,

$$\xi = [t^2 - (x-Ut)^2]^{1/2}, \quad \tanh\Theta = (x-Ut)/t\,. \tag{48}$$

The integral I^+ may then be written as

$$I^{+} = \int_{L} \frac{\exp\{i\xi[-\omega\cosh\alpha\cosh(\Theta+\alpha) + \sinh(\Theta+\alpha)((\omega\cosh\alpha)^{2} + \mu)^{1/2}]\}}{\cosh\alpha[(\omega\cosh\alpha)^{2} + \mu]^{1/2}} . \tag{49}$$

A closed form solution can then be obtained by deforming L around the branch cut. A parametric representation of ω along the branch cut is given by

$$\omega\cosh\alpha = i\mu^{1/2}\cos\psi, \quad -\pi < \psi < +\pi \tag{50}$$

with

$$[(\omega\cosh\alpha)^{2} + \mu]^{1/2} = \mu^{1/2}\sin\psi \tag{51}$$

on both sides of the cut (Figure 9). The integral I^{+} is then rearranged as

$$I^{+} = -\frac{i}{\cosh^{2}\alpha} \int_{-\pi}^{+\pi} \exp[\mu^{1/2}\xi\cos[\psi - i(\Theta + \alpha)]]\,d\psi . \tag{52}$$

The above integral is zero around the contour C sketched in Figure 11. Evaluation of the contributions along the 4 segments making up C leads to the final expression

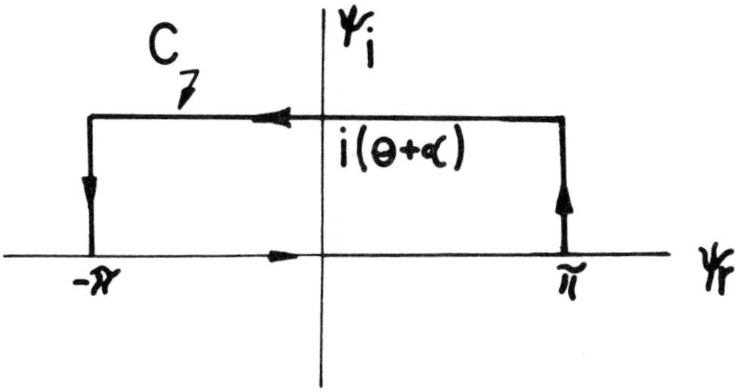

Figure 11. Contour C used to derive (53) from (52).

$$I^+ = -\frac{2i}{\cosh^2\alpha}\int_0^\pi \exp[\mu^{1/2}\xi\cos\psi]d\psi\ . \tag{53}$$

The integral representation of the modified Bessel function I_0 is recognized [19] so that

$$I^+ = -\frac{2\pi i}{\cosh^2\alpha} I_0(\mu^{1/2}\xi)\ . \tag{54}$$

Bearing in mind that $I^- = I^+$, the Green's function pertaining to the Klein-Gordon equation then reduces to

$$G(x,t) = \frac{1}{2} I_0[\mu^{1/2}\{t^2 - (x - Ut)^2\}^{1/2}]H\{t^2 - (x - Ut)^2\}. \tag{55}$$

The above result remains valid when $|U| > 1$.

In the long-time limit, one may use the asymptotic expansion of I_0 to arrive at the result one would obtain from a steepest descent calculation, namely,

$$G(x,t) \sim \frac{1}{2(2\pi)^{1/2}}\frac{\exp[\mu^{1/2}\{t^2 - (x - Ut)^2\}^{1/2}]}{\mu^{1/4}\{t^2 - (x - Ut)^2\}^{1/4}} H\{t^2 - (x - Ut)^2\}. \tag{56}$$

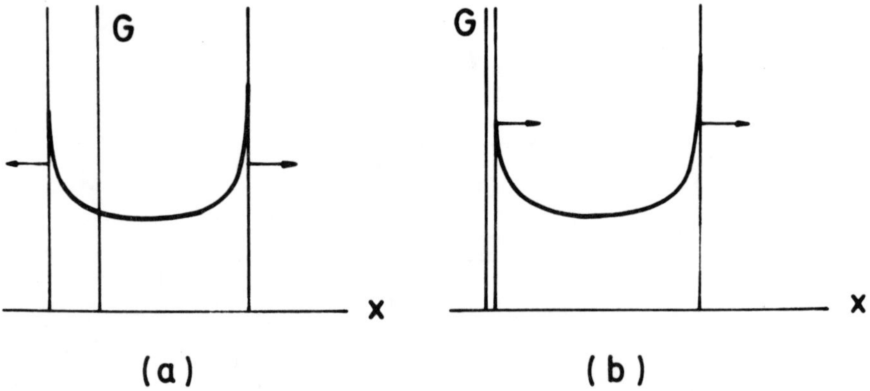

Figure 12. Green's function of the Klein-Gordon equation: (a) absolutely unstable case; (b) convectively unstable case.

The peak of the packet experiences the maximum growth rate $\mu^{1/2}$ and travels along the ray $x/t = U$. The edges move at the respective velocities $U \pm 1$. The absolute growth rate is $[\mu(1 - U^2)]^{1/2}$ when $|U| < 1$ and zero when $|U| > 1$. The impulse response is sketched in Figure 12.

In the range $|U| > 1$, spatially-developing waves can be generated with the stability characteristics sketched in Figures 13a,b. Note that both waves develop downstream of the source when $U > 1$, in agreement with a previous discussion on spatial branches. The steady-state response is of the form (22) with no terms multiplying $H(-x)$ when $U > 1$. To emphasize this feature, the spatial branches have been renamed k_1^+ and k_2^+ as in (40).

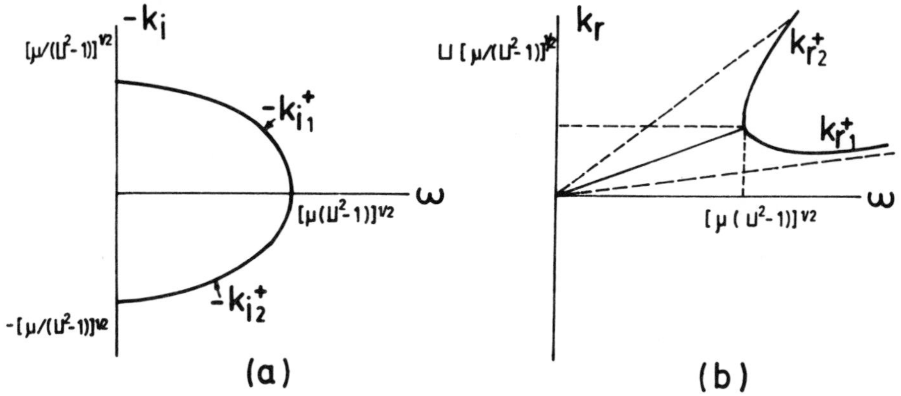

Figure 13. Spatial branches of the Klein-Gordon equation: (a) Spatial growth rate ($-k_i$) versus real frequency ω; (b) k_r versus real frequency ω. The velocity U is such that $U > 1$.

3.3. A long wavelength evolution equation

As a last example, we consider a long wavelength evolution equation pertinent to the Kelvin-Helmholtz instability in bounded mixing layers [6]. The model will serve to illustrate the type of technique which may be applied when the dispersion relation contains $|k|$ terms and is therefore not analytic. When a general forcing term $s(x,t)$ is present, the linearized version of the model reads

$$D_a\left[-i\frac{\partial}{\partial x}, i\frac{\partial}{\partial t}\right] a(x,t) \equiv \frac{1}{\pi x} * \left(\frac{\partial}{\partial t} + U\frac{\partial}{\partial x}\right) a + \mu\frac{\partial a}{\partial x} + \frac{\partial^3 a}{\partial x^3} = s(x,t), \tag{57}$$

where $a(x,t)$ and $s(x,t)$ are *real*, and

$$f * h = \int_{-\infty}^{+\infty} f(x - \xi)h(\xi)d\xi \tag{58}$$

is the spatial convolution of the functions f and h. The dispersion relation in this case takes the form

$$D_a[k,\omega] \equiv -\mathrm{sgn}k(\omega - Uk) + ik(\mu - k^2) = 0, \tag{59}$$

and there exists a single temporal mode characterized by

$$\omega_a(k) = Uk + i|k|(\mu - k^2). \tag{60}$$

The system becomes unstable when $\mu > 0$, the maximum growth rate $\omega_i(k_{max}) = 2(\mu/3)^{3/2}$ being reached at $k_{max} = (\mu/3)^{\frac{1}{2}}$, as shown in Figures 14a,b.

To facilitate the interpretation of $|k|$ in the complex plane, it is convenient to associate with the *real* signal a(x,t), the complex "*analytic signal*" A(x,t) (see, for instance, [20]) defined by

$$A(x,t) = [\delta(x) + i/(\pi x)] * a(x,t);$$
$$a(x,t) = \mathrm{Re}A(x,t). \tag{61a,b}$$

The symbol Re indicates the real part of a complex quantity. If a(k,t) and A(k,t) are the spatial Fourier transforms of a(x,t) and A(x,t) respectively, relations (61a,b) become, in wavenumber space

$$A(k,t) = 2H(k)a(k,t);$$
$$a(k,t) = \frac{1}{2}[A(k,t) + \bar{A}(-k,t)], \tag{62a,b}$$

a bar denoting the complex conjugate. Thus, the spectrum of A(x,t) is nothing but the restriction of the spectrum of a(x,t) to *positive* wavenumbers only. Hence, the analytic signal A(x,t) is merely the extension to arbitrary waveforms of the complex notation which, in linear systems, relates a monochromatic real signal $a(x) = \cos k_0 x$ and its complex equivalent $A(x) = e^{ik_0 x}$. Once A(x,t) is known, the real signal a(x,t) can be obtained by taking the real part a(x,t) = ReA(x,t), as required by (61b). All the calculations can therefore be performed on A(x,t) which is easily shown to satisfy

$$D[-i\frac{\partial}{\partial x}, i\frac{\partial}{\partial t}]A(x,t) \equiv -i\left(\frac{\partial A}{\partial t} + U\frac{\partial A}{\partial x}\right) + \mu\frac{\partial A}{\partial x} + \frac{\partial^3 A}{\partial x^3} = S(x,t). \tag{63}$$

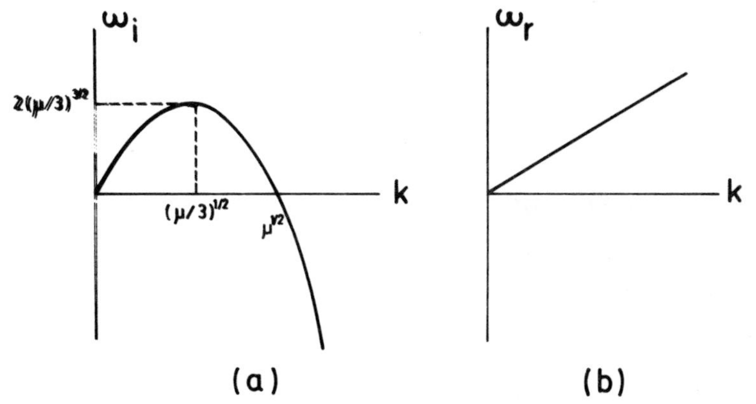

Figure 14. Temporal mode of the long wavelength equation (57): (a) Temporal growth rate ω_i versus real wavenumber k; (b) ω_r versus real wavenumber k.

The function S(x,t) is the analytic signal corresponding to s(x,t), as defined in the pair (61a,b). The new dispersion relation

$$D(k,\omega) \equiv -(\omega - Uk) + ik(\mu - k^2) = 0, \tag{64}$$

is analytic in k. It has been obtained from (59) by simply choosing sgnk = +1. The inverse of D(k,ω) defines the operator acting on the analytic signal A(x,t) in (63). Since the spectrum of A is identically zero for k < 0, the general formalism developed in section 2 holds, provided the contour F is initially taken to be the *positive* real axis instead of the entire real line. The study of the trajectories of the poles in the integrand can be *restricted to the half-plane* $k_r > 0$ and its corresponding image in the ω-plane. We note that (63) is the linearized form of the Korteweg-deVries equation with complex coefficients. The temporal mode is given by

$$\omega(k) = Uk + ik(\mu - k^2), \tag{65}$$

which can be compared with (60). The branch point of k(ω), obtained by solving for $(d\omega/dk)(k_0) = 0$, is located at

$$k_0 = [(\mu - iU)/3]^{1/2}, \qquad \omega_0 = 2ik_0^3, \tag{66}$$

with $k_{0r} > 0$. At each value of ω, there are in general 3 spatial branches, only two of which are located in the half-plane $k_r > 0$. The branch point (66) involves genuine pinching of the F contour by branches $k^+(\omega)$ and $k^-(\omega)$ originating from distinct sectors $k_i > 0$ and $k_i < 0$ as L is lowered. The absolute growth rate ω_{0i} becomes zero when $|U| = \mu\sqrt{3}$. The system is therefore absolutely (convectively) unstable when $|U| < \mu\sqrt{3}$ ($|U| > \mu\sqrt{3}$).

The Green's function $g = \text{Re}G$ is calculated by solving for

$$D\left[-i\frac{\partial}{\partial x}, i\frac{\partial}{\partial t}\right]G = \left(\delta(x) + \frac{i}{\pi x}\right)\delta(t), \tag{67}$$

equivalently,

$$G(k,\omega) = 2H(k)/D(k,\omega). \tag{68}$$

A straighforward residue evaluation in the complex ω-plane leads to the wavepacket integral

$$G(x,t) = -\frac{i}{\pi}\int_0^{+\infty}\frac{e^{i[kx-\omega(k)t]}}{(\partial D/\partial\omega)[k,\omega(k)]}\,dk, \tag{69}$$

which should be compared with the general formula (10). We note that the k-integral is limited to the range $0 < k < +\infty$. The application of the method of steepest descent involves a stationary point k^* such that $(d\omega/dk)(k^*) = x/t$, as well as a boundary point $k = 0$. Since in the particular example considered here $\omega_i(0) = 0$, the saddle point contribution at k^* will dominate the response inside the wavepacket. Rather than pursuing this argument, we choose to derive an exact representation of $G(x,t)$ when the dispersion relation $D[k,\omega]$ is given by (64). The wavepacket integral then reduces to

$$G(x,t) = \frac{i}{\pi}\int_0^{+\infty}\exp[ik(x - Ut) + k(\mu - k^2)t]\,dk\,. \tag{70}$$

Upon making the change of variable $k = \tau/(3t)^{1/3}$, $G(x,t)$ can be recast as

$$G(x,t) = \frac{i}{\pi(3t)^{1/3}}\int_0^{+\infty}\exp\{-\tau^3/3 + (3t)^{-1/3}[\mu t + i(x - Ut)]\tau\}d\tau\,, \tag{71}$$

which involves the integral representation of the function $\text{Hi}(z)$ [19]. The real Green's function takes the following final expression:

$$g(x,t) = (3t)^{-1/3}\text{Re}\{i\text{Hi}\left[\frac{\mu t + i(x - Ut)}{(3t)^{1/3}}\right]\}\,. \tag{72}$$

The asymptotic expansion of $\text{Hi}(z)$ as $|z| \to \infty$ leads to the same results as the method of steepest descent, namely

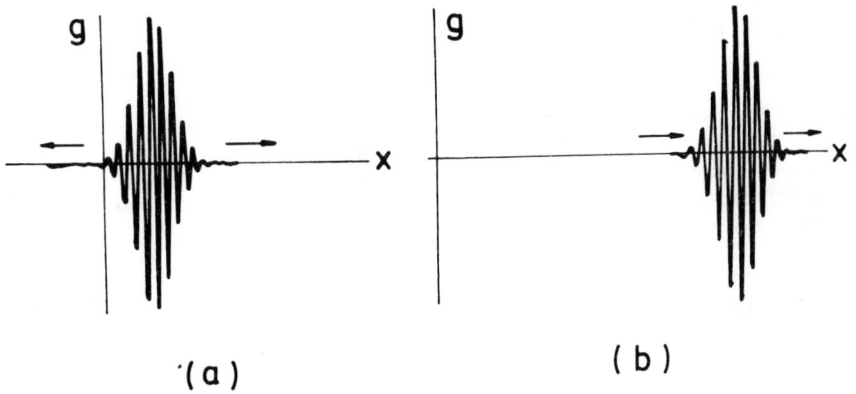

Figure 15. Green's function of the long wavelength equation (57): (a) absolutely unstable case; (b) convectively unstable case.

$$g(x,t) \sim \frac{1}{\pi^{1/2}(3t)^{1/4}} \mathrm{Re}\left[\frac{i}{[\mu t + i(x - Ut)]^{1/4}} \cdot \right. \tag{73}$$
$$\left. \cdot \exp\left\{ \frac{2}{3} \frac{[\mu t + i(x - Ut)]^{3/2}}{(3t)^{1/2}} \right\} \right]$$

when $|(x - Ut)/t| < \mu\sqrt{3}$,

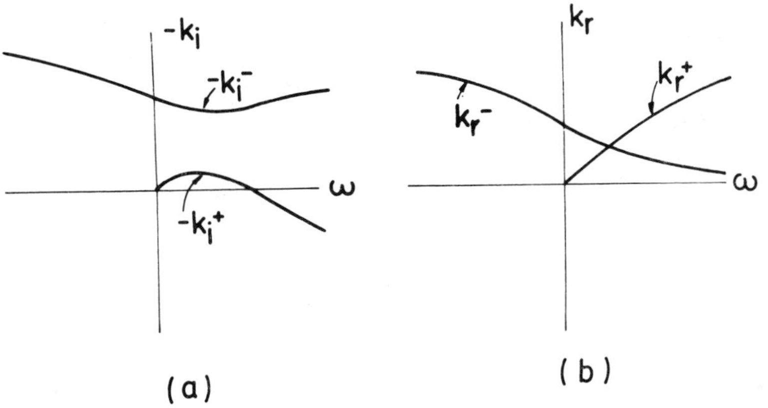

Figure 16. Spatial branches of the long wavelength equation (57): (a) spatial growth rate ($-k_i$) versus real frequency ω; (b) k_r versus real frequency ω. The velocity U is such that $U > \mu\sqrt{3}$.

and

$$g(x,t) \sim \mathrm{Re}\left[\frac{-i}{\pi[\mu t + i(x - Ut)]}\right], \text{ when } |(x - Ut)/t| > \mu\sqrt{3} . \qquad (74)$$

The edges of the wavepacket move at the respective velocities $U \pm \mu\sqrt{3}$, and the critical values $U = \pm\mu\sqrt{3}$ signal the transition from absolute to convective instability. Typical responses are shown in Figure 15.

The characteristics of spatially-evolving waves are sketched in Figure 16, when $|U| > \mu\sqrt{3}$.

4. APPLICATION TO HYDRODYNAMIC INSTABILITIES

In the sequel, we shall emphasize the physical implications of the previous mathematical development. Recall that two classes of unstable flows have been distinguished.

4.1. Convectively unstable flows

In convectively unstable systems, any initial disturbance is advected by the flow as it is amplified and the medium is ultimately left undisturbed. In this instance, solutions of the dispersion relation with ω real and k complex are physically relevant. They describe the spatial evolution of a periodic excitation applied at a fixed spatial location. As a rule, convectively unstable flows are extremely sensitive to external forcing. *Receptivity* issues therefore play a crucial role in determining the fate of infinitesimal disturbances. At the same time, intrusive measurements leave the flow relatively

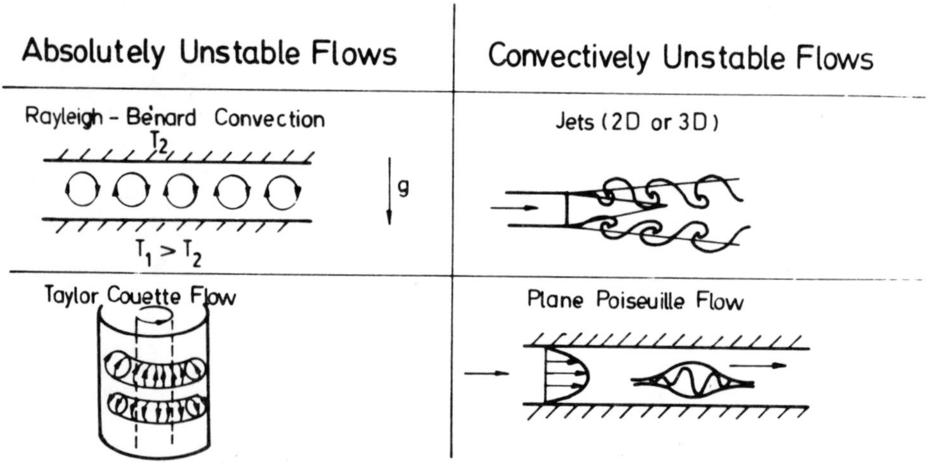

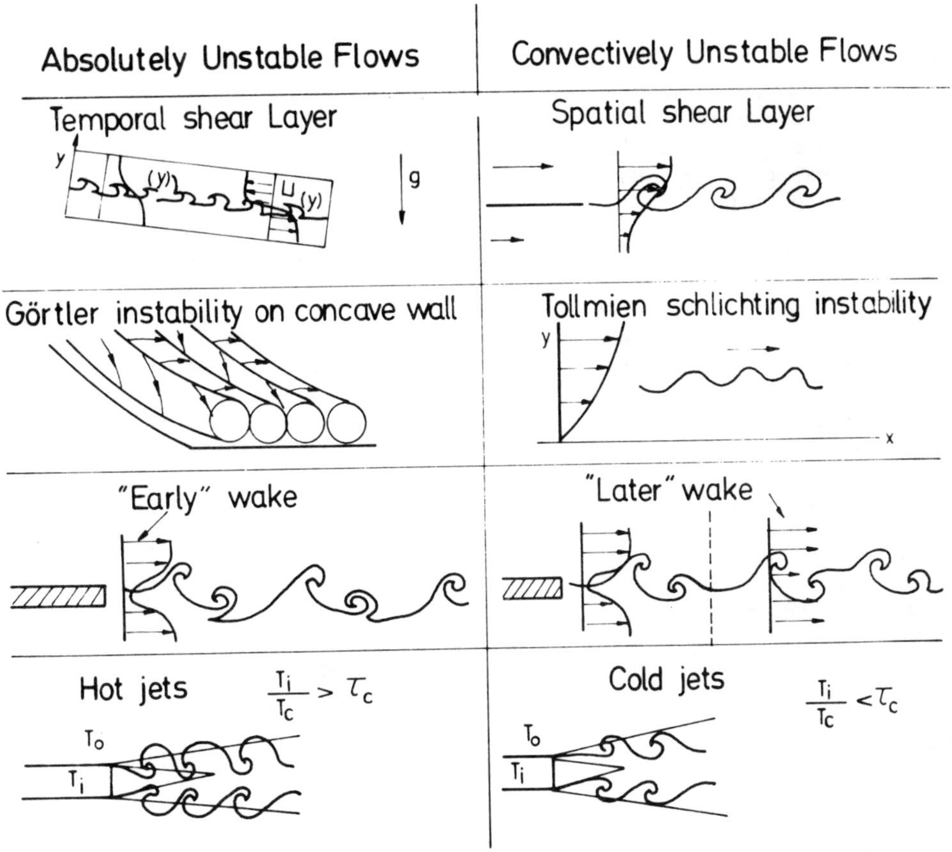

Figure 17. Tentative classification of common fluid flows into absolutely unstable flows and convectively unstable flows (see also [7]).

unaffected, provided no feedback loops or hydrodynamic resonances are willingly introduced (see later discussion). Examples of convectively unstable parallel flows are numerous in classical hydrodynamic stability theory, but few detailed calculations of the pertinent complex dispersion relations have been conducted. An incomplete, partially subjective, list is given in Figure 17. According to Huerre & Monkewitz [4], the family of parallel shear layer profiles $\bar{U}(y) = 1 + R\tanh y$ ($\bar{U}$: streamwise velocity, y: cross-stream direction, R: velocity ratio) is convectively unstable to *two-dimensional* disturbances whenever $R > 1.315$. Indeed, spatial stability theory [21,22] provides a consistent description of most instability waves in mixing layers generated downstream of a splitter plate. The *three-dimensional* wavepacket calculations of Balsa [23] further support the fact that such mixing layers are convectively unstable. A recent survey of experimental and theoretical results pertinent to shear layers has been made by Ho & Huerre [24]. Although no general results are presently available, there are very strong indications from the pioneering work of Gaster [25-28] that Tollmien-Schlichting waves in flat plate boundary layers are induced by a convective instability. It may be conjectured that plane Poiseuille flow is of a similar character. The dispersion relation associated with the Orr-Sommerfeld equation [29] is extremely rich in structure, which largely explains why no definite statement can be made regarding these viscous instabilities. Similarly, cold circular jets at low Mach numbers are probably convectively unstable and spatial instability theory [30] does represent the linear evolution of vortical disturbances in such a circular geometry. Note further that many of the above flows are far from being parallel. It is therefore implicitly assumed that the WKB formation can be applied successfully, the locally parallel basic flow being convectively unstable at all streamwise stations. Such an assumption does not necessarily take into account the effects of global pressure fluctuations, as opposed to *local vortical* fluctuations. For instance in shear layers, feedback loops between preferred downstream stations and the trailing-edge could make a locally convectively unstable medium to be globally absolutely unstable.

4.2. Absolutely unstable flows

In absolutely unstable systems, any infinitesimal disturbance contaminates the entire medium. As a result, intrusive measurements greatly affect the flow. Indeed, great care must be exercised in laboratory experiments, and non-intrusive techniques, such as Laser Doppler anemometry are often preferred. Spatial stability theory is irrelevant in absolutely unstable flows: any spatially-evolving wave is, in the course of time, overwhelmed by linear, temporally-growing fluctuations. Thus, in contrast with the previous class of instabilities, absolutely unstable flows tend to be less sensitive to infinitesimal external fluctuations. The behavior of the flow is *intrinsic* rather that *extrinsic*. Here again, one must rely on physical intuition to give fluid dynamical examples since no detailed theoretical investigations of the dispersion relations have been undertaken in the complex plane. It is probably very safe to say that a horizontal fluid layer heated from below (see reference [31] for a review) does give rise to absolute instabilities. The centrifugal Görtler instability taking place in boundary layers along concave walls does not obviously fall in one class or another. Although a recent study by Hall [32] has shed considerable light on the spatial instability problem, the convective nature of the instability has not been established. The present author's personal prejudice would tend to guess that it is absolutely unstable. Before tackling this configuration, however, one would need to extend the formalism of section 2 to highly non-parallel flows for which no separation of variables is possible.

Finally, it is interesting to note that temporally evolving mixing layers can be obtained experimentally by tilting a tank filled with a stably stratified fluid [33]. These flows are such that $R > 1.315$ for the family of profiles $\bar{U}(y) = 1 + R\tanh y$; consequently, the medium is absolutely unstable [4] in contrast with shear layers generated downstream of a splitter plate. It is found that the temporal theory of Michalke [34] does predict the main features of the linear development of the Kelvin-Helmholts vortices induced by the shear.

4.3. "Mixed flows" and Frequency Selection Mechanisms

There are also flow geometries which may give rise to a convective instability in one region of the flow and an absolute instability in another region. To our knowledge, this possibility has first been explored by Pierrehumbert [10] and Koch [35]. In such "mixed" flows, one typically assumes that the locally parallel basic flow changes only slowly in the streamwise x direction and that the WKB formalism is applicable. At each streamwise x station, the instability characteristics are, therefore, to leading-order, given by parallel flow theory. One can then easily envision physical situations in which a convection velocity, say the U parameter introduced in the models of section 3, changes slowly with x. If, at a particular station x_t, U reaches the critical value U_t where the branch point ω_0 crosses the real ω-axis, the flow will change from being locally convectively unstable to locally absolutely unstable, or vice-versa. If, more generally, several crossings take place as x is varied, the flow can be divided into several regions where the local absolute growth rate $\omega_{0i}(x)$ is positive or negative. In the border regions where $\omega_{0i}(x_t)$ vanishes, the slowly-varying analysis breaks down, and transition layers of Airy type have to be introduced, as in classical WKB theory.

The particular case of an absolutely unstable domain separating two convectively unstable regions (hereafter referred to as CU-AU-CU) was studied by Pierrehumbert [10] for a zonally-varying two-layer model of the baroclinic instability. Earlier analytical work by Thacker [18] and Merkine [36] had established that the same two-layer model in the parallel approximation underwent a transition from absolute to convective instability above a critical value of the ratio between the average speed of the mean flow and the shear. Pierrehumbert [10] demonstrated, by a combination of analytical and numerical means, that the slowly-varying baroclinic flow could support modes locally confined along the stream and growing at the *maximum absolute growth rate* $\omega_{0i\,max}$ over the entire domain $-\infty < x < \infty$. Thus Pierrehumbert's work suggests a *frequency selection criterion*, whereby, in a mixed flow situation of the type CU-AU-CU, *the dominant frequency is equal to* $\omega_{0r\,max}$, *namely the real part associated with the maximum absolute growth rate* $\omega_{0i\,max}$ *over the entire flow.*

But, in a study of two-dimensional wakes behind bluff bodies, Koch [35] proposes yet another frequency selection mechanism. According to his analysis, the developing wake is a mixed flow of the type solid body-AU-CU, a characteristic that one might have inferred from the earlier investigation of the $\mathrm{sech}^2 y$ wake by Mattingly & Criminale [3]. *The dominant frequency should*, in Koch's scenario, *lock to* $\omega_{0r}(x_t) = \omega_0(x_t)$, i.e., *to the real frequency pertaining to the transition point* x_t *separating the* AU and CU *regions.* Needless to say, $\omega_0(x_t)$ and $\omega_{0r\,max}$ are in general distinct and do not lead to the same predictions. A comparison between stability calculations and wake experiments reveals nonetheless that Koch's criterion predicts the shedding frequency of the Karman vortex street relatively well. This appears to imply that intense self-sustained oscillations occur in the AU region between the body and the transition point x_t. This *hydrodynamic resonance phenomenon* is somewhat akin to the resonances which are produced when a shear layer issuing from a streamlined or blunt body interacts with a second body placed at a finite distance downstream [37]. Note, however, that hydrodynamic resonances do not rely on the presence of a second body; the necessary streamwise length scale is generated instead by the flow itself in the form of the distance x_t. A recent analysis by Nguyen [38] further confirms Koch's results: for a two-parameter family of wake velocity profiles which closely fits experimentally-measured mean flows, it is found that the varicose mode is in general CU. The sinuous mode, however, leads to a pocket of AU within the near wake and it is responsible for the onset of self-sustained oscillations.

To close this discussion of frequency selection mechanisms in mixed flows, it is worth mentioning the interesting conjecture of Monkewitz & Sohn [5]. In the view of these authors, one should distinguish between two possible configurations and change the frequency selection criterion accordingly. If the flow is of the type solid body-AU-CU, one should follow Koch's proposition $\omega_0(x_t)$. If it is of the type solid body-CU-AU-CU, one should adopt Pierrehumbert's selection principle $\omega_{0r\,max}$. Wakes with initially thick shear layers fall within the first category [38] whereas initially very thin shear layers lead to a wake of the second category. Predictions of shedding frequency for these two families of wakes appear to follow experimental trends provided one adopts Monkewitz & Sohn's recommendation regarding the choice of selection criterion.

The reader is further referred to Monkewitz & Sohn [5] for an interesting application of absolute and convective instability concepts to the control of hot jets. Viscous liquid jets have also been examined in this light by Leib & Goldstein [39].

4.4. Absolute instabilities, convective instabilities and chaos

As emphasized by Deissler [40,41] and Deissler & Kaneko [42], the distinction between absolute and convective instabilities plays a crucial role in the search for chaos in transitional flows. Absolutely unstable flows such as Rayleigh-Bénard convection or Taylor-Couette flow can give rise, under carefully controlled conditions, to chaotic motion on a low dimensional attractor (see, for instance, Bergé, Pomeau, Vidal [43]). The motion in phase space can usually be reconstructed from a time series at a single point in physical space. Quantitative statistical measures of disorder such as the Lyapunov exponents, fractal dimension and Kolmogoroff entropy can be calculated for each attractor, independently of initial conditions within each basin of attraction. As one would expect in an absolutely unstable flow, low levels of external noise do not alter significantly statistically-averaged quantities on the attractor. In other words, chaos in absolutely unstable flows is *intrinsically* driven by temporally-developing instabilities at each spatial location. As a result, search for chaos in absolutely unstable flows has been reasonably successful, both experimentally and numerically.

Such is not the case in convectively unstable media, i.e., boundary layers, shear layers, jets, pipe flow, etc... Here, disturbances at a fixed spatial location eventually die out so that Lyapunov exponents in the laboratory frame are always negative [42]. There is, of course, plenty of experimental evidence to support that turbulence does occur in these flows sufficiently far downstream but one has not been successful in relating it to chaos produced by a deterministic mechanism (except for the Navier-Stokes equations!).

To explore possible alternative strategies, Deissler [40,41] has conducted numerical studies of the Ginzburg-Landau equation, namely,

$$\frac{\partial A}{\partial t} + U\frac{\partial A}{\partial x} - \mu A - (1 + ic_d)\frac{\partial^2 A}{\partial^2 x} + (1 + ic_n)|A|^2 A = 0\,. \qquad (75)$$

In its linearized version around $A = 0$, (75) is identical to (23), the model studied in section 3.1. The real coefficient c_n takes distinct values for different flows, as does c_d. When $U = 0$, the basic state $A = 0$ is absolutely unstable and numerical simulations indicate that, for certain values of the parameters c_d and c_n, the trajectories are confined to a low-dimensional chaotic attractor [44,45,46].

Recall from section 3.1. that the motion is convectively unstable when $U > 2[\mu(1 + c_d^2)]^{1/2}$. In this case, a single-frequency forcing applied at $x = 0$ gives rise to a spatially-developing wave in the region $x > 0$. As shown by Deissler, the characteristics of the spatial wave are well predicted by the linear model (see Figures 5a,b) sufficiently close to the source. Further downstream, the wavetrain reaches a finite-amplitude saturated state described by monochromatic nonlinear solutions of (75) at the forcing frequency. More importantly, when the single frecuency excitation is replaced by broadband random fluctuations, the external noise is *selectively* amplified by the system

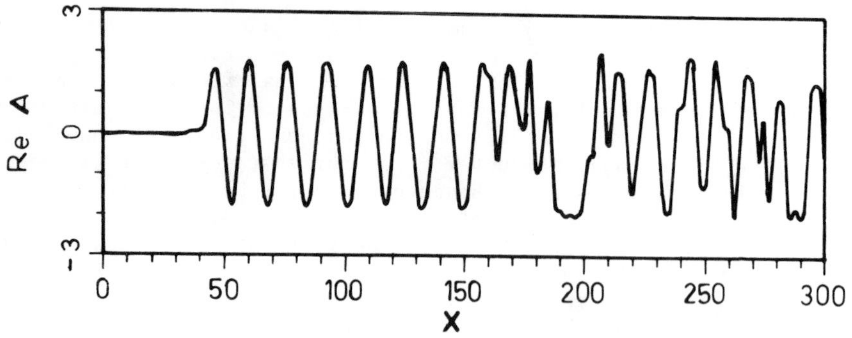

Figure 18. Plot of ReA(x,t) as a function of x for a given t. The function A(x,t) satisfies the Ginzburg-Landau equation (75). A microscopic noise is added at x = 0. Reproduced from Deissler [41].

and spatially growing waves are generated at the frequency with the maximum growth rate (see Figure 18). In other words, the initially broad spectrum at x = 0 shrinks to a narrow spectrum around the most amplified frequency, in the linear region immediately downstream of the source. Further downstream, however, in the finite-amplitude region, the nonlinear wavetrain becomes modulationally unstable to sidebands within the narrow spectrum [47]. This secondary instability brings about a break-up of the primary wave which gradually results in the generation of a broad band spectrum far downstream. As emphasized by Deissler, irregularities in the spatial wave produce random changes in the location of the break-up points, this mechanism being responsible for *intermittency*.

Again, it should be stressed that the fully-developed "turbulent flow" far downstream (Figure 18) is triggered by the external noise fed into the system at x = 0. In the absence of external fluctuations, no spatial structures can be detected, hence the name *noise-sustained structures* proposed by Deissler. This is in sharp contrast with intrinsic chaos present in absolutely unstable flows, which does not rely for its existence on the presence of external noise.

It remains to answer the following question: If "turbulent flow" in the convectively unstable Ginzburg-Landau equation is induced by external noise, can it legitimately by regarded as deterministic *intrinsic* chaos? In other words, do the usual statistical measures of chaos (dimension, Lyapunov exponents, ...) far downstream become independent of the external noise applied at x = 0? Until now, no satisfactory answer appears to have been given to this question.

But there are other problems. As noted by Deissler & Kaneko [42], the usual definition of Lyapunov exponents at a *fixed* spatial location always gives a negative number, a fact presumably related to the negative absolute growth rates prevailing in convectively unstable flows. To arrive at a meaningful measure of chaos in such flows, one must introduce a velocity-dependent largest Lyapunov exponent $\lambda(v)$, obtained by extending the usual definition to a region $\{x_1 + vt, x_2 + vt\}$ in a frame of reference moving at the velocity v. Deissler & Kaneko [42] then propose to call a convectively unstable flow chaotic if the maximum value λ_m of $\lambda(v)$ over all values of v converges to a positive value. In the case of the Ginzburg-Landau equation, a maximum positive value is obtained when v = U! For further technical details, the reader is referred to Deissler & Kaneko's paper. As previously noted, it is not clear in our view that λ_m is independent of external noise. Recent experimental research (see, for instance, [24] for a review of mixing layers) has indicated that some convectively unstable flows remain remarkably sensitive to external noise at large downstream distances, even in terms of averaged quantities. Receptivity issues and measures of chaos might have to be examined as a whole if one is to make further progress in the description of disorder in convectively unstable media.

ACKNOWLEDGMENTS

The author wishes to thank L. Brevdo, J. M. Chomaz, P. Coullet, R. J. Deissler, P. A. Monkewitz and L. G. Redekopp for helpful discussions and suggestions. This works is supported by the Air Force Office of Scientific Research under AFOSR Contract No.F49620-85-C-0080. Its presentation at the Workshop on Instabilities and Nonequilibrium Structures held in Valparaíso, Chile on December 16-21, 1985, was made possible thanks to an invitation of the Universidad Técnica Federico Santa María, Valparaíso.

REFERENCES

[1] R. J. Briggs. Electron-Stream Interaction with Plasmas, *Research Monograph* No. 29, M.I.T. Press, Cambridge, Mass., 1964.
[2] A. Bers. Linear Waves and Instabilities, in *Physique des Plasmas*, C. DeWitt & J. Peyraud (Ed.), Gordon & Breach, New York, 1975.
[3] G. E. Mattingly & W. O. Criminale. *J. Fluid Mech.*, 51, 233, 1972.
[4] P. Huerre & P. A. Monkewitz. *J. Fluid Mech.*, 159, 151, 1985.
[5] P. A. Monkewitz & K. D. Sohn. Absolute Instability in Hot Jets and Their Control, AIAA Paper No. 86-1882, 1986.
[6] P. Huerre. *J. Méca Théor. Appl.*, Special Issue on Two-Dimensional Turbulence, 121-145, 1983.
[7] D. W. Bechert. Excitation of Instability Waves, *Proc. Symposium IUTAM Aero et Hydro-Acoustique*, Lyon, 1985.
[8] N. Bleistein & R.A. Handelsman. *Asymptotic Expansions of Integrals*, Holt, Rinehart & Winston, New York, 1975.
[9] C. M. Bender & S. A. Orszag. *Advanced Mathematical Methods for Scientists and Engineers*, McGraw-Hill, New York, 1978.
[10] R. T. Pierrehumbert. *J. Atmos. Sci.*, 41, 2141, 1984.

[11] R. T. Pierrehumbert. Spatially Amplifying Modes of the Charney Baroclinic Instability Problem, Preprint, Princeton University, 1985.
[12] A. C. Newell & J. A. Whitehead. *J. Fluid Mech.*, 38, 279, 1969.
[13] L. A. Segel. *J. Fluid Mech.*, 38 203, 1969.
[14] K. Stewartson & J. T. Stuart. *J. Fluid Mech.*, 48, 529, 1971.
[15] J. Pedlosky. *J. Atmos. Sci.*, 29, 680, 1972.
[16] M. A. Weissman. *Phil. Trans. R. Soc. Lond.*, A290, 639, 1979.
[17] C. G. Lange & A. C. Newell. *SIAM J. Appl. Math.*, 21, 605, 1971.
[18] W. C. Thacker. *Geophys. Fluid Dyn.*, 7, 271, 1976.
[19] M. Abramowitz & I. A. Stegun. *Handbook of Mathematical Functions*, National Bureau of Standards, Washington D.C., 1964.
[20] F. Roddier. *Distributions et Transformation de Fourier*, McGraw-Hill, Paris, 1978.
[21] A. Michalke. *J. Fluid Mech.*, 23, 521, 1965.
[22] P. A. Monkewitz & P. Huerre. *Phys. Fluids*, 25, 1137, 1982.
[23] T. F. Balsa. Three-Dimensional Wavepackets and Instability Waves in Free Shear Layers, University of Arizona preprint, 1985.
[24] C. M. Ho & P. Huerre. *Ann. Rev. Fluid Mech.*, 16, 365, 1984.
[25] M. Gaster. *J. Fluid Mech.*, 22, 433, 1965.
[26] M. Gaster. *Phys. Fluids*, 11, 723, 1968.
[27] M. Gaster. *Proc. R. Soc. Lond.*, A347, 271, 1975.
[28] M. Gaster. The propagation of wavepackets in laminar boundary layers: asymptotic theory for non-conservative wave systems, Preprint, National Maritime Institute, Teddington, England, 1980.
[29] P. G. Drazin & W. H. Reid. *Hydrodynamic Stability*, Cambridge University Press, Cambridge, 1981.
[30] A. Michalke. *Z. Flugwiss*, 19, 319, 1971.
[31] F. H. Busse. in *Topics in Applied Physics*, 45, Springer-Verlag, Berlin, 1981.
[32] P. Hall. *J. Fluid Mech.*, 130, 41, 1983.
[33] S. A. Thorpe. *J. Fluid Mech.*, 46, 299, 1971.
[34] A. Michalke. *J. Fluid Mech.*, 19, 543, 1964.
[35] W. Koch. *J. Sound and Vib.*, 99, 53, 1985.
[36] L. O. Merkine. *Geophys. Astrophys. Fluid Dyn.*, 9, 129, 1977.
[37] D. Rockwell & E. Naudascher. *Ann. Rev. Fluid Mech.*, 11, 67, 1979.
[38] L. N. Nguyen. Frequency Selection in Jets and Wakes, Master of Science Thesis, U.C.L.A., Los Angeles, California, 1986.
[39] S. J. Leib & M. E. Goldstein, *Phys. Fluids*, 29, 952, 1986.
[40] R. J. Deissler. *J. Stat. Phys.*, 40, 371, 1985a.
[41] R. J. Deissler. Spatially Growing Waves, Intermittency and Convective Chaos in an Open-Flow System, Los Alamos Report LA-UR-85-4211, 1985b.
[42] R. J. Deissler & K. Kaneko. Velocity-Dependent Liapunov Exponents as a Measure of Chaos for Open Flow Systems, Los Alamos Report LA-UR-85-3249, 1985.
[43] P. Bergé, Y. Pomeau & C. Vidal. *L'Ordre dans le Chaos*, Herman, Paris, 1984.
[44] Y. Kuramoto. *Prog. Theor. Phys. Suppl.*, 64, 346, 1978.
[45] H. T. Moon, P. Huerre & L. G. Redekopp. *Physica D7*, 135, 1983.
[46] L. R. Keefe. *Stud. Appl. Math.*, 73, 91, 1985.

[47] J. T. Stuart & R. C. DiPrima. *Proc. R. Soc. London*, A362, 27, 1978.

MODELS OF PATTERN FORMATION FROM A SINGULARITY THEORY POINT OF VIEW

P. Coullet (*)
and
D. Repaux
Lab. de Physique Theorique, Universite de Nice
Parc Valrose, Nice Cedex 06034, France

ABSTRACT

We review simple models of one-dimensional pattern forming transitions. Elementary ideas from singularity theory are used in order to build codimension two and three transition models. As their analogs in temporal dynamics they display complicated patterns as quasiperiodic and chaotic ones.

1. INTRODUCTION

A particular simple class of patterns are described by spatially periodic solutions, static in time, of a partial differential equation

$$\partial_t U = f(U) \qquad (1)$$

where $U(x,y,z,t)$ represents a set of N fields describing a given physical system. Such patterns exist, for example in the context of fluid dynamics, chemical cinetics and many other fields [1] [2]. One can think for example of this system being a fluid layer submitted to some external field as a temperature gradient. U would then represent the velocity and the temperature field. A great simplification occurs when U is assumed to vary in one spatial dimension only. In that case a static pattern is a solution of an ordinary differential equation

$$f(U,\partial_x) = 0 \qquad (2)$$

called in the following spatial dynamical system. In the case of infinite systems all bounded solutions of equation (2) are admissible patterns. As any generic dynamical system equation (2) can have stationary solutions which describe homogeneous patterns, periodic solutions associated with periodic patterns and more complicated solutions for example quasiperiodic and aperiodic ones. Equation (1) describes physical

(*) Also Observatoire de Nice.

E. Tirapegui and D. Villarroel (eds.), Instabilities and Nonequilibrium Structures, 179–195.

systems which generally possess symmetry properties. These symmetries are reflected in the spatial dynamical system: the parity symmetry $x \to -x$ generally assumed implies that equation (2) has a conservative nature, while the translational symmetry of the initial system forces the existence of an invariant. The selection of a given pattern is a difficult question related to its dynamical stability and to the initial conditions when several patterns are found to be simultaneously stable together with the nature of the unavoidable noise present in real systems. The question of the dynamical stability can be in principle answered finding the spectrum of the Jacobian operator

$$L(\partial_x;x) = Df\Big|_{U_0(x)} \tag{3}$$

where $U_0(x)$ is a given solution of equation (2). Unfortunately this is in general technically possible only when $U_0(x)$ is a stationary solution. Other complications come from the possible multiplicity of stable patterns. Initial conditions then play an important role in the selection of a given pattern. Noise can also induce transitions from one pattern to another. A conceptual simplification arises when the evolution equation possesses a Liapunov functional F, that is when

$$f_i(U) = -\frac{\delta F}{\delta U_i} \tag{4}$$

for $i=1,..N$. In that case all initial conditions will eventually converge to stationary values since F is decreasing along all evolutions described by equation (1)

$$\partial_t F = -F^2 \tag{5}$$

This Liapunov functional acts as a thermodynamical potential when additive white noise is added to equation (1). The stationary probability of a given pattern is then given by

$$P = Z^{-1}\exp(-\beta^{-1}F) \tag{6}$$

where Z represents a normalisation factor and β measures the amplitude of the noise. Thanks to this Thermodynamical analogy the notion of metastability allows one to discuss transitions among dynamically stable patterns. Although the existence of such Liapunov functional is a rather exceptional property for evolution equations, in some circumstances equation (1) can be reduced to a simpler equation which possesses it.

The basic idea we are going to develop throughout this paper is the following: pattern transitions are related to topological changes of the corresponding solution of the spatial dynamical system. These topological changes can be studied in a prototypical way in the framework of singularity theory. This approach has been extensively used in the context of dissipative dynamical systems, where it allows in particular to reduce complicated dynamical systems to simpler ones called normal forms [3]. We intend to extend these ideas to the problem of pattern changes. As we will see it allows to rephrase in a different language the existing approaches to this problem.

This paper is organized as follows: the first part is devoted to clarify the relation between transitions from homogeneous patterns and the corresponding topological changes of the spatial dynamical system. The basic codimension one transitions are reviewed in the second part, while the third part is devoted to codimension two transitions. In the last part we eventually study a codimension three transition.

2. PATTERN TRANSITIONS AND TOPOLOGICAL CHANGES IN THE SPATIAL DYNAMICAL SYSTEM

Let U_0 be a homogeneous solution of equation (1). U_0 can be taken to be zero without loss of generality. This solution is linearly stable if all eigenvalues of the jacobian operator (3) have negative real part. Transitions occur when some eigenvalues crosses the imaginary axis. Since we are interested in describing transitions between static patterns, only the cases of crossings associated with real eigenvalues have to be consider. Let us now come to the spatial dynamical system point of view. The homogeneous pattern U_0 corresponds to a stationary solution of it. Its linearization leads to eigenvalues which characterize the topological type of this solution. Because of the conservative nature of the spatial dynamical system topological changes occur basically whenever a pair of eigenvalues meet together on the imaginary axis. In the generic situation one is left with a pair of pure imaginary eigenvalues.

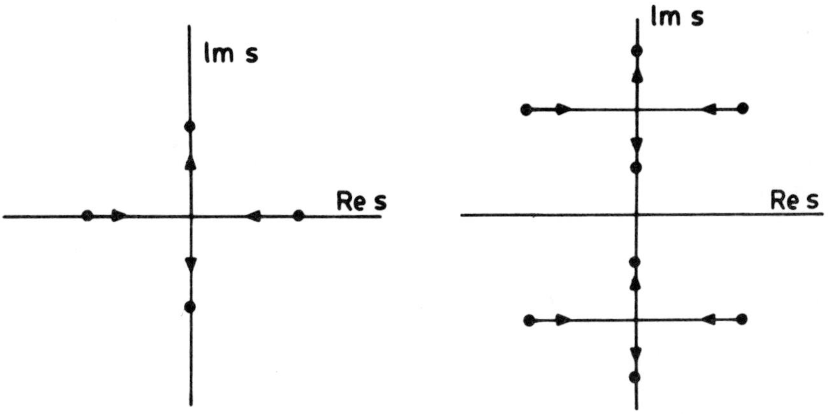

Figure 1. Motion of the eigenvalues in the complex plane when μ changes from negative to positive values. Fig. 1a corresponds to the characteristic equation (7) while fig. 1b is associated with equation (8).

The two elementary or codimension one transitions correspond to the following characteristic equations

$$s^2 + \mu = 0 \tag{7}$$

and

$$(s^2 + k_0^2)^2 + \mu = 0 \tag{8}$$

where μ represents a small bifurcation parameter and k_0 a finite one (see fig. 1)

We also assume that all other eigenvalues have finite real part. This is a necessary condition to insure the temporal dynamical stability of the homogeneous pattern described by U_0. In order to clarify this point let us come back to the initial evolution equation (1). The dynamical linear stability analysis of the solution $U_0 = 0$ proceeds as follows. Equation (1) can be split into linear and nonlinear parts

$$\partial_t U = L(\partial x)U + N(U;\partial x) \tag{9}$$

where $L(\partial x)$ is the Jacobian operator (3) evaluated for $U_0 = 0$- Thanks to the translational invariance the solutions of the linear part can be sought under the form

$$U(x,t) = U_k(0)\exp(\lambda t + ikx) \tag{10}$$

The associated characteristic equation reads

$$\lambda^N + A_{N-1}\lambda^{N-1} + \ldots + A_0 = 0 \tag{11}$$

where, thanks to the parity symmetry

$$x \to -x \ ,$$

$A_i = A_i(k^2)$, and $A_0 = (-1)^N \det(L(ik))$. The condition to have zero eigenvalues is thus

$$\det(L(ik)) = 0 \tag{12}$$

Equation (12) is nothing but the characteristic equation of the spatial dynamical system, where $s = ik$. Whenever this equation admits pure imaginary s-eigenvalues $s = \pm ik_0$, positive λ-eigenvalues are encountered in the temporal dynamical stability problem. Hence the condition to have s-eigenvalues with finite real part is necessary to insure the temporal dynamical stability of the corresponding homogeneous pattern. On the other hand change from s-eigenvalues with finite real parts to pure imaginary ones corresponds to the lost of stability of the homogeneous pattern. It is reasonable to study the corresponding topological changes in the spatial dynamical system under their prototypical forms. At the leading asymptotic order the normal forms associated with the characteristic polynomials (7) and (8) respectively reads

$$A_{XX} + \mu A - A^3 = 0 \tag{13}$$

and

$$A_{XX} + \mu A - |A|^2 A = 0 \tag{14}$$

The actual relation between U and A is the following. In the case of equation (13)

$$U(x) = A(X)\Phi + \mathcal{U}(A,\mu) \tag{15}$$

where $\mathcal{U}$ stands for linear and nonlinear corrections, and Φ is the eigenvector associated with the eigenvalue $s = 0$. In the case of equation (14)

$$U(x) = A(X,t)\Phi\exp(ik_0x) + \mathcal{U}(A,\bar{A},\mu;x) + c.c \tag{16}$$

Where $\mathcal{U}$ represents periodic linear and nonlinear corrections with period $2\pi/k_0$, and $\Phi\exp(ik_0x)$ is the eigenvector associated with the eigenvalue $s = ik_0$. In both equations (13) and (14) the sign of the nonlinear term has been chosen in order to insure the temporal dynamical stability. Moreover one has assumed in equation (13) the symmetry $A \to -A$ as often encountered in many physical examples. The spatial variable X is used instead of x to insist on the slow variation of the amplitude A. The reduction of the spatial dynamical system to its normal form is not a dynamical one, nevertheless at the leading order equation (1) can be asymptotically reduced to equations

$$\partial_t A = A_{XX} + \mu A - A^3 \tag{17}$$

or

$$\partial_t A = A_{XX} + \mu A - |A|^2 A \tag{18}$$

depending upon the type of instability. Equation (17) describes the relaxational dynamics of a ferromagnetic-like transition, while equation (18) known as the Ginzburg-Landau for superconductivity and superfluidity has been first derived in the context of pattern formation by Newell-Whitehead [4] and Segel [5]. We now review the content of these spatio-temporal dynamical systems.

3. CODIMENSION ONE PATTERN FORMING TRANSITIONS

Equations (13) and (14) are both integrable. Equation (13) describes the motion of a particle in the potential given by

$$V = \frac{1}{2}\mu A^2 - \frac{1}{4}A^4 \tag{19}$$

For $\mu < 0$ the only bounded solution of equation (13) is then $A = 0$. For $\mu > 0$, besides $A = 0$, one gets a one parameter family of periodic solutions centered on $A = 0$ which merge on heteroclinic trajectories connecting the stationary solutions $\pm\sqrt{\mu}$ (see fig. 2).

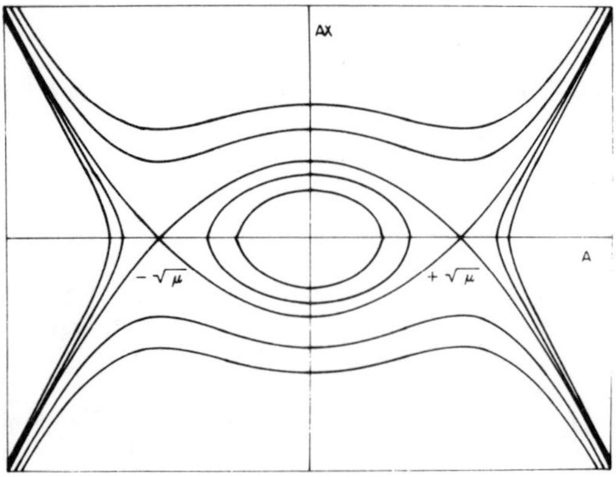

Figure 2. Phase portrait associated with equation (13).

The interpretation of these solutions in terms of patterns is the following: the periodic solutions correspond to periodic patterns, the heteroclinic trajectory represents a defect connecting the two possible homogeneous patterns $\pm\sqrt{\mu}$. This defect is called a domain wall in the context of ferromagnetic transitions. The dynamical stability of these patterns will be discussed later. In the case of equation (14) it is useful to introduce the real amplitude and phase variables. With these new variables equation (14) reads

$$R_{xx} + \mu R - R^3 - \Theta_x^2 R = 0 \tag{20}$$

and

$$R\Theta_{xx} + 2\Theta_x R_x = 0 \tag{21}$$

From equation (21) one readily deduces the following invariant

$$\Theta_x R^2 = h \tag{23}$$

Equation (20) then reads

$$R_{xx} + \frac{\partial V(R,h)}{\partial R} = 0 \tag{24}$$

where the potential V is given by

$$V(R,h) = \frac{1}{2}\mu R - \frac{1}{4}R^4 - \frac{h^2}{4R^4} \tag{25}$$

For negative μ the only bounded solution is $A = 0$. It corresponds to $h = 0$. For positive μ one gets bounded solutions only when $h < h_c = 2(\mu/3)^{3/2}$. For $h < h_c$ equation (24) has two stationary solutions R_- and R_+ and a one parameter family of periodic solutions centered around R_- which merges into a homoclinic trajectory which bi-asymptotically connect R_+ (see fig. 3).

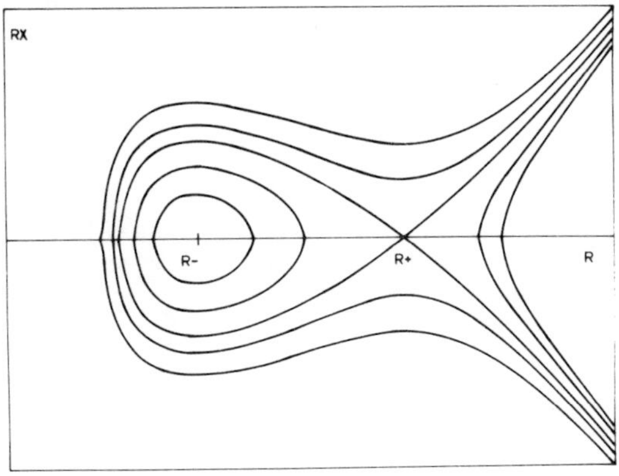

Figure 3. Phase portrait associated with equation (14).

The interpretation of these solutions in terms of patterns is the following: to stationary solutions $R_\pm$ corresponds periodic solutions $A = R_\pm \exp(ih/R_\pm^2)$ which represent periodic patterns with period $k = k_0 \pm h/R_\pm^2$. To periodic solutions correspond quasiperiodic patterns. Finally the homoclinic solution describes a single defect of a perfect pattern with amplitude R_+. The nature of the defect depends on the sign of h. For positive h it corresponds to the localized addition of an extra period, and to its substraction for negative h. The two spatial dynamical systems considered so far have in common the integrability property. As a direct consequence they cannot have complex trajectories as chaotic ones. They also have in common the fact that they are Hamiltonian systems, whose associated Lagrangian respectively reads

$$f = \frac{1}{2}A_X^2 - \frac{\mu}{2}A^2 + \frac{1}{4}A^4 \tag{26}$$

and

$$f = |A_X|^2 - \mu|A|^2 + \frac{1}{4}|A|^4 \quad . \tag{27}$$

As an immediate consequence the associated dynamical model is a gradient flow.

$$\partial_t A = -\frac{\partial F}{\partial A} = A_{XX} + \mu A - A^3 \tag{28}$$

and

$$\partial_t A = -\frac{\partial F}{\partial \bar{A}} = A_{XX} + \mu A - |A|^2 A \tag{29}$$

Where $F = \int dXf$. These equations can be deduced from equation (1) near codimension one instability of the homogeneous pattern $U_0 = 0$. More precisely let us consider the Jacobian operator

$$L = Df(U)\Big|_{U=0} \tag{30}$$

and assume that, varying some external parameter μ one finds a zero eigenvalue associated with some wavenumber. Near this parameter value, for nearby wavevectors the eigenvalues either read

$$\sigma(k) = \mu - k^2 + O(k^4) \tag{31}$$

if the instability occurs at zero wavevector, or

$$\sigma(k) = \mu - (k^2 - k_0^2)^2 + O((k^2 - k_0^2)^3) \tag{32}$$

if it occurs at finite k_0. In equations (31) and (32) the factor in front of the k variable of the eigenvalue has been chosen unity by appropriate scaling. One recognizes in the rigth hand side of equations (31) and (32) the characteristic polynomials (7) and (8) associated with the spatial dynamical system discussed so far. The reduction from equation (1) to the so called amplitude equation whose rigth hand side is given, at the leading order by the normal form of the spatial dynamical system can be done with the help of asymptotic expansions.

We now discuss the problem of dynamical stability of patterns described below. Defining stable pattern as the ones that absolutely minimizes the Liapunov functional it is readily found that it is given by $A=0$ for $\mu < 0$, $A = \pm\sqrt{\mu}$ for positive μ in the case of equation (28), while it is $A = 0$ for negative μ and $A = \sqrt{\mu}\exp i\phi$, where ϕ is an arbitrary phase for positive μ, in the case of equation (29). All the other patterns are found to be unstable or metastable. Particularly interesting are the patterns corresponding to the stationary solution R_+. These patterns are parametrized by their wavenumber or equivalently by the value of their invariant h. They can be interpreted as periodic patterns which differ by their total number of periodic cells. Since the homoclinic trajectory precisely corresponds to perfect periodic pattern with one unit cell added or substracted, the wavelength changing process involves a transition from the perfect periodic pattern to the homoclinic pattern [6]. The Liapunov functional evaluated with this homoclinic solution plays the role of an activation energy needed to initiate the transition. This activation energy is found to be maximal for $h = 0$ and decreases to vanish when $h \to h_c$. The point where $h = h_c$ is known as the Eckhaus instability transition [6],[7].

4. SOME CODIMENSION TWO PATTERN FORMING TRANSITIONS

More complicated one-dimensional patterns exist in nature. They cannot be captured by studying codimension one pattern forming transitions because of their integrability properties and their low dimensional spatial phase portrait. The situation is quite analogous as in temporal dynamical system, where only normal forms of codimension three can contain chaotic behaviors [8] [9]. Thanks to the conservative nature of the dynamical systems considered here we can already expect spatial complexity in codimension two transitions. In this section we briefly discuss some problems of codimension two. Up to now we have considered codimension one topological transitions which occur on a given stationary solution. Two cases have been looked at. In the first one a stationary solution changed its topological nature from real hyperbolic to elliptic. This case was the analog of the relaxational dynamics of a ferromagnetic transition. In the second case a stationary solution changed from complex hyperbolic to elliptic. Although the phase portrait is quadri-dimensional in that case no chaotic pattern was an admissible pattern because of the existence of two invariants, one associated with the translational invariance, the other with an invariance property of the normal form itself, namely the invariance of equation (29) under the transformation $A \to A\exp i\phi$. This last property comes from the very nature of the bifurcation. We will study later an interesting way to break this invariance. A simple example of codimension two bifurcation is associated with the following characteristic polynomial

$$-s^4 + \nu s^2 + \mu = 0 \tag{33}$$

The corresponding normal form reads at the leading order

$$-A_{xxxx} + \nu A_{xx} + \mu A - A^3 = 0 \tag{34}$$

In equation (33) and (34) the signs have been choosen in order to guarantee dynamical stability. Equation (34) can be derived from a Lagrangian f

$$f = \frac{1}{2}A_{xx}^2 + \frac{\nu}{2}A_x^2 - \frac{\mu}{2}A^2 + \frac{1}{4}A^4 \tag{35}$$

The corresponding evolution equation reads

$$\partial_t A = -\frac{\partial F}{\partial A} = -A_{xxxx} + \nu A_{xx} + \mu A - A^3 \tag{36}$$

where $F = \int dXf$. It describes the relaxational dynamics associated with a Lifchitz point [10]. In the unfolding parameter space $\mu - \nu$ the codimension one line of transitions reviewed in the first part of this paper occur. Defects can have an oscillatory asymptotic behavior due to the complex hyperbolic nature of the stationary solutions (see fig. 4).

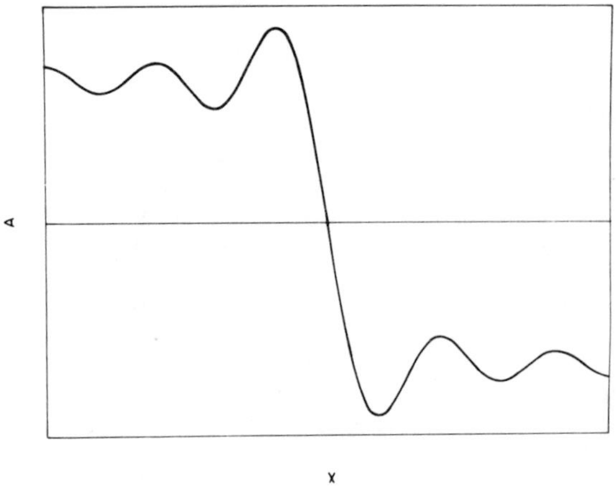

Figure 4. Oscillatory defect corresponding to an heteroclinic solution of equation (34).

A direct consequence of this oscillatory behavior is the existence of interesting chaotic patterns [11]. Although these patterns are not believed to absolutely minimize the Liapunov functional they can be metastable and then play an important role when noise is added to equation (1) [12]. Another typical example of codimension two singularity is associated with the degeneracy of the nonlinear term of equation (28). The normal form of the evolution equation reads in that case

$$\partial_t A = A_{XX} + \mu A + \nu A^3 - A^5 \tag{37}$$

It describes the relaxational dynamics near a tricritical point. The presence of a small symmetry breaking term in equation (28) also leads to a codimension two situation whose corresponding evolution equation is given by

$$\partial_t A = A_{XX} + \nu + \mu A - A^3 \tag{38}$$

The small parameter ν can be interpreted in this case as a small external magnetic field. Other codimension two transitions can be obtained from equation (18) as for example

$$\partial_t A = -A_{XXXX} + \nu A_{XX} + \mu A - |A|^2 A \tag{39}$$

It corresponds to the strongly resonant interaction between two pattern forming transitions with almost identical wavevector. Equation (39) in particular describes the appearance of stable quasiperiodic patterns.

5. AN EXAMPLE OF CODIMENSION THREE TRANSITION

An interesting example of codimension three singularity is obtained by weakly breaking the translational invariance of the initial equations [13]. The external breaking considered here will simply be a periodic one. Working near parameter values where a pattern forming transition described by equation (29) takes place, one is left with the problem of the external periodic forcing of a double non semi-simple imaginary eigenvalue singularity in a conservative system. As expected the actual value of the external frequency plays an important role since it leads to resonances. The normal form of such an instability reads, at the leading order

$$A_{XX} + 2iqA_X + (\mu - q^2)A + \alpha\bar{A}^{(n-1)} - |A|^2 A = 0 \tag{40}$$

where n is defined as $mk_1 = n(k_0 + q)$, k_1 being the wavenumber of the external excitation, q the deviation from the exact resonnance, α a small parameter which generically scale as ε^m, where ε represents the intensity of the external forcing. The actual relation between U and A is given by

$$U(x) = A(X)\Phi\exp(i(k_0 + q)x) + \mathcal{U}(A,\bar{A},\mu;x) + c.c \tag{41}$$

where $\mathcal{U}$ is a periodic function in x with period $2\pi/k_1$. The cases corresponding to $n < 4$ are termed as strongly resonant. Equation (40) posseses the discrete group of invariance $A \to A\exp(2\pi i/n)$. As a consequence one of the dynamical invariants is lost and this gives the possibility of chaotic solutions. As usual codimension three singularities contain in their unfolding some codimension two and one singularities whose nature depends on n. In both cases the codimension one transitions are of the type discussed in the first section of this paper. In the case $n = 1$ codimension two transitions occur on a line of tricritical points characterized by a infinite wavelength. In the case $n = 2$ it corresponds to a line of Lifchitz points, while in the case $n = 3$ it is a line of tricritical points associated with a zero wavelength. The dynamical model corresponding to the spatial dynamical system given by equation (40) reads [14] [15]

$$\partial_t A = \frac{\partial F}{\partial \bar{A}} = A_{XX} + 2iqA_X + (\mu - q^2)A + \alpha\bar{A}^{(n-1)} - |A|^2 A \tag{42}$$

where $F = \int dXf$ is the Lagrangian associated with equation (40)

$$f = \frac{1}{2}(R^2 - \mu)^2 + R_X^2 + R^2(\Theta_X + q)^2 - \frac{\alpha}{n}R^n\cos(n\Theta) \tag{43}$$

where $A = R\exp(i\Theta)$. This Liapunov functional turns out to be a model for the free energy charge density waves systems [16] [17]. We now focus our attention on topological changes that occur on dynamically stable stationary solutions of equation (42). The solution $A = 0$ represents the forced pattern while a non trivial solution $A = Q$ is associated with the locked pattern. The domain of stability of the forced pattern is bounded by two codimension one surfaces. On one of these surfaces the topological nature of the trivial solution changes its type from real hyperbolic to elliptic. This surface is associated with the transition from forced patterns to locked patterns. In the case $n = 1$ the corresponding transition does not really exist since the bifurcation is imperfect. It is actually described by equation (38). In the case $n = 2$ a transition described by equation (28) occurs on a surface given by

$$\mu - q^2 + \alpha = 0 \tag{44}$$

In the case $n = 3$ the bifurcation is discontinuous. The surface where metastability occurs is given by

$$\mu - q^2 + 2\frac{\alpha^2}{9} = 0 \tag{45}$$

The second surface where the forced pattern looses its stability is associated with a change from complex hyperbolic to elliptic. In both cases the transition is described by equation (29). In the case $n = 1$ the corresponding surface

$$\alpha = (5q^2 - \mu + 2\sqrt{4q^4 - \mu q^2})(4q^2 + 2\sqrt{4q^4 - \mu q^2})^{\frac{1}{2}} \tag{46}$$

ends on a line of codimension two points whose associated dynamics is described by the equation

$$\partial_t A = A_{XX} + \mu A + \nu|A|^2 A - |A|^4 A \tag{47}$$

In the case $n = 2$ the corresponding surface given by

$$\mu + \frac{\alpha}{4q^2} = 0 \tag{48}$$

ends on a line of Liftchitz points described by equation (34). In the case n = 3 it occurs on the plane

$$\mu = 0 \tag{49}$$

the corresponding surface ends on lines of codimension two

$$\mu = 0 \quad , \quad q^2 - 2\frac{\alpha^2}{9} = 0 \tag{50}$$

In each case the transition corresponds to the appearance of a stable quasiperiodic pattern [18].

In the same way the domain of stability of the locked pattern is bounded by two codimension one surfaces of transition. The first one, corresponding to the transition from locked patterns to forced ones, has been previously discussed. On the second instability surface a transition from the locked pattern to a quasiperiodic one takes place. The mechanism of the transition is of nucleation type. More precisely, when one crosses this surface by increasing the parameter $|q|$, the absolute minima of the Lagrangian changes from the locked pattern described by some solution $A = Q$ of equation (42) to an heteroclinic solution connecting asymptotically the solution Q and $Q\exp(2\pi i/n)$. This heteroclinic connection corresponds to a perfect pattern with a phase defect, called in the context of commensurate-incommensurate phase transition a discommensuration. In order to obtain this transition surface it is worthwhile to consider first the limit μ large, more precisely $\mu \gg q^2, \alpha^{2/(4-n)}$. In that limit the Lagrangian (43) becomes

$$f = \mu(\Theta_X + q)^2 - \frac{2\alpha\mu^{n/2}}{n}\cos(n\Theta) \tag{51}$$

Its associated Euler-Lagrange equations are given by

$$\Theta_{XX} - \alpha\mu^{(n-2)/2}\sin(n\Theta) = 0 \tag{52}$$

A single discommensuration is thus given by the separatrix solution of the pendulum equation (see fig. 5)

$$\Theta_S = \frac{4}{n}\tan^{-1}\exp(\sqrt{n\alpha\mu^{(n-2)/2}}X) \tag{53}$$

Evaluating the value of the Lagrangian with this solution one readily finds that the transition takes place on a surface given by

$$16\alpha\mu^{(n-2)/2} - n\pi^2q^2 = 0 \tag{54}$$

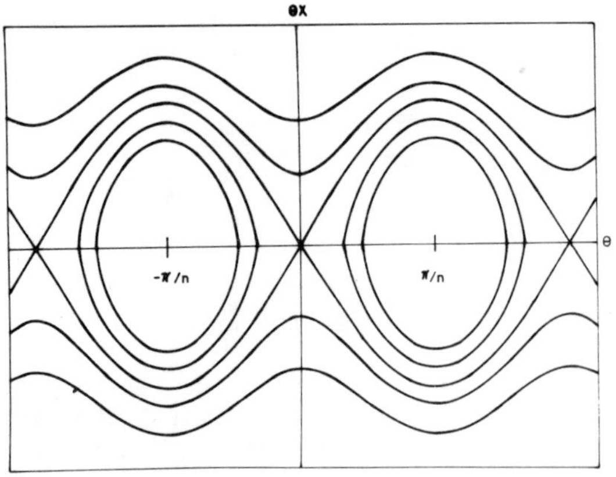

Figure 5. Phase portrait of the pendulum equation (52).

When $|q|$ is increased furthermore the absolute minima of the Lagrangian is taken on the rotating solutions of the pendulum equations. For $|q|$ close enough to $|q_c|$ these solutions are intepreted as periodic chains of discommensurations. The distance between two discommensurations obey the critical law

$$d \propto \log(q - q_c) \tag{55}$$

where q_c is defined by equation (54). The quasiperiodicity is introduced by the existence of this lattice whose period is in general incommensurate with the period of the external forcing. In the phase limit considered above the $\Theta_x - \Theta$ plane, where equation (53) lives, is an invariant manifold for the spatial dynamical system (46). The non trivial solution is real hyperbolic in the $R_x - R$ plane, characterized by strong dilatation and contraction properties. It is weakly real hyperbolic in the plane $\Theta_x - \Theta$. When one moves away from the phase approximation the "amplitude" eigenvalues decrease while the "phase" ones increase. On some surface given by

$$\mu - 2Q^2 - \frac{(\alpha(n-1)A^{(n-2)} - Q^2)^2}{4q^2} \tag{56}$$

they become equal. When this surface is crossed the topological nature of the non trivial solution changes from real to complex hyperbolic. The discommensurations then present oscillatory tails whose consequence is to make the nucleation transition discontinuous. The transition surface where the Lagrangian evaluated for a single discommensuration becomes equal to the ones evaluated for the locked solution is constructed numerically. One finds that it crosses the surface given by equation (56). In the oscillatory domain the Lagrangian of interaction between two distant discommensuration is then given by

$$F_{int} = 4\Gamma \exp(2\sigma_r d)\cos(2\sigma_1 d + \Psi) \tag{57}$$

where $\Gamma = \sigma\rho^2 - 2qQ\rho\theta - \sigma Q^2\theta^2$, $\sigma = \sigma_r + i\sigma_i$ are the eigenvalues of the spatial dynamical system linearized around the locked pattern, ρ and θ a small perturbation of this pattern and d the distance between two defects. Since F_{int} is an oscillatory function of d one readily concludes that the transition is first order. These transitions have actually been observed in a recent experiment by M. Lowe and J. Gollub [19]. The change from continuous to discontinuous when the forcing is increased is in agreement with our analysis. In order to complete the picture let us mention that the nucleation surfaces ends on the codimension two lines which terminates both transitions from forced to locked patterns and transitions from forced to quasiperiodic patterns. The various transitions are illustrated on fig. 6.

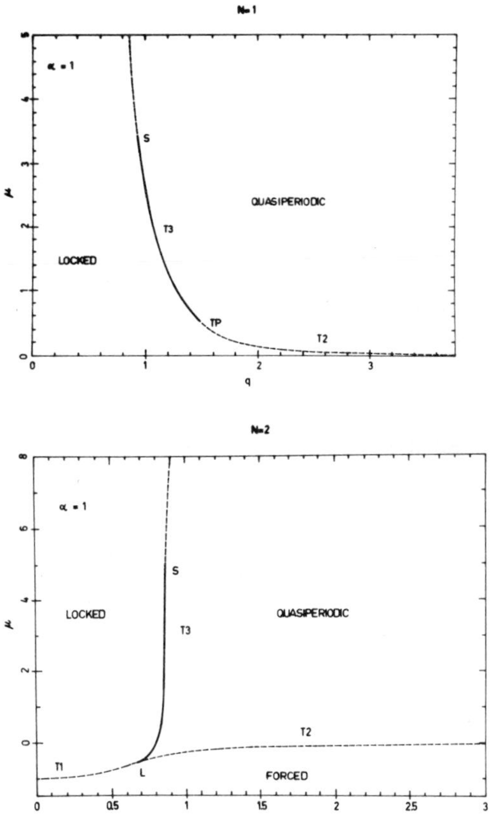

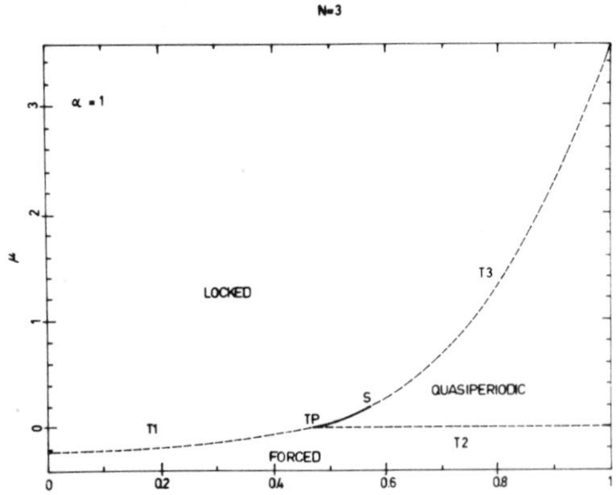

Figure 6. Phase Diagrams associated with equation (42). Transitions from the "forced patterns" to "locked patterns" are denoted as T1, transitions from "forced patterns" to "quasiperiodic" ones are denoted as T2, "nucleation" transitions are denoted as T3. In both cases dashed lines corresponds to continuous transitions while full lines are associated with discontinuous ones. The cases n = 1,2,3 respectively corresponds to figures 6a,b,c.

ACKNOWLEDGMENTS

This work has been partly supported by the C.N.R.S. through an A.T.P. and the D.R.E.T. One of us (P.C.) would like to thank the University of Arizona where part of this work has been performed under the ONR Engineering Grant N00014-85-k-0412.

REFERENCES

[1] P.C. Hohenberg and J.S. Langer, J. Stat. Phys. 28, 193 (1982) and references therein.
[2] Cellular structures in instabilities J.E. Weisfreid and S. Zaleski Eds. (Springer, New York, 1984) and references therein.
[3] See for instance J. Guckenheimer and P. Holmes, Nonlinear oscillations, dynamical systems, and bifurcation of vector fields. Springer Verlag (1984).
[4] A.C. Newell and J.A. Whitehead, J. Fluid Mech. 38, 279 (1969).
[5] L.A. Segel, J. Fluid Mech. 38, 203 (1969).
[6] L. Kramer and Zimmermann, Physica 26D, 221 (1985).
[7] W. Eckauss, Studies in Nonlinear Stability Theory, Springer Verlag (1965).
[8] A. Arneodo, P. Coullet, E.A. Spiegel and C. Tresser, Physica 14D, 327 (1984).
[9] P. Coullet, Physics Reports 103, 95 (1983).

[10] R.M. Hornreich, M. Luban and S. Shtrikman, Phys. Rev. Let. 35, 1678 (1975).
[11] R. Devaney, J. Diff. Equ. 21, 431 (1976).
[12] S. Aubry, private communication.
[13] P. Coullet and D. Repaux, "Strong resonances of periodic patterns" Preprint (1986).
[14] R.E. Kelley and D. Pal. J. Fluid Mech. 86, 433 (1978).
[15] P. Coullet, Phys. Rev. Let. 56, 724 (1986).
[16] A.E. Jacobs and B. Walker, Phys. Rev. B21, 4132 (1980).
[17] B. Schaub and D. Mukamel, J. Phys. 16C, L225 (1983).
[18] P. Coullet, D. Repaux and J.M. Vanel, "Quasiperiodic patterns" to appears in J. Am. Math. Soc. (1986).
[19] M. Lowe and J.E. Gollub, Phys. Rev. A31, 3893 (1985).

PATTERNS AND DEFECTS FAR FROM EQUILIBRIUM

Daniel Walgraef *
Service de Chimie-Physique II,
Université Libre de Bruxelles, CP 231,
B-1050 Brussels, Belgium

1. INTRODUCTION

When driven away from equilibrium by the effect of an external constraint, many nonlinear systems present spatio-temporal order beyond a threshold value for the control parameter [1]. At this point, the homogeneous steady state becomes unstable versus inhomogeneous or oscillatory perturbations and various spatial or temporal patterns may nucleate spontaneously in the medium. This has long been a puzzling phenomenon from experimental and theoretical point of views and numerous questions related to the pattern formation, selection and stability mechanisms in physical, chemical and biological systems remain still unanswered. Due to the complexity of the dynamics a complete analytical solution of the kinetic equations governing the behavior of the system is usually impossible. However near threshold great simplifications occur as a consequence of the separation between the time and space scales of the stable and unstable modes [2]. The evolution of the system may effectively be reduced to its slow mode dynamics which leads to amplitude equations for the spatio-temporal patterns. This important simplification which reduces the complete set of nonlinear kinetic equations to the kinetics of the unstable modes only is valid and may be justified by multiple scale analysis in the vicinity of the instability [3]. In this framework, various problems related to the growth rate of inhomogeneous fluctuations, wavelength selection, boundary effects, phase dynamics have been successfully described for hydrodynamical, chemical or photochemical instabilities in isotropic and anisotropic media [4].

It turns out that the linear stability analysis performed on the homogeneous reference state gives the critical wavelength and the localisation of the instability in the parameter space. Above threshold the competition between linear and nonlinear contributions to the dynamics provides further selection mechanism between the possible planforms. Boundary effects, variational principles or experimental procedures may also play a role in the pattern selection processes. Furthermore, as the

* Chercheur Qualifié au Fonds National de la Recherche Scientifique de Belgique.

E. Tirapegui and D. Villarroel (eds.), Instabilities and Nonequilibrium Structures, 197–216.

structures appear via the breaking of continuous symmetries, long range phase fluctuations may spontaneously develop in the medium and induce the nucleation of defects related to phase singularities such as vortices, dislocations or disclinations [5,6].

The question arises then to understand what happens when the intensity of the constraint or of the bifurcation parameter is increased beyond threshold. In principle the lowest order amplitude equations are no more valid and the derivation of higher order contribuyions is usually a formidable task with no guarantee on the convergence properties of the procedure. It is why model equations which correspond to extensions of the slow mode dynamics at finite distances of threshold have been used to study hydrodynamical or chemical patterns [7]. The symmetry properties of the system may also allow the derivation of the phase dynamics beyond threshold in a very powerful and elegant manner [8]. Such models have been studied analytically [9] and numerically [10] and their predictions were experimentally tested in a few convective instabilities [11]. Due to the ability of such models to describe the stability boundaries, relaxation phenomena and defect activity in various experimentally relevant situations, I would like to discuss here some examples of self-organisation phenomena in non equilibrium systems in this framework. I will especia-ly consider the case of autowaves in chemical oscillators, pattern selection and defects in anisotropic media and finally the nucleation of dislocation microstructures in fatigued metals as the result of dynamical instabilities due to the competition between the mobility and interactions of dislocations.

2. SPATIO-TEMPORAL ORGANISATION OF CHEMICAL OSCILLATORS

In many nonequilibrium systems where order may appear spontaneously, spatio-temporal organisation results from the cooperative effects within populations of individuals like electrons, molecules, cells, insects, dislocations, voids,... The instability of their homogeneous equilibrium steady states is usually due to the interplay between their mobility and local or nonlocal interactions which is modified by the external constraints driving such systems away from the thermal equilibrium. The macroscopic dynamics of the system is often deduced from phenomenological rate equations or from master equations built on elementary processes between the members of the populations under consideration [12]. Both approaches lead to dynamical systems of the type:

$$\dot{X}_i = F_i(b,\{X_j\}) + D_i\nabla^2 X_i \qquad (2.1)$$

where $\{X_i\}$ are the local concentrations of active species, D_i are the diffusion coefficients (or tensor in anisotropic media, in the case of diffusional instabilities higher order diffusion coefficients may be needed). F_i are the rates of change of X_i due to mutual interactions and b represents a set of parameters corresponding to the external constraints.

When the external constraints are increased beyond the linear regime around thermal equilibrium various instabilities may arise which are located by analyzing the linear stability of the homogeneous steady state. Among them, the Hopf bifurcation is a typical example of symmetry breaking instability in nonlinear systems [13]. Such a bifurcation occurs when two eigenvalues of the linear evolution matrix associated with the dynamical system become purely imaginary whilst the other ones have negative real parts. The corresponding eigenmodes are, hence, the two slow modes of the problem. The remaining fast modes may then be adiabatically eliminated and the evolution of the system is given on the longest time scale through its slow mode dynamics only. Hence, such systems may be described, in the vicinity of the Hopf bifurcation by the following Langevin equation for the slow modes (Σ,Σ^*) in their rotating frame ($\sigma = \Sigma \exp i\omega_0 t$ where $+(-)i\omega_0$ are the imaginary eigenvalues of the linear evolution matrix at the bifurcation point:

$$\dot{\sigma}(\vec{r},t) = [r_o + c\nabla^2]\sigma(\vec{r},t) - u|\sigma(\vec{r},t)|^2\sigma(\vec{r},t) + \eta(\vec{r},t) \tag{2.2}$$

This equation corresponds to the normal form of the problem and may be explicitely derived for concrete models.

In the ordered regime, the Langevin equation (2.2) may be written, in phase and amplitude variables ($\Sigma(\vec{r},t) = R(\vec{r},t)\cdot\exp i[\omega_0 t + \phi_0 + \phi(\vec{r},t)]$):

$$\begin{aligned}\dot{R} &= u_1(R_o^2 - R^2)R + c_1(\nabla^2 R - R(\nabla\phi)^2)\\ &\quad - c_2(2\nabla R\cdot\nabla\phi + R\nabla^2\phi) + \eta_R\\ R\dot{\phi} &= u_2(R_o^2 - R^2)R + c_2(\nabla^2 R - R(\nabla\phi)^2)\\ &\quad + c_1(2\nabla R\cdot\nabla\phi + R\nabla^2\phi) + \eta_\phi\\ R_o^2 &= r_o/u_1\end{aligned} \tag{2.3}$$

The linear stability analysis performed on this system around the limit cycle defined by $R = R_0, \phi = \phi_0$, shows that when $u_1c_1 + u_2c_2 > 0$ it is stable against small radius or phase perturbations while it is unstable in the opposite case where phase turbulence occurs [14].

It is easy to check that the eigenvalues of the linear evolution matrix associated to eq. (2.3) are well separated when the range l of the fluctuations is such that $L > L_0[(u_1c_1 + u_2c_2)/r_1u_1]^{0.5}$. Hence, inhomogeneities of range $L < L_0$ evolve on a commons time scale and are strongly coupled while for $L > L_0$ amplitude and phase fluctuations evolve on different time scales and the amplitude which is the fast variable may be adiabatically eliminated. It turns out that the asymptotic phase dynamics governs the evolution of the oscillations and is given by the stochastic Burges equation [15]:

$$\dot{\phi} = \mu\nabla^2\phi + \nu(\nabla\phi)^2 + \bar{\eta} \tag{2.4}$$

where $\mu = c_1 + u_2c_2/u_1$; $\nu = (u_1c_2 - u_2c_1)/u_1$

and

$$<\eta(\vec{r},t)\bar{\eta}(\vec{r}',t')> = \frac{2\Gamma}{R_o^2}\,\delta(\vec{r} - \vec{r}')\delta(t - t')$$

Note that the validity of this equation is not restricted to the vicinity of the Hopf bifurcation [16]. It may effectively be shown that for any stable limit cycle, the phase variable obeys a kinetic equation of the Burgers type; the coefficients μ and ν and the noise intensity depend however on the shape of the limit cycle.

The asymptotic distribution of the phase fluctuations may be deduced from (2.4) and written as:

$$P(\{\phi(\vec{r})\}) = N\exp\,[-\frac{\mu R_o^2}{2\Gamma}\int_{L_o} d\vec{r}(\vec{\nabla}\phi)^2] \qquad (2.5)$$

As a consequence large low-dimensional systems ($d \leqq 2$) cannot exhibit true long range order since the order parameter correlation function shows an algebraic decay for $d = 2$ and an exponential decay for $d = 1$ [6]. This algebraic decay is associated to quasi long range order. Hence, small systems ($L < L_0$) may oscillate synchronously while large systems ($L > L_0$) become partly desynchronized by the spontaneous nucleation of long range phase fluctuations. From the Burgers equation (2.4) we are able to estimate the probability of occurence of a phase inhomogeneity and to follow its spatio-temporal evolution.

Two types of phase fluctuations may affect the oscillatory behavior on the macroscopic level:

1) As phase and amplitude are strongly coupled on short length scales, even small disturbances may induce local frequency shifts of the oscillations. The small regions where these shifts occur oscillate with a period different than that of the bulk and radiate concentric waves of iso-concentration lines or target patterns. One finds that a fluctuation of amplitude a and rangle L_0 induces the following phase evolution far from the center:

$$\phi(\vec{r},t) = \mathrm{Max}(0,\nu k^2(t - t_0) - k|\vec{r} - \vec{r}_i|) \qquad (2.6)$$

where k, the wavenumber of the pattern increases with a. Such patterns are currently observed in various two-dimensional oscillating or excitable media (cf. Fig. 1) like the Belusov-Zhabotinskii reaction [17], chlorite oscillators [18], fungi mycelia, nerve fibers, aggregating Dictyostelium Discoideum [19], catalytic surfaces [20]... and are an intrinsic property of unstirred oscillating media.

2) Vortex-like fluctuations associated to phase singularities analogous to the vortices of XY or superfluid models may also appear in oscillating systems. Such vortices have an integer topological charge or winding number, N. Their long time evolution leads to N-armed rotating Archimedean spiral waves of isoactivity [21,22]:

$$\phi(\vec{r},t) = N \operatorname{arctg} \frac{(y-y_i)}{(x-x_i)} + \mathrm{Max}(0, \nu k^2(t - t_0) - k|\vec{r} - \vec{r}_i|) \tag{2.7}$$

The probability associated to an isolated vortex is given by:

$$P \propto \exp - \frac{2\pi\mu R_o^2}{\Gamma} \ell n(L/L_o) \tag{2.8}$$

and is vanishingly small in large systems. On the other hand, the probability of a pair of vortices of opposite winding numbers (+N, -N) located at $\vec{r}_i$ and $\vec{r}_j$ is given by:

$$P \propto \exp - \frac{2\pi\mu R_o^2}{\Gamma} \ell n \frac{|\vec{r}_i - \vec{r}_j|}{L_o} \tag{2.9}$$

and remains finite. Hence spiral waves may appear spontaneoulsy by clusters of zero total vorticity only. Up to now, no spontaneous apparition of isolated spirals has been reported.

Figure 1. Spiral waves in the BZ reaction (photograph from C. Müller, T. Plesser and B. Hess. MPI für Ernährungsphysiologie,D4600, Dortmund, FRG).

When the chemical reactions take place in fluid phases, concentration fluctuations may also result from the coupling with convective motion induced for example by local temperature gradients, surface effects like evaporative cooling and Marangoni effects or even by stirring memory. Due to the extreme sensitivity of the phase dynamics to fluctuations, this coupling is very likely to affect the overall behavior of the system and should be incorporated in its description. If we restirct ourselves to the case of passive convection, the nonlinear reaction-diffusion dynamics may then be written as:

$$\dot{X}_i = F_i(b,\{X_j\}) + D_i\nabla^2 X_i + \vec{v}\cdot\vec{\nabla}X_i \tag{2.10}$$

$\vec{v}$ being the velocity field of the solvent.

In this case the homogeneous limit cycle appearing for $b > b_c$ is not affected by the convective term which, however, affects the phase dynamics. Effectively, the adiabatic elimination of the amplitude of the oscillations leads to the following kinetic equation for the phase ϕ:

$$\dot{\phi} = \omega + \mu\nabla^2\phi + \nu(\nabla\phi)^2 + \vec{v}\cdot\vec{\nabla}\phi \tag{2.11}$$

(where $\omega = \mathrm{Im}(r_0 - uR^2)$) and the autowave propagation should hence be modified as a result of the presence of convective motion in the solution.

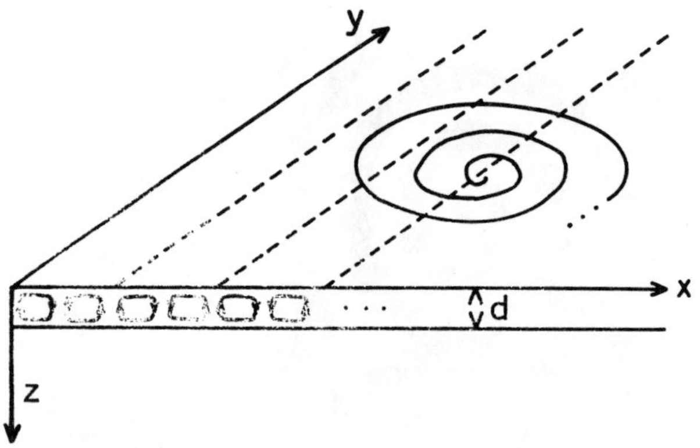

Figure 2. Sketch of the experimental situation corresponding to a spiral wave initiated in a convective layer.

Let us for example consider the simple case of a layer of solution of thickness d where convective Bénard rolls develop along the Ox direction and where a chemical spiral scroll with vertical axis is nucleated (cf. Fig. 2). The first order correction ϕ_1 to the wave shape ϕ_N ($\phi_N = N\,\mathrm{arctg}(y/x) + \mathrm{Max}(0, \nu k^2 t - k(x^2+y^2)^{\frac{1}{2}})$ is inferred from equation (2.11) and is given, far from the scroll axis, by:

$$\phi_1 = \mu\nabla^2\phi_1 + v_x\nabla_x\phi_N = \mu\nabla^2\phi_1 + 2v_x kx(x^2+y^2)^{-\frac{1}{2}} \tag{2.12}$$

In the case of free boundary conditions v_x behaves as

$$v_x = v_0 \cdot \sin(\pi z/d) \cdot \sin q_0 (x-a)$$

and ϕ_1 may be written as $\phi_1 = \sin(\pi z/d) \cdot \psi(x,y)$ leading to:

$$\dot{\psi}(x,y) = \mu((\nabla_x^2 + \nabla_y^2) - (\frac{\pi}{d})^2)\psi(x,y) + \frac{2kv_o x}{(x^2+y^2)^{\frac{1}{2}}} \sin q_0 (x-a) \tag{2.13}$$

It turns out that the propagation of the spiral wave is practically not affected near the surface or in the y direction while it is strongly modified in the center of the layer and in the x-direction and the iso-concentration lines are given at the lowest order in the velocity field by:

$$N \operatorname{arctg}(\frac{y}{x}) - k(x^2+y^2)^{\frac{1}{2}} - \frac{2kv_o}{3\mu q_0^2} \frac{x}{(x^2+y^2)^{\frac{1}{2}}} \sin q_0 (x-a) = \text{cst.} \tag{2.14}$$

leading to an irregular modulation of the wavefronts which increases with the thickness of the layer. Moreover the concentration wavefronts tend to be orthogonal to the convective rolls. The irregularity of this modulation may also be shown to be more important in the case of polygonal convective structures due to the periodicities in the x and y direction. These effects are in qualitative agreement with the observations of Krinsky et al. [23] (cf. Fig. 3).

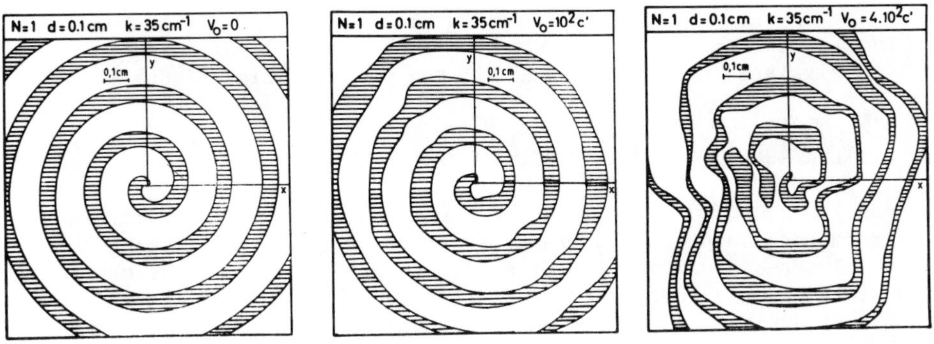

Figure 3. Deformation of a one-armed spiral wave due to the effect of convection.

When one of the reacting species shows a vertical concentration gradient like for example in the case of evaporative cooling or adsorption effects at the top surface, the phase dynamics becomes:

$$\dot{\phi} = \omega + v_z(\nabla_z g)\sin\phi + \mu\nabla^2\phi + \nu(\nabla\phi)^2 + \vec{v}\cdot\vec{\nabla}\phi \tag{2.15}$$

where g is a functional of the concentration gradients and a decreasing function of the oscillation amplitude R. If the solution of this equation is constructed as an expansion in the molecular diffusion coefficients, we have, at the lowest order:

$$\dot{\phi}_0 = \omega + (v_z\nabla_z g)\sin\phi_0 \tag{2.16}$$

The temporal oscillations may then be supressed by phase locking above a convective threshold given by $|v_z\nabla_z g| > \omega$. This implies the coexistence of regions with oscillatory or excitable behavior within the layer. The molecular diffusion mechanisms, which are very slow, are expected to alter this simple picture on long times scales only. They should then modify the zeroth order phase dynamics according to the following rate equation:

$$\dot{\phi}_1 = \mu\nabla^2\phi_1 + \nu(\nabla\phi_1)^2 + f(\vec{v},\nabla_z g,\nabla\phi_0) + h(\nabla\phi_0) + v_z k(\phi_0,\phi_1)$$
$$(\phi = \phi_0 + \phi_1) \tag{2.17}$$

At the surface, while f vanishes, h contains the print of the convective pattern (through the $(\nabla\phi_0)^2$ term for example) and hence induce spatially dependent frequency variations of the oscillations able to trigger chemical waves in the boundary layer.

In this case the coupling between chemical oscillations and convective motion may lead to a complex spatial structuration of the solution due to the extreme sensitivity of the phase dynamics to even small perturbations or fluctuations: inhomogeneous phase locking in the center of the layer, wave generation in the boundary layer,... The existence of such phenomena has already been reported [25] but more quantitative analysis is needed to check all the possibilities discussed here.

3. PATTERN SELECTION IN ANISOTROPIC MEDIA

Patterning instabilities occur in many nonlinear systems where interactions terms are coupled with transport processes and lead to macroscopic spatial structures. The linear stability analysis fixes the critical value of the control parameter and the critical wavelength of the patterns. Contrary to most hydrodynamic instabilities, the critical wavelength in reaction-diffusion systems is determined by intrinsic properties of the system and is given in terms of the two competing mechanisms (reaction rates and diffusion constants) the dynamic equilibrium of which leads to the spatial ordering of the medium. Systems with spatal dimensionalities higher than one exhibit a infinite degeneracy. In isotropic media, this degeneracy is twofold:

1) for a given wavelength, different planforms may be simultaneously stable (rolls, squares, hexagons, ...);
2) the localisation and orientation of the patterns remain undetermined in inbounded systems.

In anisotropic media, however, an intrinsic mechanism raises the orientational degeneracy by inducing preferred directions for the wave-vectors defining the structure. Close to the instability, the complete nonlinear dynamics may be reduced to that of the unstable mode only. Its linear growth rate ω_q, tends to zero when approaching the instability. In isotropic systems it depends on the length on the wavevector only while in anisotropic media it also depends on its orientation. For example in uniaxial materials we have:

$$\omega_q = \frac{b-b_c}{b_c} - D(q^2 - q_c^2)^2 - Aq^2\sin^2\varphi \tag{3.1}$$

where φ is the angle between the wavevector and the principal axis of the system, A being the anisotropy constant.

Model equations associated to reduced dynamics have been derived near Rayleigh-Bénard instabilities [7], near convective instabilities in nematic liquid crystals [4b] or in nonlinear chemical network [8]. Numerical and experimental analysis also show that such dynamical models may be valid well beyond threshold [10,11,25]. In two-dimensional systems, where most of the experimental studies are performed, these models lead to the following type of kinetics for the order parameter field $\sigma(x,y,t)$ in the case of isotropic media:

$$\dot{\sigma} = \varepsilon\sigma - \frac{\xi_0^2}{4q_0^2}(q_0^2 + \nabla^2)^2\sigma + NL(\sigma) \tag{3.2}$$

($\varepsilon = (b - b_c)/b_c$ where b is the control parameter, ξ_0 a coherence length scale, q_0 the critical wavenumber, $\nabla^2 = \nabla_x^2 + \nabla_y^2$).

The linear term defines the band of allowed wavevectors and favors the critical one far from the boundaries. The structure of the nonlinear couplings participates in the selection of the planforms. For example in the case of Rayleigh-Bénard convection for Boussinesq fluids and high Prandtl number, $NL(\sigma)_1 = -u\sigma^3$ and roll patterns are stable. When non-Boussinesq effects are present or for nonlinear reaction-diffusion networks,

$$NL(\sigma)_2 = -v\sigma^2 - u\sigma^3 \tag{3.3}$$

In this case roll patterns and hexagonal or triangular textures may be simultaneously stable. For convecting fluids between poorly conducting plates, $NL(\sigma)$ may be approximated by $u\nabla(\nabla\sigma\cdot(\nabla\sigma)^2)$ leading to the stability of square patterns [26].

A large class of systems where stable inhomogeneous structures appear spontaneoulsy are anisotropic like liquid crystals, chemically active media with electrical fields,... [4b]. As mentioned above the first manifestation of this anisotropy results in an intrinsic mechanism which raises the orientational degeneracy by inducing preferred directions for the wavevectors defining the structure. One-directional anisotropy related to an external field for example can also play an important role in the pattern selection mechanisms since it usually favors roll structures. Hence, according to its importance, anisotropy may destabilize the polygonal patterns selected by the nonlinearities of the dynamics. It may be shown in different models where roll or square patterns may appear supercritically [25,26] that close to threshold the linear anisotropic effects dominate and favor a roll structure. Let us discuss here the influence of anisotropy on the competition between the rolls and the hexagonal or triangular patterns which may appear subcritically in reaction-diffusion systems or non-Boussinesq convection. Let us consider for this purpose the following slow mode dynamics:

$$\dot{\sigma}(x,y,t) = [\varepsilon - \frac{\xi_0^2}{4q_0^2}(q_0^2 + \nabla_x^2 + \nabla_y^2)^2 + A\nabla_y^2]\sigma(x,y,t) - v\sigma^2(x,y,t) - u\sigma^3(x,y,t) \quad (3.4)$$

In the isotropic case ($A = 0$) two types of patterns may be simultaneously stable and correspond to rolls or structure built on wavevectors forming equilateral triangles (triangular or honeycomb lattices). Among them the preferred ones have critical wavevectors ($|\vec{q}| = q_0$). The absolute and relative stability of these structures may be computed via their associated Lyapunov functional. For increasing A we expect successive transitions form the homogeneous state to triangular lattices and finally to roll patterns (cf. Fig. 4).

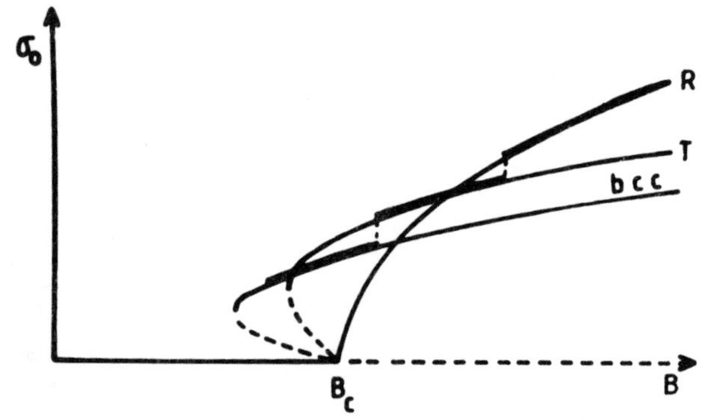

Figure 4. Bifurcation diagram associated to the dynamics given by the amplitude equation (3.2,3) for 3D systems. Heavy lines correspond to the structures minimising the associated Lyapunov functional.

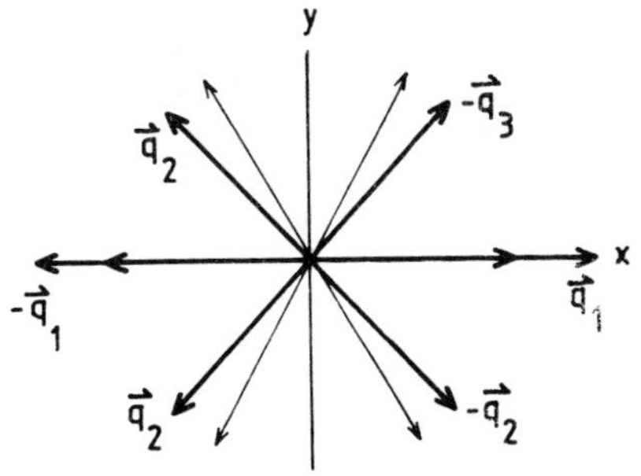

Figure 5. $\{\vec{q}_i\}, i = 1,2,3$ represent the wavevectors associated to triangular or honeycomb patterns in 2D anisotropic systems (the light vectors represent the corresponding wavevectors in the isotropic case).

In the presence of anisotropy, the instability versus one-dimensional modulations occurs first for wavevectors in the x direction for $A > 0$. Structures with triangular symmetry may still exist but are deformed by the anisotropy. Effectively they are built on three pairs of wavevectors $-\{\pm q_i\}, i = 1,2,3$-such that (Fig. 5):

$$\frac{\xi_0^2}{4q_0^2}(q_0^2 - q_1^2) = Aq_{2y}^2 = Aq_{3y}^2 \tag{3.5}$$

The amplitude of the modulations associated to this structure are given by:

$$[\varepsilon - \frac{\xi_0^2}{4q_0^2}(q_0^2 - q_1^2)^2]\sigma_T - 2v\sigma_T^2 - 15u\sigma_T^3 = 0 \tag{3.6}$$

As this structure may only be stable for:

$$\varepsilon - \frac{q_0^2\xi_0^2}{4}(1 - (\frac{q_1}{q_0})^2)^2 > -\frac{v^2}{15u} \tag{3.7}$$

it turns out from the dependance of q_1 versus A (cf. eq. (3.5)) that the anisotropy shifts the trheshold for the apparition of triangular or hexagonal patterns to higher values of the bifurcation parameter. Furthermore, once $\xi_0^2(q_0^2 - q_1^2)^2/4q_0^2$ becomes larger than $-v^2/15u$, which is easily realized for small non-Boussinesq effects in convective problems, the roll pattern is the first to appear. Hence, for a given anisotropy, experiments performed by increasing the bifurcation parameter are expected to display hexagonal or roll structures according to the intensity of the anisotropy. If, on the contrary, the experiments are performed at a given value of the bifurcation parameter, a sudden increase of the anisotropy may lead to the instability of the hexagonal pattern and the transition to a roll structure (Fig.6).

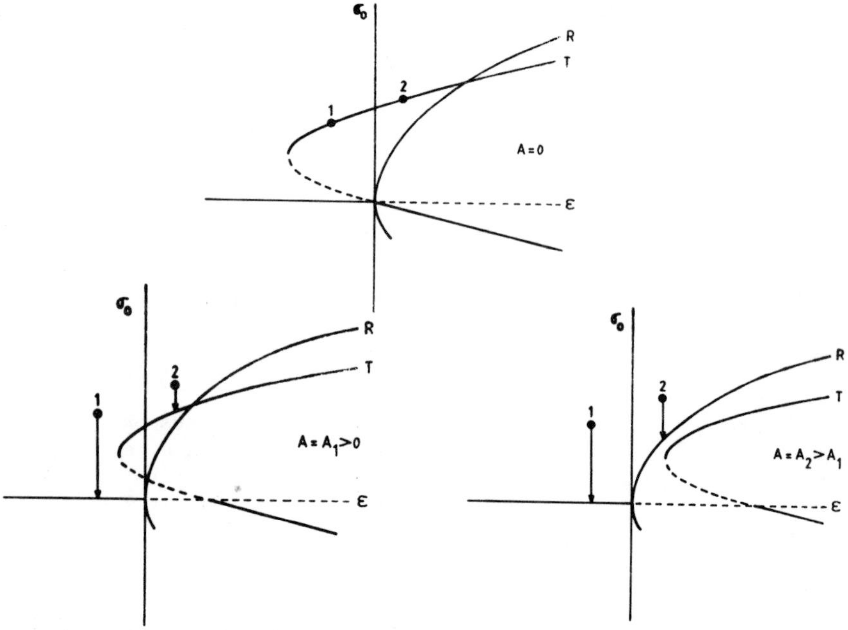

Figure 6. Bifurcation diagram associated to the amplitude equation for 2D anisotropic systems for increasing anisotropy. States 1 and 2 corresponding to stable patterns of triangular symmetry relax to new states when the anisotropy is increased.

The patterning instabilities discussed in this section are symmetry-breaking instabilities and here also the breakdown of translational and rotational symmetry generates long ranged fluctuations in the ordered region of the parameter space. These "dissipative Goldstone modes" [27] are responsible for the lack of true long range order in large low dimensional systems. These long ranged fluctuations are associated to the local changes in the position and orientation of the patterns. The corresponding dynamics is described in terms of phase variables which obey diffusion equations [28]. More precisely, in the case of a layered structure periodic along the x direction, the phase is a scalar satisfying the following equation:

$$\dot{\phi} = D_{\parallel} \nabla_x^2 \phi + D_{\perp} \nabla_y^2 \phi \tag{3.8}$$

where

$$D_{\parallel} = \xi_0^2 \cdot \frac{\varepsilon - 3\xi_0^2 (q - q_0)^2}{\varepsilon - \xi_0^2 (q - q_0)^2} \ , \ D_{\perp} = A q_0^2 + \xi_0^2 \ \frac{q - q_c}{q_c}$$

Experimental values for these coefficients have been obtained in the case of convective structures [29,30]. In isotropic systems,

$$D_{\parallel} = 2.2\cdot 10^{-3}\,cm^2 s^{-1}$$

and $D_\perp$ vanishes at threshold while in anisotropic systems $D_\perp$ is proportional to the anisotropy constant. In the (b,q) plane the curves $D_{\parallel}=0$, $D_\perp = 0$ delineate the region where the pattern is stable and one recovers the usual criteria derived for the Eckhaus and zig-zag instabilities. Similarly in the case of hexagonal or triangular patterns the phase vector obeys the following diffusion equation:

$$\dot{\phi} = D_\perp^T \nabla^2 \phi + (D_{\parallel}^T - D_\perp^T)\mathrm{grad}\cdot\mathrm{div}\phi \tag{3.9}$$

It is interesting to compare this equation with the basic equations of motion for the displacement field u of an isotropic solid:

$$\rho\ddot{u} = \mu\nabla^2 u + (\lambda + \mu)\mathrm{grad}\cdot\mathrm{div}\,u \tag{3.10}$$

where λ and μ are the Lamé coefficients. In the stationary case the diffusion coefficients $D_{\parallel}^T$ and $D_\perp^T$ play the role of the elasticity coefficients of a solid. However in nonequilibrium systems the phase dynamics is diffusive whereas the Newtonian equations for the displacement field admit elastic waves as solutions.

Describing the short range fluctuations by a Gaussian white noise it is possible to derive a potential for the long range phase fluctuations. In the case of layered structures it takes the form:

$$V = \int d\vec{r}[D_{\parallel}(\nabla_x\phi)^2 + D_\perp(\nabla_y\phi)^2] \tag{3.11}$$

while for hexagonal patterns we have:

$$V = d\vec{r}[\sum_{ij} D_\perp^T \phi_{ij}^2 + \sum_i (D_{\parallel}^T - D_\perp^T)\phi_{ii}^2] \tag{3.12}$$

where

$$\phi_{ij} = \frac{1}{2}[\partial_j\phi^i + \partial_i\phi^j]$$

Here again the analogy with the Hamiltonian of elastic solids is striking. As a result disclocations or disclinations are expected to appear spontaneously or due to experimental procedures. Moreover for increasing external constraints dislocations and grain boundaries may be expected to destroy the translational order like in the melting of 2D solids [31].

Another possible cause for the destruction of translational order well beyond threshold could be the nucleation of quasiperiodic structures via the higher order nonlinearities of the slow mode dynamics. These terms could effectively induce the formation of structures isomorphic to Penrose lattices like in "ā la Landau" approach to quasi-crystals [32,33,34].

To conclude this section let us emphasize the delicate balance between linear and nonlinear terms of the dynamics in the pattern selection processes. Moreover structures which appear to be stable in the deterministic analysis may however be strongly affected by the long range phase fluctuations induced by the breaking of translational and rotational symmetries and which lead to the spontaneous nucleation of defects in the nonequilibrium patterns. Small systems will in general not be affected by such fluctuations while in large systems they may lead to the desctruction of any long range order and to the apparition of spatial chaos.

4. SPATIO-TEMPORAL ORDER AND DISLOCATION PATTERNS IN SOLIDS

In this last section, I would like to show how dynamical nonequilibrium concepts and reaction-diffusion equations may improve the understanding of the fundamental aspects of dislocation patterning in stressed materials. Before discussing the principal aspects of a reaction-diffusion model developed for the description of self-organisation phenomena in fatigue experiments. Let me stress upon a few points in favor of the study of defect microstructures in the framework of nonlinear dynamics.

The macroscopic behavior of irradiated or mechanically or chemically driven materials is usually dictated by cooperative effects between defects like vacancies, interstitials, dislocations or voids. Moreover as such materials are driven away from thermal equilibrium, we cannot rely on traditional mechanical or thermodynamical approaches. Fortunately, elementary reactive interactions between defects and defect motion may be characterized experimentally and theoretically. As a result master equations or kinetic rate equations may be derived for the defect concentrations leading to nonlinear reaction-diffusion descriptions of their dynamics.

This is particularly interesting since such models are able to generate oscillatory or inhomogeneous solutions according to the value and variations of the control parameter and such behaviors are experimentally observed [35,36]. Furthermore theoretical modelling of yielding, spatial or temporal variation of dislocation concentration, fracture kinetics,... in the framework of contimuum mechanics [37] or irreversible thermodynamics [38] needs to incorporate the behavior of dislocation populations.

In addition to the different aspects of the instabilities discussed in the preceeding sections, specific properties of crystalline solids are expected to strongly influence the analysis. Effectively we deal here with strong anisotropies and the control parameter, say the stress or strain rates, control not only the dislocation creation and interaction rates but also the anisotropy itself via the velocity of mobile dislocations.. Moreover the main problems are related to the dynamics of the nucleation of the spatio-temporal structures since they affect the macroscopic properties of the samples during thei development. It may also happen that the failure of the sample occurs before the steady state is reached.

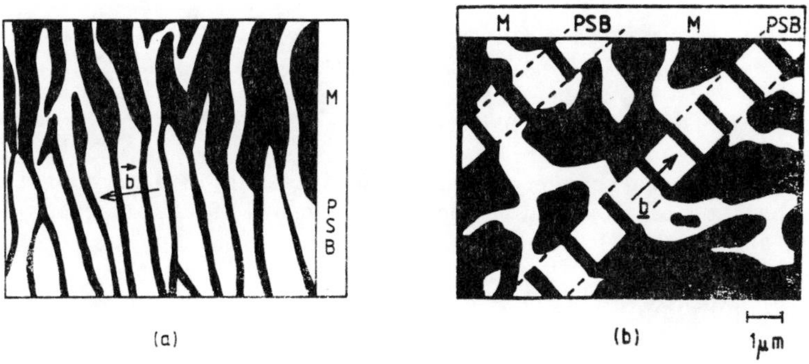

Figure 7. Vein and ladder-like dislocation patterns in fatigued copper single crystals ((a) in the primary slip planes, (b) in the cross-slip planes).

I will discuss here dislocation patterning in fatigue experiments on monocrystals of metallic compounds submitted to cyclic loading at controlled stress or strain rates. It is now well known that, according to the loading intensity, two types of dislocation patterns may appear in fatigued metals [35-38] (Fig. 7). One is the vein structure of the matrix consisting in dense multipoles of primary edge dislocations separated by dislocation poor channels. The second one is ladder-like in copper monocrystals for example and corresponds to persistent slip bands (PSB). It develops in primary slip planes and consists of regularly spaced walls of high dislocation density. The experimental observations also lead to the conclusion that the dislocation patterns within PSBs are the result of dynamic equilibrium between different processes such as creation, annihilation, pinning and motion. Hence a dynamical model may be built on the following elements:

1) The creation of dislocations under an applied stress is described by means of internal sources and the dislocation population is sufficiently high to be represented by a continuous dislocation concentration on a space scale larger than a few lattice spacings. The concentration is a local variable associated to the number of Burgers vectors originating in a space element.

2) The dynamical annihilation processes during glide are represented by nonlinear couplings in the dynamics which have the form of reactive contributions in chemical-like rate equations.

3) When the applied stress reaches threshold, i.e. when thermal or stress-induced activation dominates local energy barriers, dislocations may break free and move rapidly with a stress-dependent velocity in their glide planes [39].

Hence we divide the dislocation population in immobile or slowly moving ones (climbing, cross-slipping or pinned at obstacles like multipoles) and mobile or fast moving ones (free dislocations travelling in their glide planes according to the direction of their Burgers vectors). The local concentrations of these two dislocation populations, ρ_I and ρ_M, consequently obey the following balance equations:

$$\dot{\rho}_I + \mathrm{div}\,\vec{j}_I = c_I(\rho_I,\rho_M)$$
$$\dot{\rho}_M + \mathrm{div}\,\vec{j}_M = c_M(\rho_I,\rho_M) \qquad (4.1)$$

where c_I and c_M are source and sink terms taking into account the various mechanisms of production and annihilation of dislocations and are modelled as follows:

$$c_I(\rho_I,\rho_M) = F(\rho_I) - b\rho_I + \rho_M\rho_I^2$$
$$c_M(\rho_I,\rho_M) = b\rho_I - \rho_M\rho_I^2 \qquad (4.2)$$

where $F(\rho_I)$ represents creation and annihilation of slow dislocations, $b\rho_I$ the production rate of free dislocations and $\rho_M\rho_I^2$ corresponds to the pinning rate of free dislocations by immobile dipoles. b may be considered as proportional to the local rate of plastic deformation. A detailed analysis of the currents [40,41] leads to diffusive contributions to the kinetic equations. Effectively the current associated to the slow dislocations may be written as:

$$j_{I,i} = -[D_{ij}^{(\circ)}\nabla_j + D_{ijk}^{(1)}\nabla_j\nabla_k^2]\rho_I \qquad (4.3)$$

Due to the attractive character of the elastic interactions between dislocations of opposite Burgers vectors, $D_{ij}^{(\circ)}$ may be negative in high density regimes while $D_{ijk}^{(1)}$ remains positive. Hence a diffusive instability may occur for high creation rates or high stress levels. By defining the x axis as parallel to the resolved shear stress we also have:

$$\vec{j}_M = j_{M,x}\hat{i}_x = -D_M\nabla_x\rho_M \qquad (4.4)$$

The diffusive nature of the flux $j_{m,x}$ is supported by the fact that the stress is periodically reversed and that positive and negative dislocations move in opposite directions. It turns then out that D_m is of the order of the thermal diffusivity below the yield point and increases rapidly to its maximum value above this threshold. Effectively it may be shown that [41]:

$$D_M \cong v^2/\omega^2 \qquad (4.5)$$

where v may be related to the stress intensity via different relationships ranging from a power law to an exponential law of the type

$$v = v_0\exp(-(\tau_0/\tau)^m) \quad .$$

short range effects which occur at the atomic scale for example and which are neglected in this description may be introduced via noise terms leading to stochastic reaction-diffusion equations.

Among the various experimental situations describable by this model, let me just quote the results obtained in the simplest case, i.e. low initial densities in one slip plane. In this case the diffusion coefficients are positive and if xOy is the slip plane under consideration, the dynamical model may be reduced to:

$$\begin{aligned} \dot{\rho}_I &= F(\rho_I) + D_I^x \nabla_{xx}^2 \rho_I + D_I^y \nabla_{yy}^2 \rho_I - b\rho_I + \rho_M \rho_I^2 \\ \dot{\rho}_M &= b\rho_I - \rho_M \rho_I^2 + D_M^x \nabla_{xx}^2 \rho_M \end{aligned} \qquad (4.6)$$

where $D^x >> D^y$.

The homogeneous steady state is given by:

$$F(\rho_I^0) = 0 \qquad \rho_M^0 \rho_I^0 = b \qquad (4.7)$$

and becomes unstable versus homogeneous oscillations (Hopf bifurcation) when:

$$b \geqq b_c^0 = a + \rho_I^{0\,2} \qquad (a = -F^1(\rho_I^0)) \qquad (4.8)$$

or to inhomogeneous perturbations for:

$$b \geqq b_c = [a^{\frac{1}{2}} + \rho_I^0 (D_I^x / D_M^x)^{\frac{1}{2}}]^2 \qquad (4.9)$$

b being an increasing function of the stress or strain rates, the patterning instability is reached before the Hopf bifurcation if:

$$(D_I^x / D_M^x)^{\frac{1}{2}} < a^{\frac{1}{2}} [(1 + (\rho_I^{0\,2}/a))^{\frac{1}{2}} - 1] / \rho_I^0 \qquad (4.10)$$

This condition is satisfied for a wide range of dislocation densities when $D_I^x << D_M^x$. This is the case for sudden increases of stress intensity beyond threshold. For slow increases through the threshold region this condition may not be satisfied since D_M^x is proportional to the square velocity of mobile dislocations. In the latter case oscillations in the dislocation populations should occur which could be related to the strain bursts observed in corrresponding experimental situations [42].

When the patterning instability occurs first, the slow mode dynamics may be easily obtained showing that the preferred patterns beyond the dislocation freeing point correspond to modulations in the primary slip direction. Such dislocation patterns may be associated to the ladder-like structure of the PSBs. Moreover, the spontaneous breaking of the translational symmetry induces the development of long phase fluctuations. According to this property, a constant shift of the structure has no relaxational mechanism while phase singularities leading to layer splitting in the pattern may appear in agreement with experimental observations [43]. It is also worthnoting that the ladder-like character of the structure is induced by the strong anisotropy in the x direction. Effectively, in the case of two equivalent slip directions polygonal or labyrinth structures would be preferred. From the dynamical analysis of the slow mode amplitude equation it turns out that different nucleation mechanisms are possible for the PSBs. In the presence of inhomogeneities the pattern should propagate like shock waves from regions where the amplitude is nonzero to regions where the amplitude is zero. This type of analysis is also relevant to the description of the nucleation of PSBs from the vein structure of the matrix [44].

As a result, we see that reaction-diffusion dynamics provides a formal framework well adapted to the understanding of dislocation patterning in stressed crystals. Qualitative agreement with experimental observations is obtained but our conclusions have still to be tested by relating the various parameters to measurable quantities and also by the numerical anaysis of the model equations.

This type of approach should also be useful as a basis for the justification of the constitutive equations needed to improve the mechanical or thermodynamical approach to these phenomena. I furthermore allows to incorporate inhomogeneous effects in stability cirteria and stress-strain curves.

5. CONCLUSIONS

The apparition of spatio-temporal order is a wide-spread phenomenon in systems driven far from thermal equilibrium. The instability of the thermodynamical branch may for example lead to the nucleation of spatial patterns which behave as nonequilibrium crystals and have a profound influence on natural or technological phenomena. Having diverse origins, such structures, however, present strong similarities due to the symmetry breaking effects at the patterning instability. The spontaneous breaking of continuous symmetries also leads to the presence of long range phase fluctuations able to induce topological defects in the medium and to destroy any true long range order in low-dimensional systems. The specificities of each systems appear in the selection mechanisms (planforms, wavelengthes,...) and in the stability properties. Hence no true universal behavior may exist.

It is why we have chosen to discuss some examples of structures which appear in very different systems to show on one side the unicity of the methods used to describe them and on the other side the particular aspects related to the structure of the nonlinear couplings of the dynamics. The difficulties of the theoretical analysis lies in the complexity of the dynamics (chemical kinetics, Navier-Stokes, dislocation dynamics) and it is usually hopeless to obtain analytical solutions for these systems. Hence the approximations based on one scale separations near instabilities or on the symmetries of each problem are of great interest. In the case of patterns arising in systems with large aspects ratios, the phase dynamics governs the asymptotic dynamics and plays an essential role in the definition of the stability domains and in the behavior of topological defects.

In fact, some breakthroughs in the understanding of the spatial patterning induced by dynamical instabilities are illustrated in the examples discussed here. They suggest new experiments and more refined theoretical analysis but also show the simplicity of the concepts which govern the properties of these structures despite the complexity of the underlying dynamics. They finally contribute to fill the gaps between different domains and especially between fundamental science and technology.

ACKNOWLEDGMENTS

Part of this work was achieved in collaboration with Drs. P. Borckmans and G. Dewel at the Université Libre de Bruxelles and Prof. E.C. Aifantis at Michigan Technological University. The support of a NATO grant for international collaboration in research (082/84) is also gratefully acknowledged.

REFERENCES

[1] G. Nicolis and I. Prigogine: Self-Organisation in Non Equilibrium Systems. Wiley, New York, 1977.
[2] H. Haken: Synergetics, 2nd. ed., Springer, Berlin, 1978.
[3] A.C. Newell and J.A. Whitehead: J. Fluid. Mech. 38, 279 (1969).
[4] H.L. Swinney and J.P. Gollub: Hydrodynamic Instabilities and the Transition to Turbulence, Springer, Berlin, 1981. Cellular Structures and Instabilities, J.E. Wesfreid and S. Zaleski eds., Lecture Notes in Physics 210, Springer, Berlin, 1984.
[5] J. Toner and D. Nelson: Phys. Rev. B23, 316 (1981).
[6] D. Walgraef, G. Dewel and P. Borckmans: J. Chem. Phys. 78, 3043 (1983).
[7] R. Graham, Phys. Rev. A10, 1762 (1974). J. Swift and P.C. Hohenberg: Phys. Rev. A15, 319 (1977).
[8] D. Walgraef, G. Dewel and P. Borckmans, Adv. Chem. Phys. 49, 311 (1983).
[9] P. Coullet and S. Fauve, to appear.
[10] H.S. Greenside, W.M. Coughran and N. Schryer, Phys. Rev. Lett. 49, 726 (1982).
[11] M. Heutmaker, P. Fraenkel and J.P. Gollub, Phys. Rev. Lett. 54, 1369 (1985).
[12] M. Malek Mansour, C. Vanden Broeck, G. Nicolis and J.W. Turner, Annals of Physics 131, 283 (1981).
[13] J.E. Marsden and M. Mc Cracken, The Hopf Bifurcation and its Applications, Appl. Math. Sci. 19, Springer, Berlin, 1976.
[14] Y. Kuramoto: Progr. Theor. Phys. 56, 679 (1976).
[15] Y. Kuramoto and J. Yamada: Progr. Theor. Phys. 56, 724 (1976).
[16] J. Neu: SIAM J. Appl. Math. 36, 509 (1979); P. Hagan: Adv. Appl. Math. 2, 400 (1981).
[17] A. Pacault and C. Vidal in Evolution of Order and Chaos in Physics, Chemistry and Biology, Synergetics 17, H. Haken ed., Springer, Berlin, 1982. C. Vidal and J.M. Bodet, private communication.

[18] P. De Kepper, I.R. Epstein, K. Kustin and M. Orban, J. Phys. Chem. 86, 170 (1982).
[19] A. T. Winfree: The Geometry of Biological Time, Springer, Berlin, 1980.
[20] cf. R. Schmitz in Chemical Instabilities, Applications in Chemistry, Engineering, Geology and Materials Science, G. Nicolis and F. Baras eds., Reidel, 1983.
[21] P. Hagan: SIAM J. Appl. Math. 42, 762 (1984).
[22] T. Yamada and Y. Kuramoto: Progr. Theor. Phys. 55, 2035 (1976).
[23] G. R. Ivanitsky, V. I. Krinsky, A. N. Zaikin and A. M. Zhabotinsky, Sov. Sc. Review, D, Biology Reviews, 2, 279 (1981). K. I. Agladze and V. I. Krinsky, Nature, 269, 424 (1982).
[24] M. Orban: J. AM. Chem. Soc. 102, 4311 (1980).
[25] E. Guazzelli, G. Dewel, P. Borckmans and D. Walgraef, preprint, 1985.
[26] See for example D. Walgraef, G. Dewel and P. Borckmans: Springer Proc. Phys. 1, 50 (1984) and references therein.
[27] D. Walgraef, G. Dewel and P. Borckmans: Phys. Rev. A21, 397 (1980)
[28] Y. Pomeau and P. Manneville, J. Physique Lett. 40, L609, (1979). M. C. Cross: Phys. Rev. A25, 1065 (1982).
[29] V. Croquette and F. Schosseler, J. Physique, 43, 1183 (1982).
[30] E. Guazzelli, E. Guyo and J. E. Wesfreid in "Symmetries and Broken Symmetries in Condensed Matter Physics", N. Boccara ed., IDSET, Paris, pp. 455 - 461 (1981).
[31] D. Walgraef, G. Dewel and P. Borckmans: Z. Physik B48, 167 (1982)
[32] D. Schechtman, L. Bleck, D. Gratias and J.W. Cahn: Phys. Rev. Lett. 53, 1951 (1984).
[33] P. Bak: Phys. Rev. Lett. 54, 1517 (1985). N. D. Mermin and S. M. Troian: Phys. Rev. Lett. 54, 1524 (1985).
[34] D. Walgraef, G. Dewel and P. Borckmans, Nature, to appear.
[35] P. Neuman: Z. Metallkde, 58, 780 (1967).
[36] N. Thompson, N. J. Wadsworth and N. Louat: Phis. Mag. 1, 113 (1956).
[37] C. Caglioti in "Mechanical and Thermal Behavior of Metallic Materials", Int. School of Physcis E. Fermi LXXXII Course, Varenna 1981, C. Caglioti and A. Ferro eds., North Holland, Amsterdam, 1982.
[38] E. C. Aifantis in "On the Mechanics of Modulated Structures", NATO ASI Series, T. Tsakalakos ed., Martinus Nijhoff, 1983.
[39] D. L. Stein and J. R. Low, J. Appl. Phys. 31, 362 (1960). J. P. Hirth and W. D. Nix, Phys. Stat. Sol. 35, 177 (1969). A. S. Krausz and H. Eyring, Deformation Kinetics, Wiley, New York, 1975.
[40] D. L. Holt, J. Appl. Phys. 41, 3197 (1970).
[41] D. Walgraef and E. C. Aifantis, preprint, 1985.
[42] P. Neuman, Z. Matallkde, 59, 927 (1968).
[43] T. Tabata, H. Fujita, M. Hiraoka and K. Onishi, Phil. Mag. A47, 841 (1983).
[44] D. Walgraef and E. C. Aifantis, J. Appl. Phys. 58, 688 (1985).

NONEQUILIBRIUM CORRELATION FUNCTIONS

Charles P. Enz
Département de Physique Théorique
Université de Genève
1211 Genève 4, Switzerland

1. INTRODUCTION

Nonequilibrium phenomena occur wherever matter or energy is forced to flow, that is in physics (transport and radiation processes), in chemistry (reaction kinetics) and at various levels of biology (replication, metabolism, population dynamics). Under sufficiently strong forcing (gradients), nonequilibrium states may exhibits instabilities analogous to equilibrium phase transitions (LASER, Bénard effect, etc.). But already for small gradients that is, sufficiently far away from instabilities, nonequilibrium states show interesting correlation effects which have attracted the attention of experimenters and theorists alike.

Since nonequilibrium situations are characterized by flows, their description is always based on some form of equations of motion, either on a microscopic or on a phenomenological level. On the other hand, the calculation of correlation or response functions (transport coefficients) requires in addition some kind of averaging procedure. In such a framework, instabilities are caused by the nonlinear terms in the equations of motion (which may also give rise to chaotic behaviour) whereas for stable nonequilibrium situations these nonlinearities only dictate the deterministic flow pattern but have a negligible effect on the fluctuations.

In these lectures I restrict attention to these stable nonequilibrium situations. The problem is thus simply to obtain correlation or response functions by suitably averaging over the fluctuations. Here I will concentrate on the first type of functions, in particular the Fourier-transformed density-density correlation function or dynamic structure factor $S(\vec{q},\omega)$, the response functions being complicated by the question of the existence of a fluctuation-disipation theorem for nonequilibrium situations.

The reason why even the calculation of correlation functions may lead to severe complications is that an applied gradient in general induces spacial variations of key physical quantities which occur in the equations of motion or in the averaging procedure. These spacial variations give rise to renormalization and nonlinear effects

E. Tirapegui and D. Villarroel (eds.), Instabilities and Nonequilibrium Structures, 217–239.

in terms of the applied gradient as observed for instance in the Brillouin scattering on the every-day system of water in a temperature gradient [1,2,3]. As shown in Figure 1, the temperature gradient creates an asymmetry between the Stokes and the anti-Stokes Brillouin lines which for strong gradients deviates from a simple linear dependence. While a linear dependence has been predicted simultaneously by many authors [4-7], the problem of the nonlinear effect seems to me not to have had a fully satisfactory treatment [8].

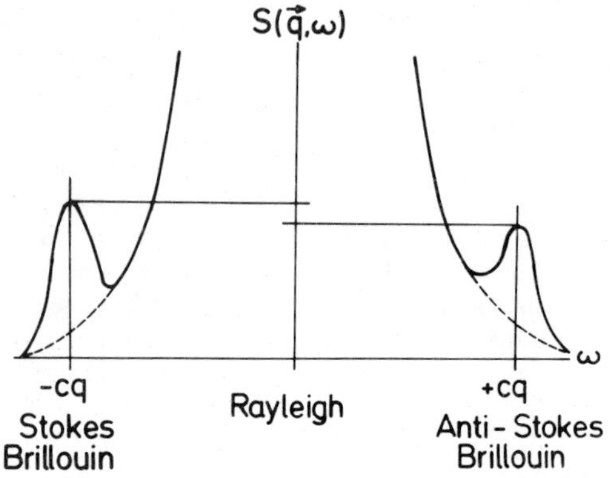

Figure 1.

2. SURVEY OF THEORETICAL METHODS AND DIFFERENT PHYSICAL SITUATIONS

The key physical quantity which in a temperature gradient acquires a spacial variation due to thermal expansion is the density. And this variation in turn affects the sound velocity and the viscosities which occur in the hydrodynamic equations. Since the latter are just the equations of motion of the phenomenological level, it is obvious that this description gives rise to the most direct method, if supplemented by the appropriate averaging procedure, namely the one obtained by the physically natural addition of Gaussian white noise terms to the deterministic hydrodynamic equations.

Correlation functions have been calculated in this way long ago by Melvin Lax [9] while the systematic application to the above situation was simultaneously done by Tremblay et al. [10] and by van der Zwan et al. [11] (see also Ref. 12). These authors did not, however, treat the problem of the nonlinear effect. The latter was qualitatively formulated within the method just described by Ueyama [13] who did not, however, work out his formulas quantitatively. A general solution of this problem has been attempted in unpublished work by the present author, part of which will be described below.

The only published calculation of the nonlinear effect is due to Kirkpatrick et al. [8] who have applied the microscopic Boltzmann equation method to the problem of a fluid in a temperature gradient [14]. Although this kinetic theory is in principle restricted to dilute gases [15], the definition of the hydrodynamic quantities occuring in the calculation is independent of this restriction and therefore makes numerical fits possible.

Since local equilibrium distributions are eigenfunctions of the Boltzmann operator with zero eigenvalue, another microscopic method to calculate nonequilibrium correlation (and response) functions is to develop the nonequilibrium average around local equilibrium. This method has been developed in a series of papers by Procaccia et al. [4,16]. Because of dissipation, local equilibrium is equivalent on the phenomenological level of a Fokker-Planck description, with the absence of flow which in turn was shown [17,18] to be equivalent with the socalled "potential conditions" [19]. As will be seen, an extension to nonequilibrium, that is to a situation with flow, is feasible provided that nonlinear terms in the equations of motion are negligible, that is, far from an instability [17, 18]. This means that the nonequilibrium stationary probability distribution (which here provides the averaging procedure) as well as the static correlation functions (given by the matrix E^{-1} in Refs 17 and 18) may be calculated. On the other hand, for the calculation of time-dependent correlation functions and, in particular the dynamic structure factor $S(\vec{q},\omega)$, the Lax-method [9] is more direct and therefore preferable for the complicated calculation of nonlinear effect of the gradient.

An obvious alternative to measuring the asymmetry of the Brillouin lines in function of the temperature gradient is to look for variations of the central or Rayleigh line (see Figure 1). The theoretical prediction here is that the Rayleigh line intensity increases with the square of the gradient [8,20,21,22]. (In Ref. 22 the question of equivalence of the thermodynamic representations is raised which, however, seems to me to be answered by an appropriate modification of Eq. (4.1b) of that reference.)

A different physical situation considered in the literature is a fluid with a velocity gradient which can either be viewed as a homogeneous shear flow [10,21,23,24,25] or as a Couette flow between two relatively rotating concentric cylinders whose radii are very large compared to their spacing [17]. While in Refs 17 and 25 only static correlation functions have been calculated (note that Eq. (14) of Ref. 25 is equivalent with the second Eq. (34) of Ref. 17) the full time dependence is given in Refs 10, 21, 23, 24. Still other situations may be found in the review by Tremblay [26].

Temperature and velocity gradients have also been considered in the vicinity of an instability namely near the critical point of an ordinary fluid or a binary mixture and the Lambda-point of a superfluid. Theoretically both situations have been investigated by Onuki [27] while experimental results on binary mixtures are contained in Ref. 2. The most striking feature here is that the anisotropy induced by the shear modifies the critical behaviour which becomes mean-field like close to the critical point.

A particularly interesting class of physical nonequilibrium situations are provided by charged fluids of metallic or superconducting kind in 2 or 3 dimensions subject to an electric field. The particularity of this class is that in general a constant electric field induces strictly no spacial variations in the system because no physical quantity depends on the vector potential (I thank Bastien Chopard for this remark). So there appears to be nonlinear effect due to the field. This is not exactly true, however, since a steady state can only be maintained by evacuating the Joule heat (for the equivalent problem in a shear flow see Ref. 21). As has been observed by Tremblay and Vidal [28] the steady state forces a temperature gradient upon the system which has to be determined self-consistently. While in Ref. 28 a metallic film is considered, the case of a superconductor near the transition temperature has been reviewed by Kawasaki [29].

To close this survey a remark on the "long time tail" of a transport coefficient is in order since in Refs 8 and 15 this problem has been associated with the correlation function in the same gradient. These long time tails are due to mode-mode coupling as are, in an appropiate definition, the transport coefficients, namely the interacting part of the corresponding response function (see, e.g., Ref. 30 for the example of the viscosity). However, in these lectures I neglect the interaction which is due to the nonlinear terms in the equations of motion and hence no long time tails are to be expected here.

3. LANGEVIN AND FOKKER-PLANCK EQUATIONS

I want to describe non-equilibrium by phenomenological equation of motion of macro-variables ψ_μ (μ a discrete continuous label) in which microcospic dynamics is taken into account through Gaussian white noise (random force) terms ξ_μ. This leads to Langevin equations

$$\dot{\psi}_\mu = f_\mu(\psi) + \xi_\mu \tag{1}$$

where the deterministic forces f_μ (drift) are functions of the ψ_μ and

$$\langle \xi_\mu(t)\xi_\nu(s)\rangle = 2C_{\mu\nu}\delta(t-s) \quad . \tag{2}$$

It may be proven [31] that the Langevin equations (1) are equivalent to the Fokker-Planck (FP) equation

$$\dot{P} + \partial_\mu J_\mu = 0 \tag{3}$$

for the probability distribution $P(\psi,t)$ to find for $\psi_\mu(t)$ the values ψ_μ at time t. Here

$$J_\mu = (f_\mu - \partial_\nu C_{\mu\nu})P \tag{4}$$

is the probability current distribution and $\partial_\mu \equiv \partial/\partial\psi_\mu$ acts on everything to the right (repeated indices are summed). The diffusion matrix in (2) and (4) is symmetric, $C_{\mu\nu} = C_{\nu\mu}$, and may depend on the ψ_μ.

The transformation properties under time reversal separate the drift into a dissipative (non-covariant, label: prime) term f'_μ and a non-dissipative (covariant, label: upper zero) term f°_μ [19,32],

$$f_\mu = f'_\mu + f^\circ_\mu \quad . \tag{5}$$

The probability current splits accordingly, the dissipative part being

$$J'_\mu = (f'_\mu - \partial_\nu C_{\mu\nu})P \quad . \tag{6}$$

In a stationary state

$$P_{st}(\psi) = Z^{-1} \exp(-F(\psi)) \quad , \tag{7}$$

defined by a generalized free energy F and normalized by the partition function $Z=\int d\psi \exp(-F)$, a vanishing J'_μ implies the potential conditions [19,32]

$$f'_\mu - \partial_\nu C_{\mu\nu} = -C_{\mu\nu}\partial_\nu F \quad . \tag{8}$$

Close to equilibrium [33], $F(\psi)$ is the second order variation of the internal energy U, divided by k_BT, expressed in terms of the variables $\psi_\mu = \Delta S, \Delta V$ (S = entropy, V = volume), or explicitly

$$k_BTF = U - T\Delta S + p\Delta V = \frac{1}{2}\left(\frac{\partial^2 U}{\partial S^2}\right)(\Delta S)^2 + \frac{\partial^2 U}{\partial S\partial V}\Delta S\Delta V + \frac{1}{2}\left(\frac{\partial^2 U}{\partial V^2}\right)(\Delta V)^2$$

$$= \frac{C_V}{2T}(\Delta T)^2 + \frac{1}{2V\gamma}\frac{C_V}{C_p}(\Delta V)^2 \quad .$$

Here $\gamma = -(\partial \log V/\partial p)_s$ is the adiabatic compressibility and C_V and C_p are the specific heats at constant volume and pressure, respectively. The stationary FP equation (3) with (7) then reduces to

$$\partial_\mu f^\circ_\mu = f^\circ_\mu \partial_\mu F \tag{9}$$

which, in principle, determines F. Therefore, Eqs (8) are consistency conditions that serve to determine the $C_{\mu\nu}$.

The meaning of these conditions becomes apparent in the case of constant $C_{\mu\nu}$ if we identify [33,34]

$$X_\mu = -k_B\partial_\mu F \tag{10}$$

as thermodynamic forces. With (10) and $\partial_\nu C_{\mu\nu} = 0$ the potential conditions (8) become the linear relations of irreversible thermodynamics [33,34],

$$f'_\mu = L_{\mu\nu}X_\nu \tag{11}$$

and $L_{\mu\nu} = C_{\mu\nu}/k_B$ (Grabert's D^{ij}, see Ref. 12, Eq. (6)) is recognized as an Onsager coefficient. Note that (11) is still correct if $C_{\mu\nu}$ depends on the ψ_μ, provided $\partial_\nu C_{\mu\nu} = 0$. The potential conditions (8) are generalized Onsager relations. To lowest order, F in (10) is a positive definite quadratic form [33,34]

$$F = \frac{1}{2}E_{\mu\nu}\psi_\mu\psi_\nu \tag{12}$$

so that in this approximation the f'_μ are linear functions,

$$f'_\mu = -\Gamma_{\mu\nu}\psi_\nu \tag{13}$$

and the $\Gamma_{\mu\nu}$ are damping constants (reciprocal mobilities).

The non-dissipative drift typically has linear and quadratic terms,

$$f^\circ_\mu = -(i\Omega_{\mu\nu} + \phi_{\mu\nu})\psi_\nu - \frac{i}{2}M_{\mu;\nu\lambda}\psi_\nu\psi_\lambda \tag{14}$$

where the $\Omega_{\mu\nu}$ are oscillation frequencies, and the $\phi_{\mu\nu}$ describe forcing by the external gradients. In linear approximation, $M_{\mu;\nu\lambda} = 0$, and in the absence of flows, $\phi_{\mu\nu} = 0$, Eqs (9) and (8) are simultaneously satisfied and take, respectively, the form

$$\Omega_{\mu\mu} = 0 \ ; \ E^{(\circ)}\Omega + \Omega^T E^{(\circ)} = 0 \tag{15}$$

and

$$\Gamma = CE^{(\circ)} \ , \tag{16}$$

the latter representing generalized Einstein relations [35]. $E^{(\circ)}$ is the trivial, i.e. diagonal, form of the matrix in (12), corresponding to a generalized Maxwell distribution (7). This is the situation close to equilibrium (upper label $(\circ)$).

With flow terms in (14) the potential conditions (8) or (11) cannot be maintained and one has to use the full stationary FP equation (3) with (7),

$$C_{\mu\nu}\partial_\mu\partial_\nu F - C_{\mu\nu}(\partial_\mu F)(\partial_\nu F) + 2(\partial_\nu C_{\mu\nu})(\partial_\mu F)$$
$$- \partial_\mu\partial_\nu C_{\mu\nu} - f_\mu\partial_\mu F + \partial_\mu f_\mu = 0 \ . \tag{17}$$

With constant $C_{\mu\nu}$ and in linear approximation, insertion of (12) - (14) leads to the conditions [9,17,18]

$$(\Lambda - CE)_{\mu\mu} = 0 \ ; \ \Lambda E^{-1} + E^{-1}\Lambda^T = 2C \tag{18}$$

where $\Lambda = \Gamma + i\Omega + \phi$ is the matrix of the total linear part of f_μ (in Ref. 9 the second condition (18) is called an Einstein relation; it is equivalent to Eq. (14) of Ref. 25 as remarked before). Eqs (18) determine the nonequilibrium matrix E^{-1} which is related to the static correlation functions according to (7) and (12),

$$\langle\psi_\mu\psi_\nu\rangle \equiv \int d\psi P_{st}(\psi)\psi_\mu\psi_\nu = (E^{-1})_{\mu\nu} \quad . \tag{19}$$

The deviation of E from the close-to-equilibrium form $E^{(\circ)}$ which is due to the external gradients will thus show up in the correlation functions, already without the nonlinear M-terms in (14). The time-dependent correlation functions may be obtained from (19) by inserting the explicity time evolution, $\psi_\mu(t) = (\exp\Lambda t)_{\mu\nu}\psi_\nu$ [32].

4. THE THERMO-HYDRODYNAMIC EQUATIONS

Phenomenological hydrodynamics is described by the conservation law for the mass density ρ,

$$\dot{\rho} + \nabla\cdot(\rho\vec{u}) = 0 \tag{20}$$

where $\vec{u}$ is the local fluid velocity, and by the balance equation for the momentum density $\rho\vec{u}$ [36],

$$\rho\dot{\vec{u}} + \rho(\vec{u}\cdot\nabla)\vec{u} + \nabla p - \sum_{i=1}^{3} \nabla_i[\eta(\nabla u_i + \nabla_i\vec{u})]$$

$$- \nabla[(\zeta - \frac{2}{3}\eta)\,\nabla\cdot\vec{u}] = \rho\vec{\xi} \tag{21}$$

where p is the pressure, η and ζ are the shear and bulk viscosities and $\vec{\xi}$ is an external force per unit mass. The variation of the local temperature is described by the balance equation for the entropy per unit mass, s, [36,37]

$$\rho T[\dot{s} + (\vec{u}\cdot\nabla)s] - \nabla\cdot[\kappa\nabla T] = C_V\xi_4$$

$$+ \frac{\eta}{2}(\nabla_i u_j + \nabla_j u_i - \frac{2}{3}\nabla\cdot\vec{u}\delta_{ij})^2 + \zeta(\nabla\cdot\vec{u})^2 \tag{22}$$

where κ is the thermal conductivity and $C_V\xi_4$ an external power density. The two positive quadratic terms in (22) constitute the Joule heat density (an additional term $(\kappa/T)(\nabla T)^2$ in Ref. 36 has been canceled by transforming $T\nabla\cdot[(\kappa/T)\nabla T]$). Note that in Eq. (20) a dissipative current density has been neglected [36]. Such a term $-D_1\nabla\mu$ where μ is the chemical potential, is considered in Section IIA of Ref. 10.

The natural dynamical variables in Eqs (20) - (22) are $\rho, \vec{u}$ and T while p and s may be eliminated with the help of the thermodynamic relations

$$\delta p = c^2\delta\rho + \frac{\alpha T\rho}{\gamma C_p}\delta s$$

$$\delta T = \frac{\alpha T c^2}{\rho C_p}\delta\rho + \frac{T\rho}{C_V}\delta s \qquad (23)$$

Here $c = (\rho\gamma)^{-\frac{1}{2}}$ is the adiabatic sound velocity and $\alpha = -(\partial\log\rho/\partial T)_p$ the thermal expansion coefficient. Considering the external variables $\vec{\xi}$ and ξ_4 as random in the sense of (2), Eqs (20) - (22) become Langevin equations.

On the other hand, the deterministic flow pattern [38] $\rho_o(\vec{r})$, $u_o(\vec{r})$, $T_o(\vec{r})$ (lower label o) is a stationary solution of equations (20) - (23) with $\vec{\xi} = 0$, $\xi_4 = 0$. (The average values of these quantities will be designated by an upper bar: $\bar{\rho}$, $\bar{\vec{u}} = 0$, $\bar{T}$). These equations are

$$\vec{u}_o\cdot\nabla\rho_o + \rho_o\nabla\cdot\vec{u}_o = 0$$

$$\rho_o(\vec{u}_o\cdot\nabla)\vec{u}_o + \nabla p_o - \sum_i \nabla_i[\eta_o(\nabla u_{oi} + \nabla_i\vec{u}_o)] - \nabla[(\zeta_o - \frac{2}{3}\eta_o)\nabla\cdot\vec{u}_o] = 0 \qquad (24)$$

$$\rho_o T_o(\vec{u}_o\cdot\nabla)s_o - \nabla\cdot[\kappa_o\nabla T_o] = \text{Joule heat terms}$$

It is easily seen that

$$\vec{u}_o = \bar{\vec{u}} = 0,\ \rho_o = \bar{\rho},\ p_o = \bar{p},\ \kappa_o\nabla T_o = \bar{\kappa}\vec{\beta}\bar{T} = \text{const} \qquad (25)$$

is a solution. The last relation shows that if the temperature dependence of the thermal conductivity is not negligible, ∇T_o is not a constant and the temperature profile across the fluid has quadratic (and higher deviations from linearity

$$T_o - \bar{T} = \bar{T}\vec{\beta}\cdot\vec{r}[1 - \frac{\vec{\beta}\cdot\vec{r}}{2\bar{\kappa}}\overline{(\frac{\partial\kappa}{\partial T})_\rho}] \quad .$$

This effect seem- to have been ignored in the literature.

Introducing velocity variables v_i, $i = 0, 1, 2, 3, 4$ for all the fluctuations, with $v_o \equiv \bar{c}(\rho-\rho_o)/\bar{\rho}$ and $v_4 \equiv \bar{c}(T-T_o)/\bar{T}$, and neglecting thermal expansion in Eqs (23), the linearized equations of motion for the fluctuations that follow from Eqs (20) - (22), (25) are

$$\dot{v}_o + \bar{c}\nabla\cdot\vec{v} = 0$$

$$\dot{\vec{v}} + c^2(\bar{p},T_o)\frac{1}{\bar{c}}\nabla v_o - D\nabla^2\vec{v} - D'\nabla\nabla\cdot\vec{v} = \vec{\xi} \tag{26}$$

$$\dot{v}_4 + \vec{v}\cdot\frac{\bar{c}}{\bar{T}}\nabla T_o - D_4\nabla^2 v_4 = \xi_4$$

Here $D \equiv \bar{\eta}/\bar{\rho}$; $D' \equiv (\bar{\zeta}+\frac{1}{3}\bar{\eta})/\bar{\rho}$; $D_4 \equiv \bar{\kappa}/\bar{C}_V$ that is, the temperature dependence has been neglected in η, ζ, κ, and C_V but not in c,

$$c_o^2 \equiv c^2(\bar{p},T_o) = c^{-2}[1+\psi(T_o-\bar{T})/\bar{T} + \ldots] \tag{27}$$

with $\psi \equiv \frac{2\bar{T}}{\bar{c}}\overline{(\frac{\partial c}{\partial T})_p}$. The justification for this is that $\chi \equiv \frac{q\bar{T}}{\bar{\rho}\bar{c}}(\frac{\partial \eta}{\partial T})_p$ where q is a typical wavenumber, is a small quantity. Indeed, for the experimental conditions valid for Ref. 1, namely water at $\bar{T} = 313$K, $\bar{c} = 1.53\times10^5$cm/sec and $2000\,\text{cm}^{-1} \leq q \lesssim 3420\,\text{cm}^{-1}$ one finds Figures 1 and 2 of Rouch et al. [39] $\psi \cong 0.73$ and $\chi \cong -0.003$. One also finds [40] $\alpha\bar{T} \cong 0.090$ wihch justifies neglecting thermal expansion effects and setting $\rho_o = \bar{\rho}$ in (25).

After Fourier transformation, $v_i(\vec{r}) \to v_{i\vec{q}}$, Eqs (26) take the form

$$\dot{v}_\mu = -\Lambda_{\mu\nu}v_\nu + \xi_\mu \tag{28}$$

with $\mu \equiv (i,\vec{q})$, $\nu \equiv (j,\vec{k})$. Writing $\Lambda_{i\vec{q},j\vec{k}} \equiv \Lambda_{ij}(\vec{q},\vec{k})$ as (i,j)-matrix one finds with Eqs (25) and (27)

$$\Lambda(\vec{q},\vec{k}) = \frac{1}{V}\int_V d^3r e^{-i\vec{q}\cdot\vec{r}}\Lambda(\vec{r})e^{i\vec{k}\cdot\vec{r}} = \begin{pmatrix} 0 & , & i\bar{c}\vec{q} \;,\; 0 \\ i\bar{c}\vec{k}(1-i\psi\vec{\beta}\cdot\frac{\partial}{\partial\vec{k}}), & & D\vec{q}^2\underline{1}+D'\vec{q}\otimes\vec{q}, \; 0 \\ 0 & , & \bar{c}\beta \;,\; D_4\vec{q}^2 \end{pmatrix}\delta_{\vec{q},\vec{k}} \tag{29}$$

V being the volume of the system.

5. THE DIFFUSION MATRIX $C_{\mu\nu}$

Close to equilibrium (upper label (°)), that is when $T_o = \vec{\beta}\bar{T} = 0$, the stationary probability distribution (7) is a Maxwell distribution in the velocity fluctuations $\vec{v}$, and in the density and temperature fluctuations it is given by the free energy indicated after Eq. (8) [33]

$$k_B\bar{T}F^{(o)} = \int_V d^3r\frac{1}{2}\frac{\bar{C}_V}{\bar{C}_p}[\frac{\bar{c}^2}{\bar{\rho}}(\delta\rho)^2 + \bar{\rho}\vec{v}^2 + \frac{\bar{C}_p}{\bar{T}}(\delta T)^2] \quad . \tag{30}$$

Written in the form (12), in terms of Fourier components of velocity variables, v_{iq}, this leads to a diagonal matrix

$$E^{(o)}_{i\vec{q},j\vec{k}} = \bar{\varepsilon}_i\delta_{ij}\delta_{\vec{q}+\vec{k},\vec{o}}$$

with

$$\bar{\varepsilon}_o = \bar{\varepsilon} = \frac{\bar{\rho}}{k_B\bar{T}}\frac{\bar{C}_V}{\bar{C}_p}, \ \bar{\varepsilon}_4 = \bar{C}_V/k_B\bar{c}^2 \quad .$$

Note that $\bar{\varepsilon}_o = \bar{\varepsilon}$ is necessary for the second condition (15) to be satisfied with Ω defined by

$$i\Omega = \Lambda\Big|_{D = D' = D_4 = o,\vec{\beta} = o}$$

in Eq. (29). Indeed, this condition says that $S \equiv E^{(o)}\Omega$ satisfies $S + S^T = 0$; here $S_{i\vec{q},j\vec{k}} = S_{ij}(\vec{q})\delta_{\vec{q}+\vec{k},\vec{o}}$ so that the skew symmetry implies $S_{ij}(\vec{q}) + S_{ji}(-\vec{q}) = 0$. Note also that $\bar{\varepsilon}_o = \bar{\varepsilon}$ determines the factor $\bar{C}_V/\bar{C}_p$ in the Maxwell-distribution term of Eq. (30).

In the nonequilibrium situation the average quantities are to be replaced by those of the flow pattern (25), the corresponding matrix E_o being

$$E_{oi\vec{q},j\vec{k}} = \varepsilon_i\delta_{ij}\delta_{\vec{q}+\vec{k},\vec{o}} \ ; \ \varepsilon_o = \varepsilon = \frac{\bar{\rho}}{k_BT_o} \ , \ \varepsilon_4 = \frac{\bar{C}_V}{k_Bc_o^2} \tag{31}$$

where we have neglected thermal expansion effects ($C_V = C_P$).

In order to determine the diffusion matrix $C_{\mu\nu}$ we use the generalization of Eq. (16)

$$2C = \Gamma E_o^{-1} + E_o^{-1}\Gamma^T \tag{32}$$

which is the extrapolation of the potential conditions to nonequilibrium. Here Γ is the dissipative part of the matrix $\Lambda = \Gamma + i\Omega + \phi$ defined in Eq. (29), $\Gamma = \Lambda\big|_{\bar{c} = o}$. The diffusion matrix $C_{i\vec{q},j\vec{k}} = C_{ij}(\vec{q},\vec{k})$ then follows from (31), (32), (27),

$$C(\vec{q},\vec{k}) = \begin{pmatrix} 0, & \vec{0} & , & 0 \\ \vec{0}, & \underline{D}q^2\bar{\varepsilon}^{-1}(1-i\beta\frac{\partial}{\partial\vec{k}}) & , & \vec{0} \\ 0, & \vec{0} & , & D_4q^2\bar{\varepsilon}_4^{-1}(1-i\psi\vec{\beta}\cdot\frac{\partial}{\partial\vec{k}}) \end{pmatrix} \delta_{\vec{q}+\vec{k},\vec{o}} \tag{33}$$

where $\underline{D} \equiv D\underline{1} + D'\hat{q}\otimes\hat{q}$.

6. THE CORRELATION FUNCTIONS

Time dependent correlation functions may be obtained from the matrix E^{-1} as determined by Eqs (18), by inserting the time evolution $v_\mu(t) = (\exp\Lambda t)_{\mu\nu}v_\nu$ into Eq. (19). However, a more direct method is the following [9,10]: Time-Fourier transformation $v_\mu(t) \to \tilde{v}_\mu(\omega)$ allows to write the linearized Langevin equations (28) in the form

$$\tilde{v}_\mu(\omega) = (\Lambda + i\omega 1)^{-1}_{\mu\nu}\tilde{\xi}_\nu(\omega) \quad . \tag{34}$$

Making use of the Fourier transformed Eq. (2) the correlation functions are now easily calculated,

$$\langle\tilde{v}_\mu(\omega)\tilde{v}_\nu(\omega')\rangle = [(\Lambda + i\omega 1)^{-1}2C\delta(\omega' + \omega)(\Lambda + \omega' 1)^{T-1}]_{\mu\nu} \quad . \tag{35}$$

Defining

$$K_{ij}(\vec{q};\omega) \equiv \sum_{\vec{k}} \int d\omega' \langle\tilde{v}_{i\vec{q}}(\omega)\tilde{v}_{j\vec{k}}(\omega')\rangle \tag{36}$$

the dynamic structure factor simply is $S(\vec{q},\omega) = \frac{\bar{\rho}^2}{c^2}K_{oo}(\vec{q};\omega)$.

From Eq.(35) one sees that the mathematical problem reduces to the construction of the inverse of $A \equiv \Lambda + i\omega 1$. Neglecting the spacial variation $T_o - \bar{T} = \bar{T}\vec{\beta}\cdot\vec{r}$ in (29) this matrix is simplified to

$$\Lambda(\vec{q},\vec{k})\Big|_{\psi=o} = \bar{\Lambda}(\vec{q})\delta_{\vec{q},\vec{k}} \quad .$$

Hence the problem reduces to diagonalizing the 5×5 matrix $\bar{\Lambda}$. If in addition $T_o - \bar{T}$ is also neglected in Eqs (31) one is led to the result linear in β, as will now be shown.

Choosing the gradient along the 3-axis, $\vec{\beta} = (0,0,\beta)$, and the direction of the wavevector $\vec{q}$ in the (1,3)-plane, $\hat{q} = (\xi,0,\zeta)$, the 2-direction may be ignored and $\bar{\Lambda}$ reduces to a 4×4 matrix. By approximately diagonalizing this matrix one finds for the inverse of $\bar{A} = \bar{\Lambda} + i\omega\bar{1}$

$$\bar{A}^{-1}(\pm\vec{q};\pm\omega) \cong \pm\frac{i}{\bar{c}qd_{\pm}}R \tag{37}$$

with

$$R(\hat{q}) = \begin{pmatrix} 1 & -\xi & -\zeta & 0 \\ -\xi & \xi^2 & \xi\zeta & 0 \\ -\zeta & \xi\zeta & \zeta^2 & 0 \\ 0 & 0 & 0 & 0 \end{pmatrix} \tag{38}$$

and

$$d_{\pm} \equiv d(q;\pm\omega) = \left(\frac{\pm\omega - i\lambda q^2}{\bar{c}q}\right)^2 - 1 \quad . \tag{39}$$

Here $\lambda \equiv \frac{1}{2}(D+D')$ and λq^2 measures the half-width of the Brillouin lines [1].

The diffusion matrix (33) simplifies to $C(\vec{q},\vec{k}) \cong C(\vec{q})(1-i\vec{\beta}\cdot\frac{\partial}{\partial\vec{k}})\delta_{\vec{q}+\vec{k},\vec{o}}$ with the 4×4 matrix

$$\bar{C}(\vec{q}) = \begin{pmatrix} 0 & 0 & 0 & 0 \\ 0 & D+D'\xi^2 & D'\xi\zeta & 0 \\ 0 & D'\xi\zeta & D+D'\zeta^2 & 0 \\ 0 & 0 & 0 & 0 \end{pmatrix} \bar{\varepsilon}^{-1}q^2 \tag{40}$$

where we have neglected the (4,4)-element $D_4\bar{\varepsilon}_4^{-1}q^2$ since it does not affect the dynamic structure factor; in addition $\bar{\varepsilon}/\bar{\varepsilon}_4 \cong 1.79$ and $D_4/\lambda \cong 0.115$ for water at 40°C [40]. Transforming the matrix (40) by (38),

$$R\bar{C}R^T = \frac{2\lambda q^2}{\bar{\varepsilon}}\bar{G} \tag{41}$$

one then finds $\bar{G} \cong R = R^T$ and, after a partial integration in $\vec{k}$ to get rid of the derivative acting on $\delta_{\vec{q}+\vec{k},\vec{o}}$ in $C(\vec{q},\vec{k})$,

$$K^{(1)}(\vec{q},\omega) \cong \frac{4\lambda}{\bar{\varepsilon}\bar{c}^2|d(q;\omega)|^2}\left(1 + i\frac{2\vec{\beta}\cdot\hat{q}}{qd^*(q;\omega)}\right)R(\hat{q}) \quad . \tag{42}$$

Finally, the integrated intensities under the Brillouin lines are

$$I_{\pm}^{(1)}(\vec{q}) = \pm\int_{o}^{\pm\infty} \frac{d\omega}{2\pi}S(\vec{q},\omega) \cong \frac{1}{2\bar{\varepsilon}}\left(1 \pm \frac{\vec{\beta}\cdot\hat{q}\bar{c}}{2\lambda q^2}\right)\frac{\bar{\rho}^2}{\bar{c}^2} \tag{43}$$

which is the well-known result linear in the gradient $\vec{\beta} = \nabla T_o/\bar{T}$. In Eqs (42) and (43) the upper indices (1) refers to this linear approximation.

7. THE NONLINEAR DEPENDENCE ON THE GRADIENT

When extending this calculation to the situation where the spacial dependence of the quantities in Eq. (29) is included, the diagonalization of Λ cannot anymore be done algebraically because of the derivative $\vec{\beta}\cdot\frac{\partial}{\partial\vec{k}}\delta_{\vec{q},\vec{k}}$. This leads to complicated differential equations which may be solved only approximately, as will be sketched presently.

The full matrix (29) may be written $\Lambda = [\bar{\Lambda}(\vec{k}) + \psi\bar{c}k\Pi(\hat{k})\vec{\beta}\cdot\frac{\partial}{\partial\vec{k}}]\,\delta_{\vec{q},\vec{k}}$ with

$$\Pi(\hat{k}) = \begin{pmatrix} 0 & \vec{0} & 0 \\ \hat{k} & \underline{0} & \vec{0} \\ 0 & \vec{0} & 0 \end{pmatrix} \quad . \tag{44}$$

Making use of Eq. (37) and of the identity $\frac{\partial}{\partial\vec{k}}\delta_{\vec{q},\vec{k}} = -\frac{\partial}{\partial\vec{q}}\delta_{\vec{q},\vec{k}}$ the matrix $A = \Lambda + i\omega 1$ then takes the form

$$A(\vec{q},\vec{k};\omega) \cong [\bar{I} - \frac{i\psi}{d(k;\omega)}P(\vec{k})\vec{\beta}\cdot\frac{\partial}{\partial\vec{q}}]\,\delta_{\vec{q},\vec{k}}\bar{A}(\vec{k};\omega) \tag{45}$$

where, in 4×4 matrix notation,

$$P(\hat{q}) \equiv -\Pi(\hat{q})\,R(\hat{q}) = \begin{pmatrix} 0 & 0 & 0 & 0 \\ -\xi & \xi^2 & \xi\zeta & 0 \\ -\zeta & \xi\zeta & \zeta^2 & 0 \\ 0 & 0 & 0 & 0 \end{pmatrix} \quad . \tag{46}$$

It is easy to verify that P is a projector, $P^2 = P$.

The form (45) of A is chosen in order that the following Ansatz suggests itself for its inverse:

$$A^{-1}(\vec{q},\vec{k};\omega) = \bar{A}^{-1}(\vec{q};\omega)[\bar{1}\delta_{\vec{q},\vec{k}} + i\psi Q(\vec{q},\vec{k};\omega)] \quad . \tag{47}$$

Freeing the δ-function in (45) from the derivative by partial integration, the condition

$$\bar{1}\delta_{\vec{q},\vec{k}} = \sum_{\vec{p}} A(\vec{q}\ \vec{p};\omega)A^{-1}(p,k;\omega)$$

leads to a differential equation for the matrix Q, namely

$$[1 - i\psi\vec{\beta}\cdot\frac{\partial}{\partial\vec{q}}\ \frac{P(\hat{q})}{d(q,\omega)}]Q(\vec{q},\vec{k};\omega) = \frac{P(\hat{k})}{d(k;\omega)}\beta\cdot\frac{\partial}{\partial\vec{q}}\delta_{\vec{q},\vec{k}} \quad . \tag{48}$$

Inserting Eqs (47) and (33) into (35), (36) one obtains by partial integrations to get rid of the derivatives acting on the δ-functions [41]

$$\begin{aligned} K(\vec{q},\omega) = \sum_{\vec{p}} & \bar{A}^{-1}(\vec{q};\omega)[\bar{1}\delta_{\vec{q}\ \vec{p}} + i\psi Q(\vec{q},\vec{p};\omega)]2\bar{C}(\vec{p}) \\ & \cdot\{(1 + i\vec{\beta}\cdot\frac{\partial}{\partial\vec{p}})\bar{A}^{T-1}(-\vec{p};-\omega) \\ & + \sum_{\vec{k}}(1 - i\vec{\beta}\cdot\frac{\partial}{\partial\vec{p}})i\psi Q^T(\vec{k},-\vec{p};-\omega)\bar{A}^{T-1}(\vec{k};-\omega)\} \quad . \end{aligned} \tag{49}$$

For the evaluation of this expression with the help of Eq. (48) the following physical argument is used: Since we are interested in the Brillouin-lines centered at $\omega = \pm\bar{c}q$, small values of denominators $d_\pm$, Eq. (39), will be important. This justifies the <u>approximation to let the derivatives</u> $\vec{\beta}\cdot\frac{\partial}{\partial\vec{q}}$ <u>act only on denominators</u> $d_\pm$ <u>and to put</u> $\omega = \pm\bar{c}q$, <u>that</u>

is $d_\pm = \mp i\nu$, in numerators, where $\nu \equiv 2\lambda q/\bar{c}$. From definition (39) it then follows that

$$-\dot{\beta}\cdot\frac{\partial}{\partial\vec{q}}d_\pm = \mu\left(\frac{\omega^2}{\bar{c}^2q^2} - \frac{\nu^2}{4}\right) \cong \mu \equiv \frac{2\vec{\beta}\cdot\hat{q}}{q} \quad . \tag{50}$$

For the experimental conditions of Ref. 1, $3.4\times10^{-4} \leq \nu \leq 4.1\times10^{-4}$ and $|\mu| \leq 1.6\times10^{-4}$. Note that because of the smallness of ν the second part of this approximation is also justified in the integration (43) of the Brillouin lines since

$$\frac{4}{\bar{c}q}\int_0^\infty \frac{d\omega}{2\pi}\,|d(q;\omega)|^{-2(\ell+1)}\left(\frac{\omega}{\bar{c}q}\right)^n = \frac{(2\ell-1)!!}{(2\ell)!!}\nu^{-2\ell-1}[1+0_n(\nu^2)];\ n\geq0 \tag{51}$$

Applied to Eq. (48) the first part of the approximation yields formally

$$Q(\vec{q},\vec{k};\omega) = \{1-\frac{i\psi}{d_+}(\frac{\mu}{d_+}+\partial_{\vec{q}})P(\hat{q})\}^{-1}\cdot\frac{P(\hat{k})}{d(k;\omega)}\partial_{\vec{q}}\delta_{\vec{q},\vec{k}} \tag{52}$$

where $\partial_{\vec{q}} \equiv \vec{\beta}\cdot\frac{\partial}{\partial\vec{q}}$. (In Ref. 22 an arbitrary assumption, Eq. (2.6), is introduced for an analogous quantity). Using again partial integrations to free the δ-functions in Q and Q^T from derivatives one obtaines with the help of Eqs (37), (42), of the above approximation and of the proyection property $P^2 = P$,

$$K(\vec{q};\omega) - K^{(1)}(\vec{q};\omega) = \frac{2\psi i}{\bar{c}^2q^2}RP\bar{C}R^T\,\frac{1}{d_+}Z_+^{-1}\partial_{\vec{q}}\,\frac{1}{d_+}(1+i\partial_{\vec{q}})\frac{1}{d_-}$$

$$+\frac{2\psi i}{\bar{c}^2q^2}R\bar{C}P^TR^T\,\frac{1}{d_+}(1-i\partial_{\vec{q}})\frac{1}{d_-}\partial_{\vec{q}}Z_-^{-1}\,\frac{1}{d_-} \tag{53}$$

$$-\frac{2\psi^2}{\bar{c}^2q^2}RP\bar{C}P^TR^T\,\frac{1}{d_+}Z_+^{-1}\partial_{\vec{q}}\frac{1}{d_+}\,(1-i\partial_{\vec{q}})\frac{1}{d_-}\partial_{\vec{q}}Z_-^{-1}\frac{1}{d_-}$$

Here $Z_\pm$ are the differential operators

$$Z_+ \equiv 1-\frac{i\psi\mu}{d_+^2}-\frac{i\psi}{d_+}\partial_{\vec{q}}\ ;\ Z_- = 1+\frac{i\psi\mu}{d_-^2}-\partial_{\vec{q}}\frac{i\psi}{d_-} \tag{54}$$

and derivatives $\partial_{\vec{q}}$ act on every denominator $d_\pm$ to the right.

For the evaluation of Eq. (52) we first calculate $f = Z_-^{-1}d_-^{-1}$ or, equivalently, we solve the differential equation $Z_-f = d_-^{-1}$, considering f as a function of d_-. Introducing variables $x_\pm \equiv \sigma d_\pm^2/2\psi\mu$ where $\sigma = \pm$ is a sign introduced to make $\mathrm{Re}\,x_\pm > 0$ and using (50) it is easy to see that

$$Z_- f(x_-) = (1 + i\sigma \frac{d}{dx_-})\, f(x_-) \quad . \tag{55}$$

In this form the differential equation is solved by Laplace transformation. Noting that

$$d_\pm^{-n} = \frac{(2\sigma\psi\mu)^{-n/2}}{\Gamma(n/2)} \int_0^\infty p^{\frac{n}{2}-1} e^{-px_\pm} dp \tag{56}$$

and $\Gamma(1/2) = \sqrt{\pi}$, the result is

$$Z_-^{-1} d_-^{-1} = (2\pi\sigma\psi\mu)^{-\frac{1}{2}} \int_0^\infty \frac{dp}{\sqrt{p}} \frac{e^{-px_-}}{1-i\sigma p} \quad . \tag{57}$$

It may also be derived by expanding f into a power series in d_-^{-1} and determining the coefficients recursively by the differential equation. With (55) this leads to the asymptotic expansion

$$f = \sum_{n=0}^\infty (2\ -1)!!\ \frac{(i\psi\mu)^n}{d_-^{2n+1}} \tag{58}$$

which may be computed with the help of the Borel summation trick [42]. (This is the method used in Ref. 8, Appendix B). It consists of using the integral representation

$$(2n-1)!! = \int_0^\infty \frac{dt}{\sqrt{(\pi t)}} e^{-t} (2t)^n \tag{59}$$

in (58) and leads immediately to the result (57).

Using the result (57) and the property RP = R together with (41), Eq. (53) takes the form

$$K(\vec{q};\omega) - K^{(1)}(\vec{q};\omega) = i\frac{4\psi\lambda}{\bar{c}^2\bar{\varepsilon}}\bar{G}\frac{1}{d_+}\{Z_+^{-1}[\frac{\mu}{d_+^2 d_-}(1+i\frac{\mu}{d_-}) + \frac{\mu}{d_+ d_-^2}(1+2i\frac{\mu}{d_-})]$$

$$+ \sigma(2\pi\sigma\mu)^{-\frac{1}{2}}\psi^{-3/2}\int_0^\infty dp\frac{e^{-px_-}}{1-i\sigma p}\sqrt{p}(1-i\frac{\sigma p}{\psi}d_-) + i\sigma Z_+^{-1}(2\pi\sigma\psi\mu)^{-\frac{1}{2}}\frac{1}{d_+}$$

$$\cdot \int_0^\infty dp\frac{e^{-px_-}}{1-i\sigma p} \cdot \sqrt{p}[\frac{\mu}{d_+}(1 - i\frac{\sigma p}{\psi}d_-) + \frac{\sigma p}{\psi}(i\mu + d_- - i\frac{\sigma p}{\psi}d_-^2)]\} \quad . \tag{60}$$

Making here the approximation to put $d_- = i\nu$ in the numerators and not to let Z_+^{-1} act on the exponentials, we are left with the problem to calculate $d_+F_{nm} = Z_+^{-1}d_+^{-n}d_-^{-m}$ for $n = 1$, $m = 0,2,3$ and $n = 2$, $m = 0,1,2$. With the first Eq. (54) it is easy to see that

$$Z_+(d_+F_{nm}) = [d_+ + i\psi\mu(\frac{\partial}{\partial d_+} + \frac{\partial}{\partial d_-})]\, F_{nm} \tag{61}$$

except for $m = 0$ in which case F is independent of d_- so that writing $F_{no} = f_n(x_+)$ we have, in analogy to (55),

$$Z_+(d_+f_n) = d_+(1 + i\sigma\frac{d}{dx_+})\, f_n \quad . \tag{62}$$

Using Laplace transforms and the indentity (56) one then finds

$$Z_+^{-1}d_+^{-n} = \frac{(2\sigma\psi\mu)^{-(n+1)/2}}{\Gamma((n+1)/2)}\, d_+ \int_0^\infty p^{\frac{n-1}{2}}\, \frac{e^{-px_-}}{1-i\sigma p}\, dp \quad . \tag{63}$$

In the case $m > 0$ Laplace transforms cease to be of any use. However, one notes that Eq. (61) becomes an ordinary differential expression in the variables $u = (d_+ + d_-)/2$, $v = (d_+ - d_-)/2$. Writing $F_{nm}(d_+,d_-) = \phi_{nm}(u,v)$, the differential equation then becomes

$$(u + v + i\psi\mu\frac{\partial}{\partial u})\, \phi_{nm} = (u + v)^{-n}(u - v)^{-m} \quad . \tag{64}$$

It may be solved with the help of the auxiliary function

$$a(u,v) = \exp\left[\frac{(u+v)^2}{2i\psi\mu}\right] \tag{65}$$

the result being

$$\phi_{nm} = (i\psi\mu a)^{-1} \int a(u + v)^{-n}(u - v)^{-m} du \quad . \tag{66}$$

With Eqs (63) and (66) the problem of calculating the expression (60) is reduced, in the mentioned approximation, to the evaluation of integrals. Finally, the integrated intensities may be obtained with the help of Eq. (51), noting that the sign σ in the definition x_+ has to be taken negative inside the Brillouin lines, $|\frac{\omega}{cq} - 1| < \frac{\nu}{2}$, and positive outside. The details of this evaluation will be given elsewhere.

8. STATIC CORRELATION FUNCTIONS NEAR FIRST INSTABILITY

Instability of a fluid at rest is the combined effect of a downward temperature gradient $\vec{\beta} = (0,0,-\beta)$ and of gravity $\vec{g} = (0,0,-g)$. Indeed, thermal expansion then causes the liquid to be lighter below than above which, for sufficiently strong gradient β, gives rise to convection with a pattern depending on the boundary conditions (rolls, hexagons).

Gravity is easily incorporated in Section 4 by adding to the external force per unit mass $\vec{\xi}$ in Eqn. (21) a destabilizing force [17,18]

$$\vec{\chi} = \frac{\delta\rho}{\rho}\vec{g} = -\alpha\vec{g}\delta T \tag{67}$$

where, as before, α is the thermal expansion coefficient [43]. Eqs (24) for the flow pattern are then modified by a term $\rho\vec{\chi}_o$ with

$$\vec{\chi}_o = -\bar{\alpha}(T_o - \bar{T})\vec{g} = -\bar{c}^2\gamma\beta\hat{n}(\hat{n}\cdot\vec{r}) \tag{68}$$

on the right of the second equation where $\gamma \equiv \bar{\alpha}\bar{T}g/\bar{c}^2$ and $\hat{n} = (0,0,1)$ is the vertical direction. Since now $\nabla p_o = \bar{\rho}\vec{\chi}_o$ the third Eq. (25) is modified as follows [43]

$$P_o = \bar{p} - \frac{\bar{\rho}}{2}\bar{c}^2\gamma\beta(\hat{n}\cdot\vec{r})^2 \quad . \tag{69}$$

The modification of Eqs (26) is a term

$$\vec{\chi} - \vec{\chi}_o = -\bar{\alpha}(T - T_o)\vec{g} = -\bar{c}\vec{\gamma}v_4 \tag{70}$$

to be added to $\vec{\xi}$ where $\vec{\gamma} = (0,0,-\gamma)$. Neglecting spatial variations due to the temperature gradient, i.e. putting $\psi = 0$ in Eqs (27) and (29), the forcing matrix modified by the destabilizing force is, in the representation (29), $\phi(\vec{q},\vec{k}) = \bar{\phi}\delta_{\vec{q},\vec{k}}$ with

$$\bar{\phi} = \begin{pmatrix} 0 & \vec{0} & 0 \\ \vec{0} & \underline{0} & \bar{c}\vec{\gamma} \\ 0 & \bar{c}\vec{\beta} & 0 \end{pmatrix} \quad . \tag{71}$$

The procedure of Section 6 could now be repeated to find the correlation function matrix $K(\vec{q};\omega)$ as modified by the destabilizing force. However, since here we are interested in the effect of the first instability on the correlation functions, the approximation of neglecting the gradient β in the diagonalization of the matrix $\bar{\Lambda}$ is not acceptable. We therefore come back to the method described at the end of Section 3 of calculating the static correlation functions.

As was done explicitly in Refs 17 and 18 for the case of an incompressible fluid, the static correlation functions (19) may be calculated directly from the second Eq. (18), using Eq. (32) to determine the diffusion matrix C and writing

$$E^{-1} = E_o^{-1} + X \quad . \tag{72}$$

Remembering that $\Lambda = \Gamma + i\Omega + \phi$ and noting that the matrix (31) also satisfies the second Eq. (15) or, equivalently,

$$\Omega E_o^{-1} + E_o^{-1}\Omega^T = 0 \tag{73}$$

one finds that X then satisfies the equation

$$\Lambda X + X\Lambda^T = -(\phi E_o^{-1} + E_o^{-1}\phi^T) \quad . \tag{74}$$

In the representation (33) we have $X(\vec{q},\vec{k}) = \bar{X}(\vec{q})\delta_{\vec{q}+\vec{k},\vec{o}}$, and the symmetry $X^T = X$ implies $\bar{X}^T(\vec{q}) = \bar{X}(-\vec{q})$ so that we may write

$$\bar{X}(\vec{q}) = \begin{pmatrix} x_o(\vec{q}) & \vec{x}(\vec{q}) & y(\vec{q}) \\ \vec{x}(-\vec{q}) & \underline{X}(\vec{q}) & \vec{y}(\vec{q}) \\ y(-\vec{q}) & \vec{y}(-\vec{q}) & x_4(\vec{q}) \end{pmatrix} \quad . \tag{75}$$

Here $x_o^{(-)} = x_4^{(-)} = 0$ where $f^{(\pm)} = \frac{1}{2}[f(\vec{q}) \pm f(-\vec{q})]$ are even and odd parts of any f.

Inserting (29) with $\psi = 0$, (75), (71) and (31) into Eq. (74) one finds the following system of equations

$$\vec{q}\cdot\vec{x}^{(-)} = 0$$

$$i\vec{c}\vec{q}\cdot\vec{y} + \vec{\beta}\cdot\vec{x} + D_4\vec{q}^2 y = 0$$

$$\vec{\beta}\cdot\vec{y}^{(+)} + D_4\vec{q}^2 x_4 = 0$$

$$i\vec{c}\vec{q}(\underline{X} - x_o\underline{1}) + \underline{D}\vec{q}^2\vec{x} + \vec{\gamma}y = \vec{0}$$

$$
\begin{aligned}
&ic\vec{q}y + \underline{\tilde{D}}\vec{q}^2\vec{y} + \vec{\gamma}x_4 + \underline{X}\vec{\beta} = -\mu \equiv -(\bar{\varepsilon}^{-1}\vec{\beta} + \bar{\varepsilon}^{-1}\vec{\gamma}) \\
&\underline{DX} + \underline{XD} = \frac{i\bar{c}}{q^2}[\vec{q}\otimes\vec{x}(\vec{q}) - \vec{x}(-\vec{q})\otimes\vec{q}] - \frac{1}{q^2}[\vec{y}(\vec{q})\otimes\vec{\gamma} + \vec{\gamma}\otimes\vec{y}(-\vec{q})]
\end{aligned}
\tag{76}
$$

where $\underline{D} = D\underline{1} + D'\hat{q}\otimes\hat{q}$, as before, and $\underline{\tilde{D}} \equiv \underline{D} + D_4\underline{1} \equiv \tilde{D}\underline{1} + D_4\hat{q}\otimes\hat{q}$. As in Section 6 we choose coordinate vectors $\hat{e}_1 = (1,0,0)$ and $\hat{e}_3 = \hat{n}$ such that $\hat{q} = (\xi,0,\zeta) \equiv \hat{e}_\ell$ and define

$$
z_{ij}^{(\pm)} \equiv (\hat{e}_i\underline{D}^{-1}\underline{X}^{(\pm)}\hat{e}_j) \; ; \; \tilde{z}_{ij}^{(\pm)} \equiv (\hat{e}_i\underline{\tilde{D}}^{-1}\underline{X}^{(\pm)}\hat{e}_j) \quad . \tag{77}
$$

We also calculate

$$
(\hat{q}\underline{D}^{-1}\hat{q}) = \frac{1}{2\lambda} \; ; \; (\hat{q}\underline{D}^{-1}\hat{n}) = \frac{\zeta}{2\lambda} \, , \; (\hat{n}\underline{D}^{-1}\hat{n}) = \frac{1+\theta\xi^2}{2\lambda} \tag{78}
$$

where $\lambda \equiv \frac{1}{2}(D + D')$, as before, and $\theta \equiv \frac{D'}{D}$, corresponding expressions being obtained by replacing $\underline{D} \to \underline{\tilde{D}}$, $D \to \tilde{D}$, $\lambda \to \tilde{\lambda}$, $\theta \to \tilde{\theta}$. The procedure now is to solve the first 5 Eqs (76) in terms of the unknowns (77). The result for the scalar quantities is

$$
\begin{aligned}
x_o &= \frac{2\lambda}{\Delta}\{\Delta' z_{\ell\ell}^{(+)} - R\zeta^2\tilde{z}_{33}^{(+)} + \frac{\Delta-\Delta'}{\zeta}(\tilde{z}_{\ell 3}^{(+)} + z_{3\ell}^{(+)}) - R\tau\frac{\mu\zeta^2}{2\lambda\beta}\} \\
x_4 &= -\frac{2\lambda\beta}{\gamma\Delta}\{\Delta_o\tilde{z}_{33}^{(+)} - \zeta(\tilde{z}_{\ell 3}^{(+)} + z_{3\ell}^{(+)}) + \zeta^2 z_{\ell\ell}^{(+)} + (\Delta + \tau\Delta_o)\frac{\mu}{2\lambda\beta}\} \\
y^{(+)} &= -i\frac{2\lambda\beta}{q\Delta_o'}(\tilde{z}_{\ell 3}^{(-)} + z_{3\ell}^{(-)}) \\
y^{(-)} &= +i\frac{2\lambda\beta}{q\Delta}\{\zeta\tilde{z}_{33}^{(+)} - \frac{\Delta+\zeta^2}{\Delta_o}(\tilde{z}_{\ell 3}^{(+)} + z_{3\ell}^{(+)} - \zeta z_{\ell\ell}^{(+)}) + \tau\frac{\mu\zeta}{2\lambda\beta}\}
\end{aligned}
\tag{79}
$$

where $R \equiv \beta\gamma/q^2$ is essentially the Rayleigh number $\mu \equiv \bar{\varepsilon}^{-1}\beta + \bar{\varepsilon}_4^{-1}\gamma$, $\tau \equiv 2\lambda D_4 q^2/R\bar{c}^2$ and

$$
\begin{aligned}
&\Delta_o \equiv 1 - R(1+\theta)\xi^2 \; ; \; \Delta_o' \equiv \Delta_o - R\zeta^2 \\
&\Delta \equiv (1 + \tilde{\theta}\xi^2 - \tau)\Delta_o - \zeta^2 \; ; \; \Delta' \equiv (1 + \tilde{\theta}\xi^2 - \tau)\Delta_o' - \zeta^2
\end{aligned}
\tag{80}
$$

Under the experimental conditions of Ref. 1, $\theta \cong 2.93$, $\tilde{\theta} \cong 2.39$ and $(\tilde{\lambda} - \lambda)/\lambda = D_4/2\lambda \cong 0.057$. The last difference was neglected as well as terms $2\lambda D_4 q^2/\bar{c}^2 \cong 2\times 10^{-8}$.

Substituting Eqs (79) into the two vector equations (76), $\vec{X}^{(\pm)}$ and $\vec{y}^{(\pm)}$ are found to be linear inhomogenous expressions of the unknowns (77) and, hence, the same follows for the right-hand side of the last equation (76) which has the form of a 3×3 matrix equation. To solve the latter we again omit the 2-component of all vectors and matrices so that $\hat{e}_1 = (1,0)$ and $\hat{e}_3 = (0,1)$ alone form a basis and $\underline{D}$, $\underline{X}$ may be expanded in terms of Pauli-matrices. The coefficients of $\underline{X}$ in this expansion are then obtained as linear homogenous functions of the right-hand side of the last Eq. (76) and, hence, as linear inhomogenous expressions of the unknowns (77). Using the definitions (77) one finally obtains a system of linear inhomogenous equations for these unknowns.

Although too complicated to be reported in detail here, this procedure is straight forward and guarantees the existence of a solution to Eq. (72). The important feature of this solution is exhibited in the denominators Δ and Δ_o' of Eqs (79) which, according to their definition (80) vanish for certain values of the control parameter R which is a measure for the distance from equilibrium. This leads to poles of the correlation functions for values of R of the order of magnitue 1. Physically, these poles are the cause for the first instability. However, it should be kept in mind that the linear theory described above cannot give the true location R_c of this instability, the latter being determined by the renormalization due to the nonlinear terms in the equations of motion (20), (21), (22).

Finally we note that Eqs. (76) to (80) are of course also valid in the limit $\gamma = 0$ where, as in Sections 4 to 7, gravity is neglected. In this case the static correlation functions (72) still contain the exact dependence on the gradient β. However, the information content is rather limited since these functions represent an integral over the Rayleigh and Brillouin lines, whereas to get the frequency dependence one cannot avoid diagonalizing the matrix Λ.

REFERENCES

[1] D. Beysens, Y. Garrabos and G. Zalczer, Phys. Rev. Lett. 45, 403 (1980).

[2] D. Beysens, Physica 118A, 250 (1983).

[3] H. Kiefte, M.J. Clouter and R. Penney, Phys. Rev. B30, 4017 (1984).

[4] I. Procaccia, D. Ronis and I. Oppenheim, Phys. Rev. Lett. 42, 287, 614E (1979); D. Ronis, I. Procaccia and I. Oppenheim, Phys. Rev. A19, 1324 (1979); I. Procaccia, D. Ronis, M.A. Collins, J. Ross and I. Oppenheim, Phys. Rev. A19, 1290 (1979).

[5] T, Kirkpatrick, E.G.D. Cohen and J.R. Dorfman, Phys. Rev. Lett. 42, 862 (1979).

[6] G. van der Zwan and P. Mazur, Phys. Lett. 75A, 370 (1980).

[7] A.-M.S. Tremblay, E.D. Siggia and M.R. Arai, Phys. Lett. 76A, 57 (1980).

[8] T.R. Kirkpatrick, E.G.D. Cohen and J.R. Dorfman, Phys. Rev. A26, 995 (1982).

[9] M. Lax, Rev. Mod. Phys. 32, 25 (1960).
[10] A.-M.S. Tremblay, M. Arai and E.D. Siggia, Phys. Rev. A23, 1451 (1981).
[11] G. van der Zwan, D. Bedeaux and P. Mazur, Physica 107A, 491 (1981).
[12] H. Grabert, J. Statist. Phys. 26, 113 (1981).
[13] H. Ueyama, J. Phys. Soc. Japan 51, 3443 (1982).
[14] T.R. Kirkpatrick, E.G.D. Cohen and J.R. Dorfman, Phys. Rev. A26, 950, 972 (1982).
[15] E.G.D. Cohen, Physics Today, January 1984, p. 64.
[16] D. Ronis, I. Procaccia and I. Oppenheim, Phys. Rev. A19, 1307 (1979) I. Procaccia, D. Ronis and I. Oppenheim, Phys. Rev. A20, 2533 (1979); D. Ronis, I. Procaccia and J. Machta, Phys. Rev. 22, 714 (1980).
[17] C.P. Enz, in Field Theory, Quantization and Statistical Physics, ed. E. Tirapegui (Reidel, Dordrecht, 1981), p. 263.
[18] C.P. Enz, J. Statist. Phys. 24, 109 (1981).
[]9] R. Graham, Springer Tracts in Modern Physics (Springer, Berlin, 1973), Vol. 66; H. Haken, Rev. Mod. Phys. 47, 67 (1975).
[20] D. Ronis and I. Procaccia, Phys. Rev. A26, 1812 (1982).
[21] L.S. Garcia-Colin and R.M. Velasco, Phys. Rev. A26, 2187 (1982).
[22] R. Garibay-Jiménez and L.S. Garcia-Colin, Physica 130A, 616 (1985).
[23] J. Machta, I. Oppenheim and I. Procaccia, Phys. Rev. Lett. 42, 1368 (1979).
[24] T, Kirkpatrick, E.G.D. Cohen and J.R. Dorfam, Phys. Rev. Lett. 44, 472 (1980).
[25] E. Peacock-López and J. Keizer, Phys. Lett. 108A, 85 (1985).
[26] A.-M.S. Tremblay, in Recent Developments in Nonequilibrium Thermodynamics, ed. J. Casas-Vásquez, D. Jou and G. Lebon (springer, Berlin, 1984) Lecture Notes in Physics, Vol. 199. p. 267.
[27] A. Onuki and K. Kawasaki, Ann. Phys. (NY) 121, 456 (1979); A. Onuki, J. Low, Temp. Phys. 50, 433 (1983).
[28] A.-M.S. Tremblay and F. Vidal, Phys. Rev. B25, 7562 (1982).
[29] K. Kawasaki and A. Onuki, in Dynamical Critical Phenomena and Related Topics, ed. C.P. Enz (Springer, Berlin, 1979) Lecture Notes in Physics, Vol. 104, p. 337.
[30] C.P. Enz, Physica 94A, 20 (1978).
[31] L. Arnold, Stochastic Differential Equations: Theory and Applications (Wiley-Interscience, New York, 1974), Sec. 2.6.
[32] C.P. Enz, Physica 89A, 1 (1977).
[33] L.D. Landau and E.M. Lifshitz, Statistical Physics, (Pergamon, Oxford, 1958), Sec. 114.
[34] L. Onsanger and S. Machlup, Phys. Rev. 91, 1505, 1512 (1953).
[35] R. Kubo, J. Phys. Soc. Japan 12, 570 (1957).
[36] C.P. Enz, Rev. Mod. Phys. 46, 705 (1974).
[37] L.D. Landau and E.M. Lifshitz, Fluid Mechanics (Pergamon, Oxford, 1959), Sec. 49.
[38] S. Chandrasekhar, Hydrodynamic and Hydromagnetic Stability (Clarendon, Oxford, 1961).
[39] J. Rouch, C.C. Lai and S.H. Chen, J. Chem. Phys. 65, 4016 (1976).
[40] Handbook of Chemistry and Physics, ed. R.C. Weast (CRC Press, Boca Raton, Florida, any edition).

[41] For an earlier application of this technique see C.P. Enz, Helv. Phys. Acta 33, 89 (1960).
[42] See, e.g., R.B. Dingle, Aymptotic Expansions: Their Derivation and Interpretation (Academic, New York, 1973), Sec. 21.4.
[43] Note the errors in the corresponding equations of Refs. 17 and 18.

FLUCTUATIONS IN UNSTABLE SYSTEMS

N. G. van Kampen
Institute for Theoretical Physics
Princetonplein 5, P. O. Box 80.006
3508 TA Utrecht, The Netherlands

ABSTRACT. This review is concerned with systems described by a Fokker-Planck equation. Such systems may escape from a locally stable state through thermal fluctuations. The mean escape time may be identified with a mean first-passage time, which can be computed by means of singular perturbation theory. A system initially at or near an unstable state moves to one or the other adjacent stable states with probabilities determined by the initial fluctuations. This splitting probability can be computed and also the evolution of the probability density itself. All calculations are approximations for small fluctuations, that is, for low temperature.

1. THE FOKKER-PLANCK EQUATION

A stochastic process or random function is determined by the infinite sequence of joint probability densities

$$P_n(x_1,t_1;x_2,t_2;\dots;x_n,t_n) \;, \quad (n = 1,2,3,\dots) \;.$$

For instance, $P_1(x_1,t_1;x_2,t_2)dx_1dx_2$ is the probability that the function takes at t_1 a value between x_1 and $x_1 + dx_1$ and at t_2 a value between x_2 and $x_2 + dx_2$. If the value x_1 at t_1 is known, the conditional probability that the value at t_2 lies between x_2 and $x_2 + dx_2$ is

$$\frac{P_2(x_1,t_1;x_2,t_2)dx_2}{P_1(x_1,t_1)} = T(x_2,t_2|x_1,t_1)dx_2 .$$

$T(x_2,t_2|x_1,t_1)$ is the transition probability density from x_1 to x_2.

A Markov process is characterized by the fact that all P_n can be expressed in terms of P_1 and T in the following way. If $t_1 < t_2 < \dots t_n$ one has

E. Tirapegui and D. Villarroel (eds.), Instabilities and Nonequilibrium Structures, 241–270.

$$P_n(x_1,t_1;x_2,t_2;\ldots;x_n,t_n) = P_1(x_1,t_1)T(x_2,t_2|x_1,t_1)\ldots T(x_n,t_n|x_{n-1},t_{n-1}).$$

One easily finds (*) the additional properties

$$P_1(x_2,t_2) = \int T(x_2,t_2|x_1,t_1)P_1(x_1,t_1)dx_1$$

$$T(x_3,t_3|x_1,t_1) = \int T(x_3,t_3|x_2,t_2)T(x_2,t_2|x_1,t_1)dx_2 \qquad (t_1 < t_2 < t_3).$$

The latter one is called the Chapman-Kolmogorov equation and expresses the semi-group nature of the Markov process.

A process is stationary if all these functions depend on time differences alone, i.e., they are invariant for time translations. Then $P_1(x_1)$ cannot involve time. Also $T(x_2,t_2|x_1,t_1)$ depends on time only through $t_2 - t_1 = \tau$ and will be written $T_\tau(x_2|x_1)$. This transition probability is obviously non-negative and has the properties

$$T_0(x_2|x_1) = \delta(x_2 - x_1). \qquad \int T_\tau(x_2|x_1)dx_2 = 1$$

$$P_1(x_2) = \int T_\tau(x_2|x_1)P_1(x_1)dx_1 \qquad (\text{all } \tau).$$

From now on we are only concerned with stationary Markov processes.

For small $\Delta t > 0$ one has for a stationary Markov process

$$T_{\Delta t}(x_2|x_1) = \delta(x_2 - x_1) + \Delta t \underline{W}(x_2|x_1) + o(\Delta t)$$

Substitute in the Chapman-Kolmogorov equation $t_3 = t_2 + \Delta t$ and go to the limit $\Delta t \to 0$:

$$\frac{\partial T_\tau(x|x_0)}{\partial \tau} = \int \underline{W}(x|x')T_\tau(x'|x_0)dx' \quad (\text{all } \tau > 0).$$

(*) Word sequences such as this one indicate exercises. One may also resort to standard textbooks [1,2,3] or reviews [4].

This is the master equation. After integration over x the left hand side must vanish; hence one may also write

$$\frac{\partial T_\tau(x|x_0)}{\partial \tau} = \int \{W(x|x')T_\tau(x'|x_0) - W(x'|x)T_\tau(x|x_0)\}dx',$$

which has the form of a gain-loss equation for the probability. $W(x|x')$ is the transition probability per unit time and is nonnegative. (The W in the previous equation still contained a delta function with a negative sign).

We emphasize that the master equation is an equation for the transition probability. In the following we shall abide by the custom to write it as an equation for a probability $P(x,t)$, but it should be understood to mean: There is a stationary Markov process such that for any x_0, t_0 the solution of the master equation determined by the initial condition

$$P(x,t_0) = \delta(x-x_0)$$

is the transition probability $T_{t-t_0}(x|x_0)$ of the process.

It may happen that the linear operator W is a second order differential operator. Then our master equation takes the form of a Fokker-Planck equation

$$\frac{\partial P(x,t)}{\partial \tau} = -\frac{\partial}{\partial x} A(x)P(x,t) + \frac{1}{2}\frac{\partial^2}{\partial x^2} B(x)P(x,t).$$

where the functions $A(x)$, $B(x)$ are properties of the system, and $B(x) \geq 0$. (This is a small subclass of all Markov processes. It can be shown that higher order differential operators cannot occur [5], whereas a first order equation is just a Liouville equation, belonging to the deterministic equation $\dot{x} = A(x)$.)

The Fokker-Planck equation has been widely used for describing physical, chemical, biological and sociological systems in which random fluctuations (or 'noise') are of importance. Sometimes it can be shown to be a valid approximation [2,6,7], but often it is merely based on the hope that at least the qualitative features of the actual fluctuating system are reproduced correctly. (That is called 'modeling' the system.) The literature is so overwhelming that we can only refer the reader to the general reviews mentioned already [1-4] and to books and reviews on special applications [8].

For our purpose it is sufficient to visualize the equation as diffusion of a particle in an inhomogeneous medium (i.e. B depends on x) and subject to an external force (to account for A). We restrict ourselves to one dimension and allow x to range from $-\infty$ to $+\infty$. According to this picture the equation contains both drift and diffusion, but these two effects cannot be identified with the two terms respectively. In fact, one could as well write the equation alternatively as a conservation law for the probability, with a probability flow $J(x,t)$,

$$\frac{\partial P}{\partial t} = - \frac{\partial J}{\partial x}, \qquad J = \left\{A + \frac{1}{2}\frac{dB}{dx}\right\}P - \frac{1}{2} B \frac{\partial P}{\partial x}.$$

One would then be inclined to identify the drift with the term proportional to P itself and the diffusion with the term proportional to the gradient. However, neither choice for the identification is invariant for transformations of the variables.

Our task is to solve the Fokker-Planck equation with the above initial condition. An important role is played by the time-independent or stationary solution $P^s(x)$, determined by the equation

$$0 = - \frac{d}{dx} A(x)P^s(x) + \frac{1}{2} \frac{d^2}{dx^2} B(x)P^s(x) \equiv - \frac{dJ}{dx}.$$

This equation states the physically obvious fact that J must be constant. If this constant differs from zero there is a flow entering at $x = -\infty$ and leaving at $x = +\infty$ (or vice versa). For closed systems this should not occur, hence we take $J = 0$ and find

$$P^s(x) = \frac{C}{B(x)} \exp\{2 \int_0^x \frac{A(x')}{B(x')} dx'\}.$$

The normalization constant C is determined by

$$\frac{1}{C} = \int_{-\infty}^{\infty} \frac{dx}{B(x)} \exp\{2 \int_0^x \frac{A(x')}{B(x')} dx'\}.$$

When the integral does not converge there is no normalizable stationary solution, as in the case of ordinary diffusion ($A = 0$, $B = \text{const.}$). When, however, the normalizing integral converges a stationary distribution $P^s(x)$ exists. Sufficient conditions are that $B(x)$ has a positive lower bound and that $A(x)$ tends to $\mp\infty$ sufficiently rapidly as $x \to \pm\infty$. The latter can always be achieved by a modification of $A(x)$ far away, or by adding reflecting walls at a large distance.

Incidentally, in closed, isolated physical systems one knows that the Boltzmann-Gibbs distribution $P^e(x)$ is a solution. Hence in these cases one knows a priori $P^s(x) = P^e(x)$, which leads to an identity relating A and B. This is the Einstein relation [9], which is a prototype of the fluctuation-dissipation theorem [10]. However, in most cases the FP equation is merely an approximation and one cannot therefore identify its P^s with the exact P^e. The resulting relation between A and B can be trusted only if B is constant, and A linear in x. Such FP equations we shall call linear (bearing in mind that all FP equations - and in fact all master equations - are linear in the unknown P).

2. SMALL FLUCTUATION EXPANSION FOR STABLE SYSTEMS

For linear FP equations the explicit solution can easily be found in the form of a Gaussian. For other $A(x)$, $B(x)$ an explicit solution exists only in rare cases. We are therefore concerned with approximate methods. For a systematic approximation scheme it is necessary to have a small parameter. As in most practical cases the fluctuations are small, we suppose that B is small and indicate this by inserting a small parameter θ :

$$\frac{\partial P(x,t)}{\partial t} = -\frac{\partial}{\partial x} A(x)P + \frac{1}{2}\,\theta\,\frac{\partial^2}{\partial x^2} B(x)P.$$

In the case of diffusion θ is the absolute temperature, and the same is true in many other cases. (Thermal fluctuations are small at low temperatures).

The equations cannot be solved by ordinary perturbation theory, however, because the small θ multiplies the highest derivatives. Such cases have to be treated by means of singular perturbation theory [11], which is less straightforward and requires some ingenuity. The reason is that for small temperature θ the P will be sharply peaked, so that its second derivative is large. Both effects have to balance, which is done in the following way [12].

Decompose x in a macroscopic part and a fluctuating part (Fig. 1),

$$x = \varphi(t) + \theta^{\frac{1}{2}}\,\xi\ ,$$

where $\varphi(t)$ is a solution of the deterministic equation

$$\dot{\varphi}\,(t) = A(\varphi).$$

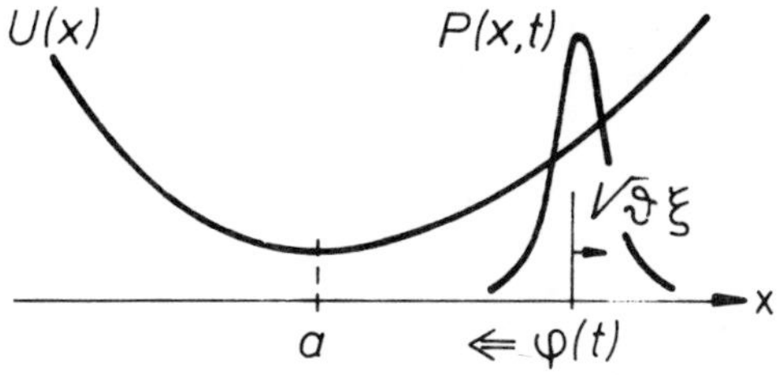

Figure 1. The peaked distribution P in a potential field U.

Transform the probability P into a probability Π for the new variable ξ,

$$P(\varphi(t) + \theta^{\frac{1}{2}}\xi, t) = \Pi(\xi, t)$$

The transformed FP equation is

$$\frac{\partial\Pi}{\partial t} - \theta^{-\frac{1}{2}} \frac{d\varphi}{dt}\frac{\partial\Pi}{\partial\xi} = -\theta^{-\frac{1}{2}} \frac{\partial}{\partial\xi}A(\varphi + \theta^{\frac{1}{2}}\xi)\Pi + \frac{1}{2}\frac{\partial^2}{\partial\xi^2}B(\varphi + \theta^{\frac{1}{2}}\xi)\Pi \quad .$$

The terms of order $\theta^{-1/2}$ cancel and there remains

$$\frac{\partial\Pi(\xi,t)}{\partial t} = -A'(\varphi)\frac{\partial}{\partial\xi}\xi\Pi + \frac{1}{2}B(\varphi)\frac{\partial^2\Pi}{\partial\xi^2} + O(\theta^{1/2})$$

To lowest order this is a <u>linear</u> FP equation, albeit with coefficients $A'(\varphi)$, $B(\varphi)$ that depend on time. Yet its solution can again be found in the form of a Gaussian. The Gaussian is fully determined by its first and second moments. These can be obtained directly from the equation by multiplying it with ξ and ξ^2 respectively and integrating:

$$\partial_t \langle\xi\rangle_t = A'(\varphi)\langle\xi\rangle_t$$

$$\partial_t \langle\xi^2\rangle_t = 2A'(\varphi)\langle\xi^2\rangle_t + B(\varphi).$$

If the function $\varphi(t)$ is known these two linear equations can be solved so as to obtain $\langle\xi\rangle_t$ and $\langle\xi^2\rangle_t$ as functions of t. (Exercise: Even if $\varphi(t)$ is not known explicitly one can still find $\langle\xi\rangle_t$ and $\langle\xi^2\rangle_t$ as functions of φ).

We now have the ingredients for constructing the solution $P(x,t)$ that reduces to $\delta(x - x_0)$ at $t = t_0$. First choose for $\varphi(t)$ that solution of the deterministic equation for which $\varphi(t_0) = x_0$. Then

$$\langle\xi\rangle_{t_0} = \langle\xi^2\rangle_{t_0} = 0.$$

From the homogeneous equation for $\langle\xi\rangle_t$ follows $\langle\xi\rangle_t = 0$ for all $t > t_0$, but $\langle\xi^2\rangle_t$ has to be found by solving its equation. Having done this one knows the Gaussian Π and therefore also

$$P(x,t) \equiv T(x,t|x_0,t_0) = \Pi\left(\frac{x - \varphi(t)}{\theta^{1/2}}, t\right).$$

Thus <u>the fluctuations have a Gaussian distribution centered at the macroscopic value $\varphi(t)$</u>.

A stochastic process is called <u>Gaussian</u> if all P_n mentioned in section 1 are (multivariate) Gaussian distributions. Then also $T(x_2,t_2|x_1,t_1)$ is Gaussian. For Markov processes the converse holds: if P_1 and T are Gaussian all P_n are Gaussian. A famous theorem of Doob [14] states that the only stationary, Gaussian Markov process is the Ornstein-Uhlenbeck process. Our underlying stationary Markov process is <u>not</u> Gaussian (unless the FP equation happens to be linear). Our present approximation is Gaussian, but not stationary, owing to the imposition of an initial value. In higher order of θ the fluctuations are no longer Gaussian and the average no longer coincides with φ .

I add one remark. It is possible to reduce B(x) to a constant by transforming x. Take as new variable.

$$\bar{x} = \int_0^x \sqrt{\frac{2}{B(x')}}\, dx'.$$

which has again the range $(-\infty,\infty)$ provided that B(x) lies between two positive bounds. The new probability density

$$\bar{P}(\bar{x}) = P(x)\,\frac{dx}{d\bar{x}} = \sqrt{B(x)/2}\,P(x)$$

obeys the transformed equation

$$\frac{\partial\bar{P}}{\partial t} = -\frac{\partial}{\partial\bar{x}}\bar{A}(\bar{x})\bar{P} + \theta\,\frac{\partial^2\bar{P}}{\partial\bar{x}^2}$$

$$\bar{A}(\bar{x}) = \sqrt{\frac{2}{B(x)}}\,\{A(x) - \frac{\theta}{4}\frac{dB}{dx}\}.$$

For simplicity we use in the following this form of the FP equation, omitting the overbars. In the multivariate case, however, such a reduction is not generally possible [15].

3. THE ONE-DIMENSIONAL BISTABLE CASE

The approximation in the previous section is based on the smallness of the fluctuations. The question is therefore whether $<\xi^2>_t$ remains finite or grows in time. To investigate this we use the reduced form of the FP equation

$$\frac{\partial P(x,t)}{\partial t} = \frac{\partial}{\partial x}U'(x)P + \theta\,\frac{\partial^2 P}{\partial x^2},$$

where we also set $A(x) = -U'(x)$. One then has

$$P^s(x) = C\,e^{-U(x)/\theta}, \qquad C^{-1} = \int_{-\infty}^{\infty} e^{-U(x)/\theta}dx.$$

The relevant equations take the form

$$\partial_t\varphi = -U'(\varphi), \qquad \partial_t<\xi^2>_t = -2U''(\varphi)<\xi^2>_t + 2.$$

If $U(\varphi)$ is convex, as in Fig. 1, the solution $\varphi(t)$ tends to the minimum at a and $<\xi^2>_t$ remains finite and tends to

$$<\xi^2>_\infty = 1/U''(a).$$

Thus $P(x,t)$ tends to $P^s(x)$, which in our approximation is the Gaussian

$$P^s(x) = \sqrt{\frac{U''(a)}{2\pi\theta}} \exp - \{\frac{U''(a)}{2\theta} (x-a)^2\}.$$

The relaxation time is of order $1/U''(a)$.

Now suppose $U(x)$ is bistable, as in Fig. 2. When the starting point x_0 is somewhere in the trough around a the evolution of $P(x,t)$ will be practically the same as before, tending to the same Gaussian as above. The only difference is that there is a very small probability per unit time, say $1/\tau_{ca}$, to fluctuate across the barrier at b into the trough around c. Thus the local equilibrium around a slowly leaks away with a decay time τ_{ca}. This escape time τ_{ca} will involve the Arrhenius factor $\exp[\{U(b) - U(a)\}/\theta]$, which for small θ is very large. This factor is suggested by the transition state theory of chemical reactions, which is based on the idea that the escape occurs via an intermediate 'activated complex' with energy $U(b)$ [16]. In addition there is a prefactor and our aim is to compute it.

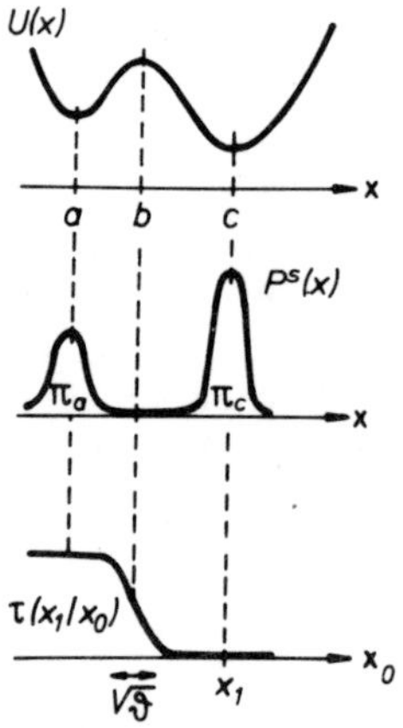

Figure 2. A bistable potential with its two-peaked stationary distribution and the escape time τ as a function of the starting point x_0.

The leakage has the effect that a probability peak slowly builds up around c, which is again an almost Gaussian local equilibrium distribution. As a consequence there is a probability $1/\tau_{ac}$ per unit time to leak back into the trough at a. Let $\pi_a(t)$ be the total probability to be near a and $\pi_c(t)$ to be near c; then (neglecting the very small probability to be in the no-man's-land near b) one has $\pi_a(t) + \pi_c(t) = 1$. These probabilities vary in time according to

$$\frac{d\pi_a}{dt} = -\frac{\pi_a}{\tau_{ca}} + \frac{\pi_c}{\tau_{ac}} = -\frac{d\pi_c}{dt}.$$

After many aeons of order τ_{ca} or τ_{ac} they reach their stationary values

$$\pi_a^s = \frac{\tau_{ac}}{\tau_{ca} + \tau_{ac}}, \qquad \pi_c^s = \frac{\tau_{ca}}{\tau_{ca} + \tau_{ac}}.$$

Then $P(x,t)$ has reached its equilibrium value $P^s(x)$, which consists of two peaks near a and c. The integral of each is π_a^s, π_c^s so that approximately

$$P^s(x) = \pi_a^s \sqrt{\frac{U''(a)}{2\pi\theta}} \exp\left\{-\frac{U''(a)}{2\theta}(x-a)^2\right\} + \pi_c^s \sqrt{\frac{U''(c)}{2\pi\theta}} \exp\left\{-\frac{U''(c)}{2\theta}(x-c)^2\right\}.$$

The ratio π_a^s/π_c^s is also the ratio of the times the particle spends near a and c; it has been called the 'relative stability' [17] of a, c. It is equal to τ_{ac}/τ_{ca} and the Arrhenius factors provide the rough estimate

$$\frac{\pi_a^s}{\pi_c^s} = \frac{\tau_{ac}}{\tau_{ca}} \sim \exp\left[-\frac{U(a) - U(c)}{\theta}\right].$$

On the other hand, this ratio is given by $P^s(x)$ and therefore the prefactor is also known:

$$\frac{\pi_a^s}{\pi_c^s} = \left[\frac{U''(c)}{U''(a)}\right]^{1/2} \exp\left[-\frac{U(a) - U(c)}{\theta}\right].$$

This cannot give us the value of τ_{ca} itself, however. (In fact, $P^s(x)$ contains no times scale). One sees that unless $U(a) - U(c)$ is zero or very small the equilibrium probability is almost totally confined to the lowest energy trough.

All these considerations suppose that τ_{ca} and τ_{ac} are much larger than the relaxation times $1/U''(a)$, $1/U''(c)$ inside each separate trough, that is, that the Arrhenius factor is large, θ small. If that is not so, no clear definition of escape time is possible, because the time to reach from a starting point x_0 near a a final point x_1 near c depends appreciably on the precise choices of x_0 and x_1. The very definition of escape time has a margin of indeterminacy of the order of the relaxation times in the separate troughs, that is, of relative order of the reciprocal Arrhenius factor.

The escape time τ_{ca} was computed by Becker and Döring in connection with nucleation [18], by Kramers for unimolecular chemical reactions [19], by Landauer for the tunnel diode [20], by Griffiths et al. for spin flips [21], all using similar ideas. Another approach is based on the eigenfunction expansion of the FP equation in a bistable potential; the lowest eigenvalue is zero and the (almost degenerate) next eigenvalues is $1/\tau_{ca}$. It can be computed by WKB [22], singular perturbation theory [23], or some other approximation method [24].

We shall treat the problem by means of the general theory of first-passage times: A particle that starts at t_0 in x_0 will arrive at a given point x_1 for the first time at some time t_1. The time interval $t_1 - t_0$ depends on the path the particle happens to take and is therefore a random quantity. Its average is the mean first passage time and will be denoted $\tau(x_1|x_0)$. Clearly $\tau(x_1|x_1) = 0$ (Appendix). If one chooses x_0 near a and x_1 near c, $\tau(x_1|x_0)$ is equal to the escape time τ_{ca} (within the margin of indeterminacy of the latter's definition). One may take $x_0 = a$, $x_1 = c$. Alternatively one may take $x_0 = a$, $x_1 = b$, but in that case one has $\tau_{ca} = 2\tau(b|a)$, because once arrived at b the particle has equal chances to escape or return. Before making a special choice, however, we shall derive an equation for $\tau(x_1|x_0)$ as a function of the starting point x_0.

4. CALCULATION OF THE ESCAPE TIME

A particle starting at x_0 arrives after a short time Δt in a point x' with probability $T_{\Delta t}(x'|x_0)$. Hence

$$\tau(x_1|x_0) = \Delta t + \int T_{\Delta t}(x'|x_0)\tau(x_1|x')dx' .$$

Substitute for $T_{\Delta t}$ the expression in section 1 and go to the limit:

$$\int \tau(x_1|x') \underline{W}(x'|x_0)dx' = -1$$

Thus τ as a function of the starting point obeys an equation in which the transposed of the kernel $\underline{W}$ occurs. For our FP equation the result reads

$$-U'(x_0)\frac{d\tau(x_1|x_0)}{dx_0} + \theta\frac{d^2\tau(x_1|x_0)}{dx_0^2} = -1 .$$

This is called the Dynkin equation [25]. It has to be solved for

$x_0 < x_1$ with the boundary condition $\tau(x_1|x_1) = 0$. (The boundary condition at $x_0 = -\infty$ is determined by the requirement that the operator is the adjoint of the one in the FP equation, that is, that the partial integrations leave no boundary term. This is easily seen to amount to $\exp(-U(x_0)/\theta)(d\tau/dx_0) = 0$ at $-\infty$, which is just what is needed in the following calculation).

It is not hard (in this one-variable case) to solve the equation for $\tau(x_1|x_0)$ explicitly. If one then takes $x_0 = a$, $x_1 = c$ one obtains:

$$\tau_{ca} = \frac{1}{\theta}\int_a^c e^{U(x')/\theta}dx' \int_{-\infty}^{x'} e^{-U(x'')/\theta}dx''.$$

Taking into account that the first exponential has a sharp maximum at b and the second at a one obtains

$$\tau_{ca} = \frac{2\pi}{\sqrt{U''(a)\,|U''(b)|}} \exp\left[\frac{U(b) - U(a)}{\theta}\right] .$$

One recognizes the Arrhenius factor and its prefactor.

This exact method is possible only in the one-variable case. We shall now give an approximate method, due to Schuss and Matkowski [26], which can also be used for more variables (section 8). It is based on the fact that $\tau(x_1|x_0)$ (as a function of x_0, with x_1 fixed and closed to c) has the following behavior. When x_0 is in the a-trough, $\tau(x_1|x_0)$ is practically equal to a very large constant, namely the τ_{ca} that has to be calculated. For x_0 in the c-trough, $\tau(x_1|x_0)$ is practically zero; more precisely: of order $1/U''(c)$ for any θ. In between, $\tau(x_1|x_0)$ falls off from the large value to zero in a narrow interval $(b-\Delta, b+\Delta)$, see Fig. 2. The behavior in this interval is crucial for the calculation of τ_{ca}. The point is that Δ turns out to be of order $\theta^{1/2}$, so that in the crucial interval $U(x)$ may be replaced by a parabola,

$$U(x) = U(b) - \tfrac{1}{2}|U''(b)|\,(x-b)^2 + \dots .$$

This makes the solution of the Dynkin equation in this interval possible. We shall now carry out this program.

Set $\tau(x_1|x_0) = \tau_{ca}w(x_0)$ so that

$$w(x_0) \approx 1 \qquad (x_0 < b - \Delta); \qquad w(x_0) \approx 0 \qquad (x_0 > b + \Delta)$$

The equation reduces to

$$-U'(x_0)\frac{dw}{dx_0} + \theta\frac{d^2w}{dx_0^2} = 1/\tau_{ca} \approx 0.$$

As this equation has no sensible limit as $\theta \to 0$ one applies a stretching transformation

$$x_0 = b + \theta^{1/2}\xi,$$

so that, to lowest order in θ.

$$- U''(b)\xi \frac{\partial w}{\partial \xi} + \frac{\partial^2 w}{\partial \xi^2} = 0.$$

This equation is of second order and can therefore obey both boundary conditions. On the other hand it is simple enough to be solved, thanks to the fact that the omission of higher orders of θ amounts to linearizing $U'(x)$.

The general solution is

$$w = \alpha \int_0^{\xi} d\xi' \exp[-\tfrac{1}{2}|U''(b)|\xi'^2] + \beta.$$

The integration constants α, β are fixed by the prescribed values for $\xi \to \infty$ and $\xi \to -\infty$:

$$w = \sqrt{\frac{|U''(b)|}{2\pi}} \int_{\xi}^{\infty} d\xi' \exp[-\frac{1}{2}|U''(b)|\xi'^2]$$

$$= \sqrt{\frac{|U''(b)|}{2\pi\theta}} \int_{x}^{\infty} dx' \exp[-\frac{|U''(b)|}{2\theta}(x'-b)^2] .$$

Note that the required limiting values are reached for $|x - b| \sim \theta^{1/2}$, so that indeed $\Delta \sim \theta^{1/2}$.

But we still have the unknown τ_{ca} as a factor. The reason is that the overall magnitude of $\tau(x_1|x_0)$ is fixed by the term -1 in the Dynkin equation, which disappeared in our approximation. We therefore return to that equation, write it in the form

$$\frac{d}{dx_0} e^{-U(x_0)/\theta} \frac{d\tau(x_1|x_0)}{dx_0} = -\frac{1}{\theta} e^{-U(x_0)/\theta} ,$$

and integrate from $-\infty$ up to b:

$$e^{-U(b)/\theta}\left[\frac{d\tau(x_1|x_0)}{dx_0}\right]_b = -\frac{1}{\theta}\int_{-\infty}^{b} e^{-U(x_0)/\theta}\, dx_0.$$

The derivative of τ has been found by our calculation of w:

$$\left[\frac{d\tau(x_1|x_0)}{dx_0}\right]_b = \tau_{ca}\left[\frac{dw}{dx_0}\right]_b = -\tau_{ca}\sqrt{\frac{|U''(b)|}{2\pi\theta}}\,.$$

Hence we find

$$\tau_{ca} = e^{U(b)/\theta}\sqrt{\frac{2\pi}{\theta|U''(b)|}}\int_{-\infty}^{b} e^{-U(x_0)/\theta}\,dx_0.$$

Evaluation of the remaining integral around its peak at $x_0 = a$ leads to the same result as before. (A more exact evaluation of the integral would be spurious in view of the indeterminacy inherent in the definition of τ_{ca}).

5. OTHER SHAPES OF THE POTENTIAL BARRIER

So far it has been tacitly assumed that the potential barrier has a smooth maximum with a second derivative . Other shapes may occur [19, 27], for instance the one in Fig. 3 :

$$U(x) = U(b) - \gamma|x - b| + \mathcal{O}(x - b)^2, \quad \gamma > 0.$$

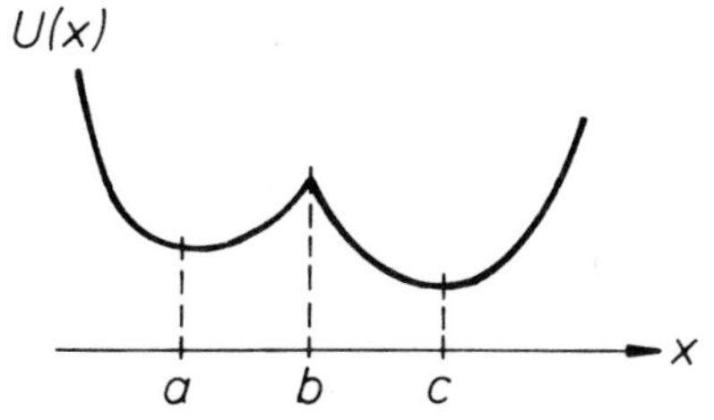

Figure 3. Bistable potential whose barrier is not smooth.

The equation for w in the crucial interval is now

$$\{\gamma\,\mathrm{sgn}(x_0 - b) + \ldots\}\frac{dw}{dx_0} + \theta\frac{d^2w}{dx_0^2} = 0.$$

The proper stretching transformation is in this case

$$x_0 = b + \theta\xi.$$

The equation can again be solved with the conditions $w(\xi\to\infty) = 0$ and $w(\xi\to-\infty) = 1$

$$w = \frac{1}{2}\, e^{-\gamma\xi} \qquad (\xi > 0); \qquad w = 1 - \frac{1}{2}\, e^{\gamma\xi} \qquad (\xi < 0).$$

The normal derivative at b is $[dw/dx_0]_b = -\gamma/2\theta$ and one finds

$$\tau_{ca} = \frac{2}{\gamma}\, e^{U(b)/\theta} \int_{-\infty}^{b} e^{-U(x_0)/\theta}\, dx_0$$

$$= \frac{2}{\gamma} \sqrt{\frac{2\pi\theta}{U''(a)}}\, \exp\left[\frac{U(b) - U(a)}{\theta}\right]$$

The main difference with the smooth barrier is the novel factor $\theta^{1/2}$. It shows that small fluctuations have more trouble overcoming a sharp ridge than a smooth hill. It is left to the reader to investigate potentials of the form

$$U(x) = U(b) - \frac{1}{2}\, \gamma |x - b|^{\delta} \qquad (\gamma, \delta > 0),$$

or with different values for γ on either side of b.

As an extreme case of a very flat potential top we consider the example in Fig. 4. (The letters a, b, c have a slightly different meaning than hitherto). The figure represents the following diffusion problem.

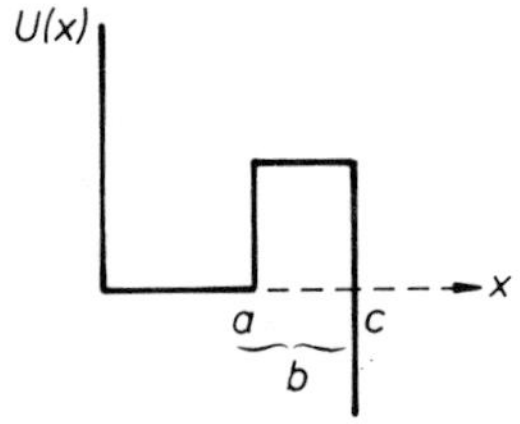

Figure 4. An explicitly solvable case.

$$\frac{\partial P(x,t)}{\partial t} = \theta \frac{\partial^2 P}{\partial x^2} \quad \text{for} \quad 0 < x < a \quad \text{and} \quad a < x < c$$

$$P(a+,t) = e^{-W/\theta}\, P(a-,t), \qquad [\partial P/\partial x]_{a+} = [\partial P/\partial x]_{a-}$$

$$[\partial P/\partial x]_{x=0} = 0, \qquad P(c,t) = 0.$$

(The origin is reflecting and c is absorbing).

There is no narrow interval where everything happens and which can be treated approximately. On the other hand, this can readily be treated exactly. In order to construct the Dynkin equation we have to find the adjoint operator, that is the operator $\underline{W}^+$ such that

$$\int_0^c \tau(x)\, \underline{W}P(x)dx = \int_0^c P(x)\, \underline{W}^+\tau(x)dx.$$

It is easily verified that this identity holds when

$$\underline{W}^+\tau(x_0) = \theta \frac{d^2\tau}{dx_0^2} \qquad (0 < x_0 < a \text{ and } a < x_0 < c)$$

$$\tau(a+) = \tau(a-), \qquad [d\tau/dx_0]_{a+} = e^{W/\theta} \left[\frac{d\tau}{dx_0}\right]_{a-}$$

$$[d\tau/dx_0]_0 = 0, \quad \tau(c) = 0.$$

Hence we have to solve

$$\theta \frac{d^2\tau}{dx_0^2} = -1 \quad \text{for} \quad 0 < x_0 < a \quad \text{and} \quad a < x_0 < c$$

with the above jump and boundary conditions to obtain the first passage time $\tau(c|x_0)$.

The solution is elementary and one finds for $0 < x_0 < a$

$$\tau(c|x_0) = \frac{c^2 - x_0^2}{2\theta} + \frac{a(c-a)}{\theta}(e^{W/\theta} - 1).$$

The first term depends on the precise starting point in the trough but is small. For the escape time from the trough one has therefore (within the margin with which it is defined)

$$\tau_{ca} = \theta^{-1}a(c-a)e^{W/\theta}.$$

An alternative calculation, not using the Dynkin equation but solving the equation for P directly, leads to the same result [28].

6. SPLITTING PROBABILITY

Consider again the bistable potential of Fig. 2 and let x_0 lie near the top b. Then the probability density, which start as a delta distribution at x_0 will split into two parts, which end up as peaks near a and c.

The total probabilities in each of these parts will be called the splitting probabilities $\pi_a(x_0)$ and $\pi_c(x_0)$ and of course $\pi_a(x_0) + \pi_c(x_0) = 1$. They are well-defined on the scale of the relaxation times in each through. On the much longer time scale computed in section 3 they exchange probability and approach their ultimate values π_a^s, π_c^s, see section 3. We compute the $\pi_a(x_0)$, $\pi_c(x_0)$ arising from the original splitting before this slow exchange is effective. It tells us how the fluctuations in an initial unstable state are magnified by the diverging of the subsequent deterministic trajectories and grow into a macroscopic effect. They give rise to a probabilistic description even for an observer who can only distinguish between the states *a* and *c* and does not see the fluctuations themselves.

The definition of these splitting probabilities has a margin of uncertainty of the reciprocal Arrhenius factor. Within this margin it is possible (in the present one-variable case) to find them exactly, but we prefer again an approximate method based on small θ. However, the limiting scenario must be formulated with some care. For $x_0 = b$ one will find of course in the limit $\theta \to 0$ that $\pi_a(b) = \pi_c(b) = 1/2$, because $U(x)$ is in first approximation a symmetric parabola. For $x_0 > b$ one will find in this limit $\pi_a(x_0) = 0$, $\pi_c(x_0) = 1$, because that is the deterministic result. To obtain anything of interest we must take $x_0 - b \sim \theta^{1/2}$, that is, we must allow x_0 to move towards *b* while θ decreases.

In the same way as for the Dynkin equation we argue that

$$\pi_a(x_0) = \int \pi_a(x')T_{\Delta t}(x'|x_0)dx'$$

and conclude

$$0 = \int \pi_a(x') \, \underline{W}(x'|x_0)dx'$$

$$= -U'(x_0)\frac{d\pi_a}{dx_0} + \theta\frac{d^2\pi_a}{dx_0^2} .$$

This is the same equation as previously used for W, but it now applies to $a < x_0 < c$ with boundary conditions

$$\pi_a(a) = 1 , \qquad \pi_a(c) = 0.$$

Our approximate solution is again based on the knowledge that $\pi_a(x_0)$ varies only in an interval $(b - \Delta, b + \Delta)$ whose width 2Δ is of order $\theta^{1/2}$. In that interval the solution is

$$\pi_a(x_0) = \alpha\int_0^{(x_0-b)/\theta^{1/2}} d\xi' \exp[-\tfrac{1}{2}|U''(b)|\xi'^2] + \beta .$$

The integration constants α, β follow from the requirements that $\pi_a(x_0) = 0$ for $x_0 - b >> \theta^{1/2}$ and $\pi_a(x_0) = 1$ for $x_0 - b << -\theta^{1/2}$; one finds

$$\pi_a(x_0) = \sqrt{\frac{|U''(b)|}{2\pi}} \int_{(x_0-b)/\theta^{\frac{1}{2}}}^{\infty} d\xi' \exp[-\frac{1}{2}|U''(b)|\xi'^2] .$$

Clearly this result reduces to the trivial macroscopic one for $|x - b| >> \theta^{1/2}$. It is easy to verify it by means of the exact calculation.

7. THE EVOLUTION OF THE DISTRIBUTION

The more ambitious problem of finding the distribution $P(x,t)$ itself for an initial x_0 close to b has been attacked by Suzuki [29] and others [30, 31, 32]. The basic idea is again that one first treats the neighborhood of the instability in linear approximation and then smoothly joins the result with the solution for the nonlinear region found in section 2. It is clear that the latter solution could not be used exclusively, because it cannot reproduce a splitting of the probability into two peaks. Another approach makes use of the eigenfunctions, but it is clear that many eigenfunctions are needed to represent a sharp peak in P [32, 22].

For simplicity we take the unstable point b as origin and adjust the time scale in such a way that

$$U(x) = -\frac{1}{2} x^2 + \mathcal{O}(x^3).$$

Setting (*) $x = \theta^{1/2}\eta$ one has to lowest order

$$\frac{\partial P}{\partial t} = -\frac{\partial}{\partial \eta} \eta P + \frac{\partial^2 P}{\partial \eta^2} .$$

Of course this equation can be used only when x_0 is of order $\theta^{1/2}$ as well, say $x_0 = \theta^{1/2}\eta_0$ with $\eta_0 > 0$. The solution is

$$P(x,t) = [2\pi\theta(e^{2t} - 1)]^{-\frac{1}{2}} \exp[-\frac{(x - \theta^{\frac{1}{2}}\eta_0 e^t)^2}{2\theta(e^{2t} - 1)}] .$$

Its average and variance are

$$\langle x \rangle_t = \theta^{1/2}\eta_0 e^t, \qquad \langle x^2 \rangle_t - \langle x \rangle_t^2 = \theta(e^{2t} - 1).$$

(*) We here use η rather than ξ, as ξ will be needed later.

As long as $\theta^{1/2}e^t \ll 1$ they are small and it is justified to omit the higher orders in $U(x)$.

Although this distribution is a single-peaked Gaussian it does achieve the splitting of the probability. In fact, the total probability on the right is

$$\int_0^\infty P(x,t)dx = \frac{1}{\sqrt{2\pi}}\int_y^\infty e^{-(1/2)z^2}dz, \qquad y = -\eta_0 \frac{e^t}{\sqrt{e^{2t}-1}} .$$

For $e^t \gg 1$ the integration limit y is constant, so that the total probability is permanently split into

$$\pi_c(x_0) = \frac{1}{\sqrt{2\pi}}\int_{-\eta_0}^\infty e^{-(1/2)z^2}dz, \quad \pi_a(x_0) = \frac{1}{\sqrt{2\pi}}\int_{-\infty}^{-\eta_0} e^{-1/2z^2}\,dz.$$

This is the same result as in the preceding section.

In order to find the distribution for longer times, when the non-linear terms in $U'(x)$ are important, we link up this linearized solution with the one found in section 2. Chapman-Kolmogorov tells us

$$T(x,t|\theta^{1/2}\eta_0,0) = \int_{-\infty}^\infty T(x,t|x_1,t_1)dx_1 T(x_1,t_1|\theta^{1/2}\eta_0,0).$$

Choose t_1 such that $\theta^{1/2}e^{t_1} < 1 < e^{t_1}$. Then the last factor is the solution we just obtained. We are interested in $x > 0$; as the splitting has been completed at t_1 there are no longer transitions across the barrier and $T(x,t|x_1,t_1)$ vanishes for $x_1 < 0$. To find it for $x_1 > 0$ one first has to obtain the solution of $\dot\varphi = -U'(\varphi)$ with initial condition $\varphi(t_1) = x_1$, which we denote $\varphi(t - t_1|x_1)$. Next obtain the variance from

$$\partial_t\langle\xi^2\rangle_t = -2U'(\varphi)\langle\xi^2\rangle_t + 2, \qquad \langle\xi^2\rangle_{t_1} = 0$$

and denote it $\sigma^2(t|t_1)$. Then $T(x,t|x_1,t_1)$ for $t > t_1$, $x > 0$ is approximated by the Gaussian with mean $\varphi(t|x_1,t_1)$ and variance $\theta\sigma(t|t_1)$.

This solves our problem. The explicit result is

$$P(x,t) = \frac{e^{-t_1}}{2\pi\theta\sigma(t|t_1)}\int_0^\infty \exp\left[-\frac{\{x-\varphi(t-t_1|x_1)\}^2}{2\theta\sigma^2(t|t_1)}\right]dx_1 \cdot$$

$$\cdot \exp\left[-\frac{\{x_1 - x_0e^{t_1}\}^2}{2\theta e^{2t_1}}\right].$$

It is understood that $x_0 = \theta^{1/2}\eta_0$ is of order $\theta^{1/2}$. The result is independent of the precise location of t_1 (inside its permitted interval); this is evident from the Chapman-Kolmogorov equation, and can also be verified from the explicit expression. However, our result is not very simple or elegant, but a more satisfactory formulation is given in an Appendix.

Another simplification can be obtained by the following additional approximation [30]. As mentioned before, the spreading of $P(x,t)$ is mainly due to the fluctuations in the vicinity of *b*, from where the deterministic trajectories fan out. That suggests that the subsequent fluctuations may be ignored, that is $T(x,t|x_1,t_1)$ is replaced with a delta function:

$$P(x,t) = \frac{e^{-t_1}}{\sqrt{2\pi\theta}} \int_0^\infty \delta\{x - \varphi(t-t_1|x_1)\}dx_1 \exp\left[-\frac{\{x_1 - x_0 e^{t_1}\}^2}{2\theta e^{2t_1}}\right].$$

This amounts to computing linearly the distribution of x_1 at t_1 and subsequently mapping it into a distribution of x at t using the transformation $x = \varphi(t - t_1|x_1)$. With the aid of the inverse transformation $x_1 = \varphi(t_1 - t\,|x\,)$ the result may also be written

$$P(x,t) = \frac{e^{-t_1}}{\sqrt{2\pi\theta}} \frac{\partial\varphi(t_1 - t|x)}{\partial x} \exp\left[-\frac{\{\varphi(t_1 - t|x) - x_0 e^{t_1}\}^2}{2\theta e^{2t_1}}\right].$$

Unfortunately this result is not independent of the precise choice of t_1, because that is the time at which the fluctuations have wilfully been amputated. The result in the Appendix overcomes this drawback.

8. SEVERAL VARIABLES

The general FP equations with n variables x_i reads

$$\frac{\partial P(x,t)}{\partial t} = -\sum_i \frac{\partial}{\partial x_i} A_i(x)P + \frac{1}{2}\sum_{ij}\frac{\partial^2}{\partial x_i \partial x_j} B_{ij}(x)P \quad .$$

We consider the more special case that $B_{ij}(x)$ is just a scalar constant θ. Nonetheless there are complications that do not exist for one variable. To minimize them we assume in this section that the vector $A_i(x)$ is the gradient of a potential $U(x)$, so that one has again $P^s(x) = C \exp[-U(x)/\theta]$. Moreover we take $n = 2$ and use vector notation:

$$\frac{\partial P(\vec{r},t)}{\partial t} = \vec{\nabla}\cdot(\vec{\nabla}U)P + \theta\vec{\nabla}^2 P.$$

Let Ω be a region bounded by the curve $\partial\Omega$. For any starting point $\vec{r} \in \Omega$ the average time $\tau(\vec{r})$ for arriving at $\partial\Omega$ for the first time is again given by the Dynkin equation

$$-\vec{\nabla}U(\vec{r}) \cdot \vec{\nabla}\tau(\vec{r}) + \theta\nabla^2\tau(\vec{r}) = -1 \quad (r \in \Omega); \quad \tau(\vec{r}) = 0 \quad (\vec{r} \in \partial\Omega).$$

Let U have the shape of a crater as in Fig. 5. We ask for the

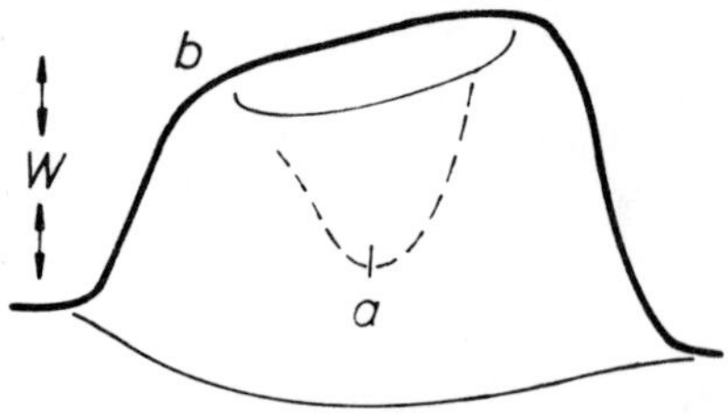

Figure 5. A crater-like potential in two dimensions.

escape time τ_a from the trough around $\vec{a}$, which may be identified with $2\tau(\vec{a})$. The crater ridge has a lowest point $\vec{b}$ of height $W = U(\vec{b}) - U(\vec{a})$. The escape time will contain the Arrhenius factor $e^{W/\theta}$; escape across any higher point of the ridge will be exponentially less likely. We set again $\tau(\vec{r}) = \tau(\vec{a})w(\vec{r})$ so that

$$-\vec{\nabla}U(\vec{r}) \cdot \nabla w(\vec{r}) + \theta\vec{\nabla}^2 w(\vec{r}) \approx 0.$$

Moreover $w(\vec{r}) = 0$ on $\partial\Omega$ and $w(\vec{r}) = 1$ in Ω excepting a narrow zone along $\partial\Omega$.

To compute $w(\vec{r})$ near the pass $\vec{b}$ we first introduce suitable local variables as in Fig. 6. Then, to sufficient order,

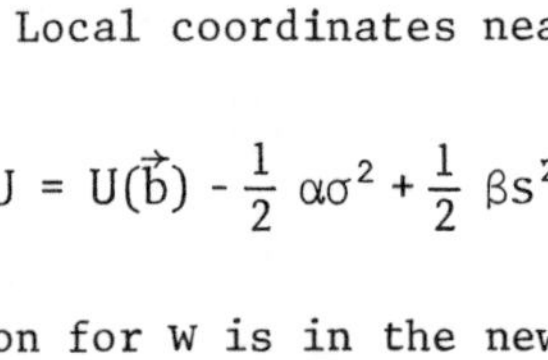

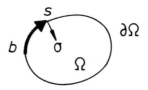

Figure 6. Local coordinates near the mountain pass.

$$U = U(\vec{b}) - \frac{1}{2}\alpha\sigma^2 + \frac{1}{2}\beta s^2 \quad \text{with} \quad \alpha,\beta > 0.$$

The equation for w is in the new variables

$$-\beta s \frac{\partial w}{\partial s} + \alpha\sigma\frac{\partial w}{\partial \sigma} + \theta\left(\frac{\partial^2 w}{\partial s^2} + \frac{\partial^2 w}{\partial \sigma^2}\right) = 0.$$

Again some terms have been omitted which will disappear in the next step anyway. The next step consists of the stretching transformation $\sigma = \theta^{1/2}\xi$, $s = \theta^{1/2}\eta$:

$$- \beta\eta \frac{\partial w}{\partial \eta} + \alpha\xi \frac{\partial w}{\partial \xi} + \frac{\partial^2 w}{\partial \eta^2} + \frac{\partial^2 w}{\partial \xi^2} = 0.$$

This equation is separable; set $w = X(\xi)Y(\eta)$,

$$X'' + \alpha\xi X' + \lambda X = 0 \qquad Y'' - \beta\eta Y' - \lambda Y = 0.$$

Here a new integration constant λ has entered, but it has to vanish in order that $X(\xi)$ approaches 1 for $\xi\to\infty$. In fact, integrate the equation for X :

$$X'(\xi) - X'(0) + \alpha[\xi' X(\xi')]_0^\xi + (\lambda - \alpha) \int_0^\xi X(\xi')d\xi' = 0.$$

If $\xi\to\infty$ this gives

$$\alpha\xi + (\lambda - \alpha)\xi + \text{finite terms} = 0.$$

Hence $\lambda = 0$.

As also $X(0) = 0$ one is left with

$$X(\xi) = \sqrt{\frac{2\alpha}{\pi}} \int_0^\xi e^{-\frac{1}{2}\alpha\xi'^2} d\xi' \quad .$$

Moreover, $Y(\eta)$ must be equal to unity in order that w approaches 1 for each η. Thus

$$w = \sqrt{\frac{2\alpha}{\pi}} \int_0^{\sigma/\sqrt{\theta}} e^{-\frac{1}{2}\alpha\xi'^2} d\xi' \quad .$$

Finally return to the Dynkin equation, multiply it with $e^{-U/\theta}$ and integrate over Ω :

$$\oint_{\partial\Omega} e^{-U/\theta} \frac{\partial\tau}{\partial\sigma} ds = \int_\Omega e^{-U/\theta} d^2\vec{r}.$$

The main contribution to the left-hand side comes from the minimum of U at $\vec{b}$ and gives

$$e^{-U(\vec{b})/\theta} \sqrt{\frac{2\pi\theta}{\beta}} \left(\frac{\partial\tau}{\partial\sigma}\right)_b = 2e^{-U(b)/\theta}\tau(a)\sqrt{\frac{\alpha}{\beta}}.$$

Collecting results we have for the escape time from the potential trough near $\vec{a}$

$$\tau_a = 2\tau(a) = \sqrt{\frac{\beta}{\alpha}}\, e^{W/\theta} \int_\Omega e^{-[U(\vec{r})-U(\vec{a})]/\theta} d^2r$$

$$= e^{W/\theta}\left[\frac{\partial^2 U}{\partial s^2}\right]^{\frac{1}{2}}_{\vec{b}} \cdot \left[-\frac{\partial^2 U}{\partial \sigma^2}\right]^{\frac{1}{2}}_{\vec{b}} \cdot 2\pi\theta \left[\frac{\partial^2 U}{\partial x^2}\frac{\partial^2 U}{\partial y^2} - \left(\frac{\partial^2 U}{\partial x \partial y}\right)^2\right]^{-\frac{1}{2}}_{\vec{a}} .$$

9. SEVERAL VARIABLES, WITHOUT POTENTIAL

In closed isolated physical systems it is usually true that A_i is the gradient of a potential U, as assumed so far. In open or driven systems, and also in nonphysical systems, it may well happen that there exists no such potential. Yet the equations $\dot{x}_i = A_i(x)$ may have stable equilibrium points (point attractors), in which the system can reside for a long time, in spite of fluctuations. One may again ask for the escape time from one attraction basin to another. A generalization of Kramers'method becomes rather awkward [33], but the theory of section 8 can easily be adapted. It has been applied to optical stability [34].

We take again a two-variable FP equation

$$\frac{\partial P(\vec{r},t)}{\partial t} = -\nabla\cdot\vec{A}(\vec{r})P + \theta\nabla^2 P.$$

Under suitable conditions on the behavior of $\vec{A}(\vec{r})$ at large $|\vec{r}|$ there is one stationary solution $P^s(\vec{r})$. We must regard it as known, although it can often not be found except numerically. We suppose that the deterministic equation $\dot{\vec{r}} = \vec{A}(\vec{r})$ has just two attracting points $\vec{a}$, $\vec{c}$ with a separatrix between them, on which lies one unstable equilibrium $\vec{b}$, as in Fig. 7. Obviously P^s must have two peaks at $\vec{a}$ and $\vec{c}$ and for small θ will be very small on the separatrix. Yet its values along the separatrix have a maximum at $\vec{b}$ because on both sides the flow is directed into $\vec{b}$. Thus P^s has a saddle point at $\vec{b}$, see Fig. 8.

Let Ω be the basin of attraction of $\vec{a}$ and $\partial\Omega$ the separatrix. For any $\vec{r}\in\Omega$ let $\tau(\vec{r})$ be the average time of first arrival at $\partial\Omega$. Then

$$\vec{A}\cdot\nabla\tau + \theta\nabla^2\tau = -1 \qquad (r\in\Omega); \qquad \tau(r) = 0 \qquad (r\in\partial\Omega).$$

Moreover we expect τ to be large and almost constant inside Ω excepting

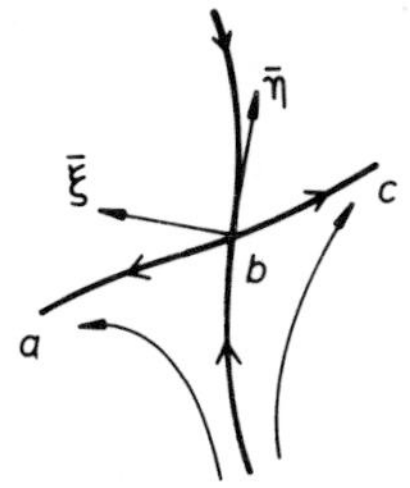

Figure 7. The saddle point on the separatrix, with local coordinates.

a narrow zone along $\partial\Omega$. We have to compute how τ deviates from this constant in the neighbourhood of $\vec{b}$, since that is where the action is.

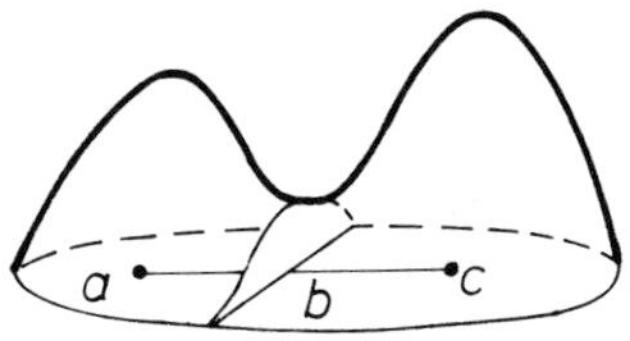

Figure 8. Stationary distribution in two dimensions when there are two attractors.

As before it will be necessary, in order to take the - 1 into account, to utilize the identity

$$\theta \int_{\partial\Omega} P^s \nabla\tau \cdot d\vec{s} = \int_{\Omega} P^s d^2\vec{r}.$$

The integral on the right is regarded as known and in fact is by definition π_a^s. The one on the left receives its main contribution from the neighborhood of $\vec{b}$ and may be approximated by

$$\theta\left(\frac{\partial\tau}{\partial n}\right)_{\vec{b}} \int_{\partial\Omega} P^s ds,$$

where n is the normal to $\partial\Omega$ directed into Ω. We therefore again have to find only the normal derivative of τ at $\vec{b}$.

Assume again $\tau(\vec{r}) = \tau(\vec{a})w(\vec{r})$. Then

$$\vec{A}(\vec{r}) \cdot \nabla w(\vec{r}) + \theta \nabla^2 w(\vec{r}) \approx 0$$

and one expects $w(\vec{r}) = 1$ inside Ω. To solve this equation near $\vec{b}$ set $x = b_x + \theta^{1/2}\xi$, $y = b_y + \theta^{1/2}\eta$, and expand :

$$A_x = \theta^{1/2}(a_{11}\xi + a_{12}\eta) + \mathcal{O}(\theta)$$

$$A_y = \theta^{1/2}(a_{21}\xi + a_{22}\eta) + \mathcal{O}(\theta).$$

The equation for w reduces to

$$(a_{11}\xi + a_{12}\eta)\frac{\partial w}{\partial \xi} + (a_{21}\xi + a_{22}\eta)\frac{\partial w}{\partial \eta} + \frac{\partial^2 w}{\partial \xi^2} + \frac{\partial^2 w}{\partial \eta^2} = 0.$$

To solve this equation we utilize the freedom to rotate the axes, so that ξ, η are transformed into $\bar{\xi}$, $\bar{\eta}$; the sum of both second derivatives is invariant.

If A were a gradient the matrix a_{ij} would be symmetric, and by choosing its eigenvectors as new axes we could transform the equation for w into the same separable form as in section 8. In the present case, however, the eigenvectors need not be orthogonal to one another. One of them is tangent to the separatrix and has an eigenvalue $\lambda_1 < 0$. The other determines the unstable direction (called the 'relevant' eigenvector in renormalization group theory) and has an eigenvalue $\lambda_2 > 0$. We take the former as $\bar{\eta}$-axis, and the normal to the separatrix, directed towards $\vec{a}$, as $\bar{\xi}$-axis, as in Fig. 7. Elementary linear algebra yields the transformed matrix

$$\bar{a}_{11} = \lambda_2; \quad \bar{a}_{12} = 0$$

$$\bar{a}_{21} = a_{21} - a_{12}; \quad \bar{a}_{22} = \lambda_1.$$

Hence the transformed equation for w reads

$$\lambda_2\bar{\xi}\frac{\partial w}{\partial \bar{\xi}} + [(a_{21} - a_{12})\bar{\xi} + \lambda_1\bar{\eta}]\frac{\partial w}{\partial \bar{\eta}} + \frac{\partial^2 w}{\partial \bar{\xi}^2} + \frac{\partial w}{\partial \bar{\eta}^2} = 0.$$

The general solution of this equation is complicated, but the following particular solution obeys all requirements

$$w = \sqrt{\frac{2\lambda_2}{\pi}} \int_0^{\bar{\xi}} e^{-\frac{1}{2}\lambda_2\xi'^2} d\xi' .$$

It vanishes on $\partial\Omega$ and tends to 1 at a distance of order $\theta^{1/2}$ into Ω.

Its normal derivative on $\partial\Omega$ is

$$\frac{\partial w}{\partial n}\sqrt{\frac{2\lambda_2}{\pi\theta}}\,.$$

This was the quantity we need in the equation

$$\pi_a^s = \theta\left(\frac{\partial\tau}{\partial n}\right)_{\vec{b}} \int_{\partial\Omega} P^s ds = \tau(\vec{a})\sqrt{\frac{2\theta\lambda_2}{\pi}}\int_{\partial\Omega} P^s ds.$$

Now the mean first passage time has been expressed in known quantities. Finally the escape time from $\vec{a}$ is

$$\tau_{ca} = 2\tau(\vec{a}) = \pi_a^s \sqrt{\frac{2\pi}{\theta\lambda_2}} \Big/ \int_{\partial\Omega} P^s ds.$$

Notice that the expression for τ_{ac} differs from it only in that π_c^s occurs instead of π_a^s. Thus one has the identity $\pi_a^s/\tau_{ca} = \pi_c^s/\tau_{ac}$, which expresses the obvious fact that in the stationary state the number of hops from $\vec{a}$ to $\vec{c}$ per unit time must be equal to the reverse hops.

10. CONCLUSION AND SUMMARY

This review was confined to random systems that are described by a Fokker-Planck equation with constant scalar diffusion coefficient θ. The equation describes the overdamped motion of a Brownian particle in an external field. If the motion is not overdamped its velocity or momentum has to be included [19, 35] and the problem becomes essentially more difficult [36].

Unless the FP equation can be solved exactly one must resort to an approximation method. For a systematic approximation a small parameter is needed. Physically the most natural parameter is the temperature θ, which governs the size of the fluctuations, keeping the external field constant. Other limiting scenarios, in which the external field varies with θ, lead to other results [37].

The limit $\theta = 0$ yields a deterministic equation. If this equation is stable (has a single point attractor) a singular perturbation expansion in $\theta^{1/2}$ yields a description of the fluctuations, which to lowest order are Gaussian.

If the deterministic equation is bistable the escape time (or transmission rate) could be found. For one-dimensional system the splitting probability and the probability distribution itself were also computed. Each time the essential step is the decomposition of the total range into a linearized region around the instability and a nonlinear region were the instability no longer matters.

There are of course many other problems. Other instabilities are possible for instance limit cycles [38, 13]. There are critical points, where on varying a control parameter the instability ceases to exist [7, 39]. And there are systems that cannot be described by our FP equation. Yet discrete one-step processes (also called birth-and death processes) can be treated in much the same way [18, 20, 40].

APPENDIX TO SECTION 3

The reason is that the paths of a Markov process described by a FP equation are continuous with probability 1. (That is why mathematicians call such processes somewhat confusingly 'continuous Markov processes'). We give an example to show that for other processes it is not true. Take a process whose transition probability per unit time is an integral kernel $W(x|x')$. The process is confined to the range $0 < x < \infty$ with a reflecting boundary at $x = 0$. The master equation is

$$\frac{\partial P(x,t)}{\partial t} = \int_0^\infty W(x|x')P(x')dx' + \int_0^\infty W(-x|x')dx' - P(x)\int_{-\infty}^\infty W(x'|x)dx'.$$

The first passage time $\tau(x_1|x_0)$ obeys in $0 < x_0 < x_1$

$$-1 = \int_0^{x_1} \tau(x_1|x_0')W(x_0'|x_0)dx_0' + \int_0^{x_1} \tau(x_1|x_0')W(-x_0'|x_0)dx_0' -$$

$$- \tau(x_1|x_0)\int_{-\infty}^\infty W(x'|x_0)dx'.$$

This integral equation determines $\tau(x_1|x_0)$ without additional boundary conditions.

For the special choice $W(x|x') = e^{-|x-x'|}$ it can be solved explicitly thanks to the identity

$$\left(\frac{d^2}{dx^2} - 1\right)e^{-|x-x'|} = -2\delta(x-x').$$

Applying this differential operator to the equation one gets

$$\left(\frac{d^2}{dx_0^2} - 1\right)\{-1 + 2\tau(x_1|x_0)\} = -2\tau(x_1|x_0).$$

Hence $\tau(x_1|x_0) = -\frac{1}{4}x_0^2 + ax_0 + b$, with two integration constants a, b, which will depend on x_1. They are found by substituting this $\tau(x_1|x_0)$

back into the integral equation; the result is that the equation is satisfied provided that $a = 0$, $b = \frac{1}{4}x_1^2 + \frac{1}{2}x_1 + \frac{1}{2}$, so that

$$\tau(x_1|x_0) = \frac{1}{4}(x_1^2 - x_0^2) + \frac{1}{2}(x_1 + 1).$$

This does not vanish for $x_0 = x_1$. The reason is that a particle starting at x_1 has a non-zero probability to jump back into the interval. If one takes

$$W(x|x') = \lambda^3 e^{-\lambda|x-x'|}$$

the equation can be reduced to the previous one, but one can now go to the limit $\lambda \to \infty$. The master equation then reduces to the standard diffusion equation while in the same limit the second term in $\tau(x_1|x_0)$ vanishes. Thus the boundary condition $\tau(x_1|x_1) = 0$ applies to the case of diffusion only.

APPENDIX TO SECTION 7

The probability distribution in the c-trough $(x > 0)$ at t_1 is proportional to $T(x,t_1|\theta^{1/2}\eta_0,0)$, but has to be normalized by dividing by $\pi_c(x_0)$.

$$P_c(x,t) = \frac{e^{-t_1}}{\pi_c\sqrt{2\pi\theta}} \exp\left[-\frac{\{x-x_0e^{t_1}\}^2}{2\theta e^{2t_1}}\right]\Theta(x),$$

where Θ is the Heaviside step function. Its average is

$$\langle x\rangle_c = e^{t_1}\left[x_0 + \frac{\theta^{1/2}e^{-(1/2)\eta_0^2}}{\pi_c\sqrt{2\pi}}\right].$$

The second term in the brackets, Δx_0, is due to the fact that the negative values of x have been cut off. The variance of the distribution is

$$\langle x^2\rangle_c - \langle x\rangle_c^2 = \theta e^{2t_1}\left[1 - \eta_0\frac{e^{-\eta_0^2/2}}{\pi_c\sqrt{2\pi}} - \left\{\frac{e^{-\eta_0^2/2}}{\pi_c\sqrt{2\pi}}\right\}^2\right],$$

which of course has been reduced by the cutting off. (It can be verified algebraically that this expression is positive).

The point is that $\langle x\rangle_c$ as a function of t_1 is again a solution of the deterministic equation (in the linear regime). Its initial value at $t_1 = 0$, however, is not the original x_0, but $x_0 + \Delta x_0$. Similarly the variance is again a solution of the equation for $\langle\xi^2\rangle_t$, but its initial value is not zero. Hence it is permissible to utilize for $t > t_1$ the same solution as in section 2, provided one determines φ and $<\xi^2>_t$ with these modified initial values. In that way one avoids the explicit use of the Chapman-Kolmogorov equation, while the result is insensitive to the precise choice of t_1.

REFERENCES

[1] R. L. Stratonovich, Topics in the Theory of Random Noise I (Gordon and Breach, New York 1963); D. R. Cox and H. D. Miller, The Theory of Stochastic Processes (Methuen, London 1965).
[2] N. G. van Kampen, Stochastic Processes in Physics and Chemistry (North-Holland, Amsterdam 1981).
[3] C. W. Gardiner, Handbook of Stochastic Methods (Springer, Berlin 1983); H. Risken, The Fokker-Planck Equation (Springer, Berlin 1984).
[4] S. Chandrasekhar, Rev. Mod. Phys. 15, 1 (1943); M. C. Wang and G. E. Uhlenbeck, Rev. Mod. Phys. 17, 323 (1945). (Both article are reprinted in: N. Wax ed., Noise and Stochastic Processes, Dover, New York 1954)). H. Haken, Rev. Mod. Phys. 47, 67 (1975); I. Oppenheim, K. E. Shuler, and G. H. Weiss, Stochastic Processes in Chemical Physics: The Master Equation (M.I.T. Press, Cambridge, Mass. 1977); P. Hänggi and H. Thomas, Physics Reports 88, 207 (1982).
[5] R. F. Pawula, Phys. Rev. 162, 186 (1967).
[6] N. G. van Kampen, Can. J. Phys. 39, 551 (1961) and in: Adv. Chemical Physics 34 (Wiley, New York 1976); D. R. McNeil, Biometrika 59, 494 (1972); T. G. Kurtz, J. Chem. Phys. 57, 2976 (1972).
[7] R. Kubo, K. Matsuo, and K. Kitahara, J. Stat. Phys. 9, 51 (1973).
[8] N. S. Goel, S. C. Maitra, and E. W. Montroll, Rev. Mod. Phys. 43, 231 (1971); N. S. Goel and N. Richter-Dyn, Stochastic Models in Biology (Acad. Press, New York 1974); E. W. Montroll and W. W. Badger, Introduction to Quantitative Aspects of Social Phenomena (Gordon and Breach, New York 1974); G. Nicolis and I. Prigogine, Self-organization in Nonequilibrium System (Wiley, New York 1977); H. Haken, Synergetics (Springer, Berlin 1977); W. Weidlich and C. Haag, Concepts and Models of a Quantitative Sociology (Springer Berlin 1983).
[9] A. Einstein, Ann. Physik [4] 17, 549 (1905); 19 371 (1906). Translations in: A. Einstein, Investigations on the Theory of the Brownian Movement (Methuen, London 1926).
[10] H. B. Callen and R. F. Greene, Phys. Rev. 86, 702 and 88, 1387 (1952); S. R. de Groot and P. Mazur, Non-Equilibrium Thermodynamics (North-Holland, Amsterdam 1962); L.E. Reichl, A Modern Course in Statistical Physics (Univ. of Texas, Austin, Texas 1980).
[11] R. E. O'Malley Jr., Introduction to Singular Perturbations (Acad. Press, New York 1974); W. Eckhaus, Matched Asymptotic Expansions and Singular Perturbations (North-Holland, Amsterdam 1973).
[12] R. F. Rodriguez and N. G. van Kampen, Physica 85A, 347 (1976); H. Grabert and M. S. Green, Phys. Rev. A19, 1747 (1979); H. Grabert, R. Graham, and M. S. Green, Phys. Rev. A21, 2136 (1980).

[13] N. G. van Kampen, in: Adv. Chem. Phys. 34 (Wiley, New York 1976).
[14] J. L. Doob, Annals Math 43, 351 (1942). (Also reprinted in Wax, see ref [4]). R. F. Fox and G. E. Uhlenbeck, Phys. Fluids 13, 1893 (1970).
[15] M. San Miguel, Z. Phys. B33, 307 (1979).
[16] H. Eyring, J. Chem. Phys. 3, 107 (1935).
[17] I. Procaccia and J. Ross, Suppl. Prog. Theor. Phys. 64, 244 (1978).
[18] R. Becker and W. Döring, Ann. der Phys. [5] 24, 719 (1935); J. S. Langer, Annals Phys. 54, 258 (1969); J. L. Katz and M. D. Donohue, in: Adv. Chem. Physics 40 (Wiley, New York 1979).
[19] H. A. Kramers, Physica 7, 284 (1940).
[20] R. Landauer, J. Appl. Phys. 33, 2209 (1962).
[21] G. B. Griffiths, C. Y. Wang, and J. S. Langer, Phys. Rev. 149, 301 (1966).
[22] B. Caroli, C. Caroli, and B. Roulet, J. Stat. Phys. 21 415 (1979).
[23] R. S. Larson and M. D. Kostin, J. Chem. Phys. 69, 4821 (1978).
[24] N. G. van Kampen, in: Fundamental Problems in Statistical Mechanics IV (E. G. D. Cohen and W. Fiszdon eds., Nauka, Warszawa 1978); K. Voigtlaender and H. Risken, J. Stat. Phys. 40, 397 (1985).
[25] E. B. Dynkin and A. A. Juschkewitz, Sätze und Aufgaben über Markoffsche Prozesse (Springer, Berlin 1969).
[26] Z. Schuss and B. J. Matkovsky, SIAM J. Appl. Math. 35, 604 (1979); Z. Schuss, Theory and Applications of Stochastic Differential Equations (Wiley, New York 1980). Also: P. Talkner and D. Ryter, Phys. Letters 88A, 162 (1982).
[27] H. Brand and A. Schenzle, Phys. Lett. 68A 427 (1978).
[28] N. G. van Kampen, Helv. Phys. Acta (to be published).
[29] M. Suzuki, J. Stat. Phys. 16, 11 (1977) and in: Adv. Chem. Physics 46 (Wiley, New York 1981).
[30] F. Haake, Phys. Rev. Letters 41, 1685 (1978); F. de Pasquale, P. Tartaglia, and P. Tombesi, Physica 98A, 581 (1979).
[31] Y. Saito, J. Phys. Soc. Jap. 41, 388 (1976); G. Hu and Q. Zheng, Phys. Letters 110A, 68 (1985).
[32] H. Risken and D. Wollmer, Z. Phys. 204, 240 (1967); H. Tomita, A. Ito, and H. Kidachi, Prog. Theor. Phys. 56, 786 (1976).
[33] C. W. Gardiner, J. Stat. Phys. 30, 157 (1983).
[34] P. Talkner and P. Hänggi, Phys. Rev. A29, 768 (1984).
[35] O. Klein, Arkiv Mat. Astr. Pys. 16, no. 5 (1922); J. Meixner, Z. Phys. 149, 624 (1957).
[36] R. Landauer and J. A. Swanson, Phys. Rev. 121, 1668 (1961); C. Cercignani, Annals Phys. 20, 219 (1962); M. A. Burschka and U. M. Titulaer, Physica 112A, 315 (1982); W. Renz, Z. Phys. B59, 91 (1985); W. Renz and F. Marchesoni, Phys. Letters 112A, 124 (1985).
[37] N. G. van Kampen, J. Stat. Phys. 17, 71 (1977); C. Chaturvedi and F. Shibata, Z. Phys. B35 297 (1979); F. Haake, Z. Phys. B48, 31 (1982).
[38] R. Lefever and G. Nicolis, J. Theor. Biol. 30, 267 (1971); K. Tomita, T. Ohta, and H. Tomita, Prog. Theor. Phys. 52 1744 (1974).
[39] H. Dekker and N. G. van Kampen, Phys. Letters 73A, 374 (1979); H. Dekker, Physica 103A, 55 and 80 (1980).
[40] N. G. van Kampen, Suppl. Prog. Theor. Phys. 64, 389 (1978);

P. Hänggi, H. Grabert, P. Talkner, and H. Thomas, Phys. Rev. A29, 371 (1984).

WEAK NOISE LIMIT AND NONEQUILIBRIUM POTENTIALS OF DISSIPATIVE DYNAMICAL SYSTEMS

R. Graham
Fachbereich Physik, Universität - GHS, Essen
West Germany

ABSTRACT

The possibility of generalizing thermodynamic potentials to non-equilibrium steady states of dissipative dynamical systems subject to weak stochastic perturbations is discussed. In the first part the general formal approach is presented and illustrated by a number of examples where a rather straightforward definition of non-equilibrium potentials is possible. In the second part a number of generic properties of non-equilibrium potentials is discussed, such as only piecewise differentiability and extremum properties. We conclude with a nontrivial example involving the coexistence of an attracting fixed point and a limit cycle.

1. GENERAL DEFINITION, SPECIAL EXAMPLES

1.1 Introduction

There is a deep connection in physics between fluctuation phenomena on one side and extremum principles on the other. One may go even to the extreme and state that all fundamental extremum principles in physics have their origin and find their explanation in underlying fluctuation phenomena. E.g. extremum principles of equilibrium thermodynamics, which are all different versions of the statement that the state of thermodynamic equilibrium in a closed system corresponds to the maximum of entropy, are explained by the nature of thermodynamic fluctuations and express the fact that in a system with small relative fluctuations the equilibrium state is the one which maximizes the probability. Or Hamilton's principle of extreme action for conservative classical systems is understood in the framework of the path integral description to result from quantum fluctuations probing an infinitesimal neighborhood of the classical path.

Large classes of systems exist in physics for which extremum principles are not commonly available. An important class is defined by classical dissipative (i.e. non-conservative) dynamical systems in steady states outside the realm of thermodynamic equilibrium. Here, neither the

E. Tirapegui and D. Villarroel (eds.), Instabilities and Nonequilibrium Structures, 271–290.

extremum principles of equilibrium thermodynamics, nor the extremum principles of conservative Hamiltonian dynamics apply.

However, intrinsic fluctuations exist also in such systems. E.g. a fluid in a convecting state is subject to thermodynamic fluctuations just as a fluid at rest. Or an externally pumped laser in CW-operation is subject to quantum fluctuations (i.e. spontaneous emission) just as a thermal light source. Also, these fluctuations are often only very minor perturbations, just as in the examples of thermodynamics or classical systems mentioned before. Therefore if seems plausible that one should be able to derive extremum principles also for such systems. They should characterize, e.g.- stable attractors of a system as extrema of a potential, similar to the entropy in thermodynamic equilibrium, and should provide Lyapunov functions for analyzing their stability. In these lectures I plan to review work which we have done over the last years in this direction and present principal results together with simple applications (see also [1,2])

1.2 Stochastic equilibrium thermodynamics

It is useful to begin by recalling the basic structure of equilibrium thermodynamics for a discrete system described by the macroscopic variables $q^\nu(t)$, $\nu = 1,\ldots n$ [3-7]. The macroscopic equations with thermodynamic fluctuations are written in the form

$$\dot{q}^\nu = K^\nu(q) + g_i^\nu \xi^i(t) \tag{1.1}$$

with the Gaussian white noise

$$<\xi^i(t)> = 0, \ <\xi^i(t)\xi^j(t')> = \eta\delta^{ij}\delta(t-t') \tag{1.2}$$

We interprete (1.1) in the sense of Ito, for concreteness. The drift $K^\nu(q)$, in thermodynamics, can be split into a reversible part $r^\nu(q)$ and a dissipative part according to

$$K^\nu = r^\nu - \frac{1}{2}Q^{\nu\mu}\frac{\partial\phi}{\partial q^\mu} \tag{1.3}$$

Here

$$Q^{\nu\mu} = \sum_i g_i^\nu g_i^\mu \tag{1.4}$$

is a non-negative, symmetric transport matrix and $\phi(q)$ is a thermodynamic potential with q as variables. (Strictly, $\phi(q)$ is a "coarse-grained" thermodynamic potential, since for its statistical mechanical definition the q must be considered fixed, while all other fluctuations are averaged over). The reversible drift conserves, the dissipative drift decreases the potential

$$r^{\nu} \frac{\partial \phi}{\partial q^{\nu}} = 0; \quad -\frac{1}{2} Q^{\nu\mu} \frac{\partial \phi}{\partial q^{\nu}} \frac{\partial \phi}{\partial q^{\mu}} \leqq 0 \tag{1.5}$$

An example is furnished by the equations of motion of a simple incompressible fluid (in Fourier representation, e.g. with periodic boundary conditions, in order to have a discrete system, even though $n \to \infty$). ϕ is given by the quadratic expression in the velocity field giving the kinetic energy, the convection and pressure terms yield r^{ν} and satisfy (1.5), and the viscosity term is expressible as a velocity gradient of the kinetic energy and satisfies (1.5).

The basic idea of the work to be described is to generalize this structure of the equilibrium theory for non-equilibrium systems [5].

1.3 Non-equilibrium potentials, statistical significance

We now consider a general dynamical system of the form (1.1), whose drift K^{ν} and diffusion $Q^{\nu\mu}$ are no longer restricted by the requirements of equilibrium thermodynamics. Thus, the form (1.3) can no longer be achieved by simply splitting K^{ν} into its reversible and irreversible part. However, we can still carry over eq. (1.3) to non-equilibrium systems by considering this equation, together with eq. (1.5), as providing a definition of r^{ν} and ϕ. Thus, we assume $K^{\nu}(q)$ and $Q^{\nu\mu}$ to be given by (1.1) and look for a potential ϕ which satisfies [1,8-12]

$$\frac{1}{2} Q^{\nu\mu} \frac{\partial \phi}{\partial q^{\nu}} \frac{\partial \phi}{\partial q^{\mu}} + K^{\nu} \frac{\partial \phi}{\partial q^{\nu}} = 0 \tag{1.6}$$

which follows from (1.3) and (1.5). If we interpret q^{ν} as generalized coordinates, ϕ as minimizing action and $p_{\nu} = \partial\phi/\partial q^{\nu}$ as the canonically conjugate momenta we have an equivalent mechanical interpretation of (1.6) as a Hamilton Jacobi equation corresponding to the Hamiltonian

$$H(q,p) = \frac{1}{2} Q^{\nu\mu} p_{\nu} p_{\mu} + K^{\nu} p_{\nu} \tag{1.7}$$

and the canonical equations

$$\dot{p}_{\nu} = -\frac{\partial H}{\partial q^{\nu}}, \quad \dot{q}^{\nu} = \frac{\partial H}{\partial p_{\nu}} \tag{1.8}$$

forming the characteristics of eq. (1.6).

It follows from (1.5) that we want a solution ϕ of (1.6) which is minimal in attractors. Reversing time in (1.1) we can argue similarly that ϕ should be maximal in repellors and stationary in saddles. These requirements are boundary conditions which select ϕ from the manifold of solutions of (1.6). Eq. (1.6) has a simple statistical interpretation as the weak noise limit of the Fokker Planck equation for the probability density $P(q,\eta)$, corresponding to eq. (1.1). Eq. (1.6) is obtained from that Fokker Planck equation by the ansatz (for $\eta \to 0$)

$$P_\infty \sim \exp(-\frac{\phi(q)}{\eta}) \tag{1.9}$$

for the probability density in the steady state, $P_\infty(q,\eta)$, and taking the limit $\eta \to 0$

$$\phi(q) = -\lim_{\eta\to 0} \eta \ln P_\infty(q,\eta) \tag{1.10}$$

In the same manner a statistical interpretation of the quantity r^ν of eq. (1.3) is obtained, namely

$$r^\nu(q) = \lim_{\eta\to 0} \frac{g^\nu(q,\eta)}{P_\infty(q,\eta)} \tag{1.11}$$

where $g^\nu(q,\eta)$ is the probability current density corresponding to (1.1) in the statistical steady state. It is clear from all these properties that $\phi(q)$ represents a genuine generalization of a thermodynamic potential to non-equilibrium steady states.

1.4 Neighborhood of a monostable steady state

Let us assume, as a simplest example, that $q^\nu = 0$, $\nu = 1,\dots,n$ is a stable fixed point of the deterministic ($\eta = 0$) version of eq. (1.1), $\dot{q}^\nu = K^\nu(q)$, and let us expand around this point in order to investigate its neighborhood [1],

$$\begin{aligned} K^\nu(q) &= A^\nu{}_\mu q^\mu + \dots \\ Q^{\nu\mu}(q) &= Q^{\nu\mu} + \dots \end{aligned} \tag{1.12}$$

The solutions of eq. (1.6) with eq. (1.12) satisfiying the boundary conditions $p_\nu = \partial\phi/\partial q^\nu = 0$ for $q^\nu = 0$ is given by

$$\phi = \frac{1}{2}(\sigma^{-1})_{\nu\mu} q^\nu q^\mu + \dots \tag{1.13}$$

where $\sigma^{\nu\mu}$ satisfies

$$A^\alpha{}_\mu \sigma^{\mu\beta} + A^\beta{}_\mu \sigma^{\mu\alpha} = Q^{\alpha\beta} \tag{1.14}$$

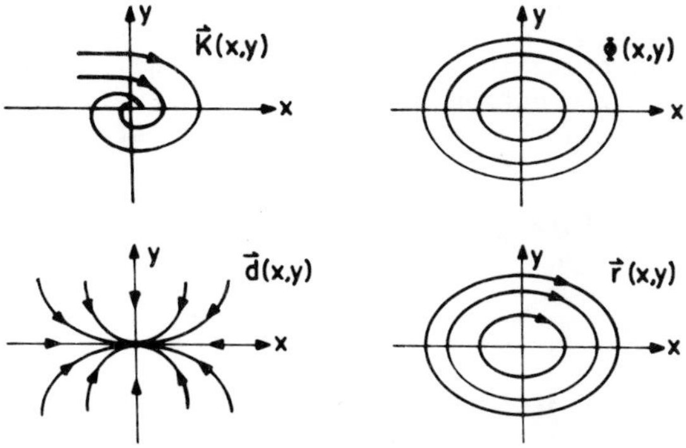

Figure 1. The drift $\vec{K}$, and its two components $\vec{r}$ and $\vec{d}$ together with the equipotential lines of ϕ near a stable fixed point. (Schematic plot from [1]). (With permission from Springer Verlag)

Fig. 1 shows, for an n = 2 dimensional example with the coordinates x, y, the field-lines of $\vec{K}(x,y)$, $\vec{r}(x,y)$ and of

$$\vec{d}(x,y) = -\frac{1}{2} Q\vec{\nabla}\phi(x,y) \tag{1.15}$$

together with the contours $\phi(x,y)$ = const. The properties (1.5) are apparent.

1.5 Application to Hopf and pitchfork bifurcations

Let us consider a system in the neighborhood of a supercritical Hopf bifurcation. A standard example is a single-mode laser near threshold [13,15]. We make use of two important results of center manifold theory [16] which state, in physical terms

(i) On a sufficiently long time-scale $\Delta t = O(\frac{1}{\varepsilon})$ (where $\varepsilon = |\mu-\mu_c|/\mu_c \ll 1$, μ is the control parameter and μ_c the critical value where Hopf bifurcation occurs) a reduced description of the system in a 2-dimensional configuration space (center manifold) is possible.

(ii) The dynamics on the 2-dimensional center manifold is governed by an amplitude equation (normal form)

$$\dot{\beta} = ((\mu-\mu_c) - i\omega)\beta - (b-i\Omega)|\beta|^2\beta + \sqrt{\eta Q}\,\xi(t) \tag{1.16}$$

Here β is a complex amplitude, ω, b, Ω, Q are real parameters (following from a microscopic description of the system) and $\xi(t)$ is a complex Gaussian white noise. Sufficiently close to the bifurcation point ($\varepsilon \ll 1$) it is sufficient to take (cf. [37])

$$\langle\xi(t)\rangle = 0 \;,\; \langle\xi^*(t)\xi(t')\rangle = \delta(t-t') \tag{1.17}$$

all other second order correlation functions vanishing.

Supercritical Hopf bifurcation require $b > 0$. Eq. (1.6) has a straightforward solution, in the case of (1.16), satisfying $\partial\phi/\partial\beta = 0 = \partial\phi/\partial\beta^*$,

$$\phi = \frac{1}{Q}(-(\mu-\mu_c)|\beta|^2 + \frac{b}{2}|\beta|^4) \tag{1.18}$$

and we find from (1,3)

$$r_\beta = -i\omega\beta + i\Omega|\beta|^2\beta \tag{1.19}$$

The properties (1.5) are again apparent. The phase-degeneracy of (1.18) and its Landau form put in evidence the close relationship between phase transitions of mean-field type, like in superconductors, and Hopf bifurcations in non-equilibrium steady states. The laser serves here as the standard example, where this analogy has been first observed [17,18].

A very similar analysis applies to pitchfork bifurcations [16]. The appearance of convection rolls at the Bénard instability of thermal convection serves here as a standard example [19,20,21]. In spatially confined systems the only difference to eqs. (1.16) and (1.18) is that the center manifold is 1-dimensional, i.e. β in (1.16) becomes real, and $\omega = 0 = \Omega$ must vanish. A more interesting situation occurs in large aspect ratio systems, where the amplitude becomes spatially varying. This case was considered from the point of view of non-equilibrium potentials in ref. [22], based on earlier work by Segel [21] and Newell and Whitehead [20]. Subsequently non-equilibrium potentials in such systems have been considered and used by a number of authors, e.g. [23-25]. Particularly interesting is the case where the structure appearing at supercriticality contains defects. A non-equilibrium potential for this case was first considered in [26] and was partially derived and extended in [25]. E.g. let us consider a plane convecting fluid layer with a lattice of rolls containing dislocations. The vertical component of the streaming velocity in such case can be written as

$$V_z = W(x,y,t)e^{ik_c\varphi(x,y,t)}f(z) + c.c. \tag{1.20}$$

where x,y are horizontal coordinates in the layer, k_c is a critical wave number, $f(z)$ a function determined by boundary conditions, W is a slowly varying complex amplitude, and $\varphi(x,y,t)$ is a slowly varying phase function which determines the local normal $\vec{n}$ of the rolls

$$\vec{n} = \vec{\nabla}\varphi(x,y) \;,\; |\vec{n}| = 1 \tag{1.21}$$

It follows that $\vec{n}$ satisfies locally

$$\vec{\nabla} \times \vec{n} = 0 \tag{1.22}$$

but

$$\oint \vec{n} \cdot d\vec{s} = \frac{2\pi}{k_c} m \tag{1.23}$$

where the integer m denotes the number of enclosed dislocations. Sufficiently close to the bifurcation point a non-equilibrium potential can again be given in terms of W(x,y) and $\vec{n}(x,y)$ [25,26]

$$\phi = \frac{1}{Q} \int d^2x \{ -\varepsilon |W(x,y)|^2 + \frac{b}{2} |W(x,y)|^4 + \xi_0^2 |(\vec{n} \cdot \vec{\nabla} + \frac{1}{2}(\vec{\nabla} \cdot \vec{n}) - \frac{i}{2k_c}(\vec{n} \times \vec{\nabla})^2) W|^2 \} \tag{1.24}$$

A discussion of defects, surface effects, and textures in non-equilibrium systems is possible, using this potential just like a free energy. An analysis along these lines was given by Cross [25].

1.6 Optical bistability

Another physical example to which the present formalism has been applied is optical bistability [27]. This is an interesting example, also from the general point of view adopted in these lectures, because global aspects play a role in determining the relative stability of the two competing attractors, in this case, beyond the purely local analysis which was sufficient for the examples of the preceding sections. For a full treatment of this example I refer to the literature [27]. Here it is sufficient to state the dynamical equations

$$\dot{E} = K(E) + \sqrt{\eta Q}\, \xi(t)$$
$$K(E) = -(1 + i\delta)E - \Gamma^2 \frac{1 + i\Delta}{1 + |E|^2} E + E_0 \tag{1.25}$$

where E is a complex field amplitude, $\xi(t)$ is Gaussian white noise as in (1.17), Q, δ, Γ, Δ are real parameters, E_0 is a real externally applied field amplitude serving as a control parameter. The Hamilton Jacobi equation

$$K(E)\frac{\partial \phi}{\partial E} + K^*(E)\frac{\partial \phi}{\partial E^*} + Q\frac{\partial \phi}{\partial E}\frac{\partial \phi}{\partial E^*} = 0 \tag{1.26}$$

is here easily solved for $E_0 = 0$, where the solution satisfying the desired boundary conditions is given by

$$\phi_0 = \frac{1}{Q}(|E|^2 + \Gamma^2 \ell n(1+|E|^2)) \tag{1.27}$$

For $E_0 \neq 0$ an exact solution

$$\phi = \phi_0 + \frac{E_0}{Q\sqrt{1+\delta^2}}(E\exp(i\arctan\Delta) + c.c.) \tag{1.28}$$

is still obtained for $\delta = \Delta = 0$ and for $\delta = \Delta$ [27]. Otherwise various perturbative solutions of eq. (1.26) have been derived in [27]. In fig. 2 the field-lines of

$$\vec{r} = \begin{pmatrix} \delta(E_y + \frac{\delta E_0}{1+\delta^2}) + \delta\Gamma^2 \frac{E_y}{1+|E|^2} \\ \\ -\delta(E_x - \frac{E_0}{1+\delta^2}) - \delta\Gamma^2 \frac{E_x}{1+|E|^2} \end{pmatrix} \tag{1.29}$$

and of

$$\vec{d} = -\frac{1}{2}Q \begin{pmatrix} \frac{\partial\phi}{\partial E_x} \\ \\ \frac{\partial\phi}{\partial E_y} \end{pmatrix} \tag{1.30}$$

are sketched, together with the equipotential contours of ϕ, for a bistable situation. The relative depth of the two valleys of ϕ corresponding to the two coexisting attractors, indicate their relative stability. in a situation where the potential minima are sufficiently shallow for transitions from one attractor to the other one to occur in reasonable time intervals. The two potential valleys are joined continuously at the saddle point in between. It should be noted that this continuity of the potential at the saddle point fixes the potential difference in the two attractors. The competition of the two attractors expressed by this potential difference is a global property of bistable system. Computations of first passage times from one attractor to the other one can be based on the results for the nonequilibrium potential. For the present example this has been done in [28].

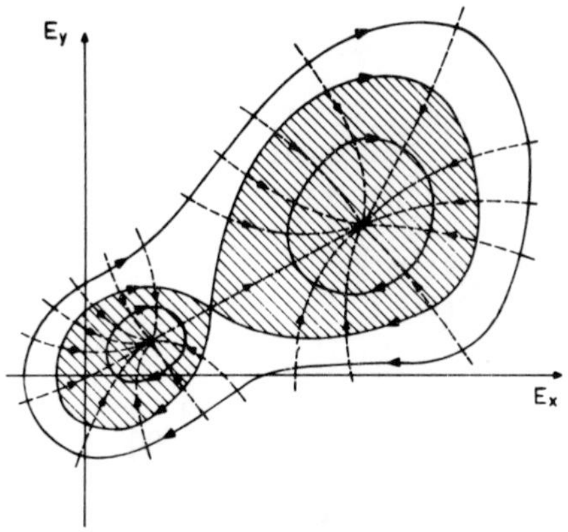

Figure 2. Equipotential lines of (1.28) with field lines of $\vec{r}$ (1.29) (full lines) and $\vec{d}$ (1.30) (dashed lines). Regions where the dynamics is trapped by the two attractors are hatched. Outside the hatched regions the two basins of attractions are interlaced in a complicated way (Schematic plot).

2. PIECEWISE DIFFERENTIABLE NON-EQUILIBRIUM POTENTIALS

2.1 Phase space structure of non-equilibrium potentials

After having seen some simple examples of non-equilibrium potentials in section 1 let us now return to general considerations, and investigate the phase space structure of the zero-energy surface $H \equiv 0$ of the Hamiltonian (1.7) [9,10]. In fig. 3 we consider a schematic plot of a $(2n-2)$-dimensional Poincaré cross section of $H \equiv 0$ at a fixed value of the coordinate q_n.

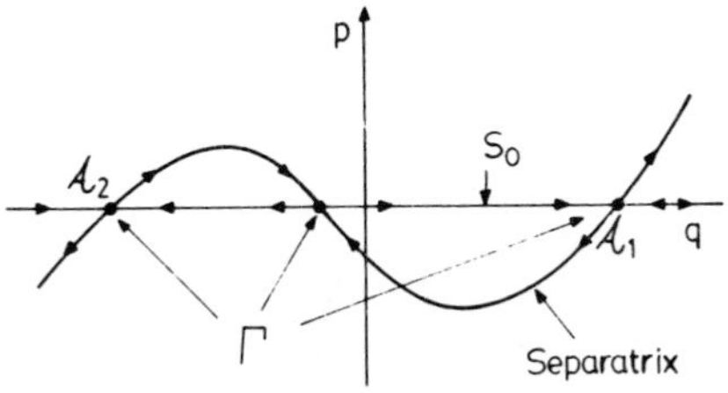

Figure 3. Phase space structure of the Hamiltonian (1.7) for the case where a smooth non-equilibrium potential exist. (Schematically, after [2]).

Schematically, we consider the case where two attractors (not necesarily fixed points) of the deterministic system $\dot{q}^\nu = K^\nu(q)$ coexist. Clearly, the dynamics $\dot{q}^\nu = K^\nu$ is restricted, in the Hamiltonian system, to the surface $S_0 : p_\nu = 0, \nu = 1, \ldots, n$, and all its limit sets Γ therefore are on this surface. In the $2n$-dimensional phase space of the Hamiltonian system the limit sets Γ are hyperbolic objects which are connected not only by the n-dimensional surface S_0 but also by n-dimensional invariant manifolds transverse to S_0 forming separatrices

$$p_\nu = p_\nu(q)$$

As these separatrices, by construction, pass through the limit sets Γ they satisfy automatically $p_\nu(\Gamma) = 0$. Identifying $\partial\phi/\partial q^\nu = p_\nu(q)$ we therefore define a function $\phi(q)$, solving the Hamilton Jacobi equation, which satisfies the required boundary condition that its first derivatives vanish in the limit sets. A smooth, differentiable potential therefore exists, if there is a smooth separatrix of the Hamiltonian system besides S_0 which connects all the limit sets of $\dot{q}^\nu = K^\nu$. This is the case drawn schematically in fig. 3. Smooth separatrices are a typical feature of integrable Hamiltonian systems. Thus, smooth non-equilibrium potentials necessarily exist if the Hamiltonian (1.7) is integrable at $H = 0$. However, smooth separatrices are structurally unstable against small perturbations $K^\nu \to K^\nu + \delta K^\nu$, or $Q^{\nu\mu} \to Q^{\nu\mu} + \delta Q^{\nu\mu}$. In thermodynamic equilibrium only special perturbations of K^ν and $Q^{\nu\mu}$ are allowed satisfying microscopic reversibility, and it turned out that they preserve the smoothness of the separatrix [9]. Hence, there are smooth thermodynamic potentials in equilibrium. In non-equilibrium systems similar restrictions do not apply and smooth separatrices are exceptional. The typical phase space structure of a Hamiltonian of the form (1.7) therefore looks like fig. 4 where the separatrices emanating from the limit sets Γ split up and oscillate wildly around each other.

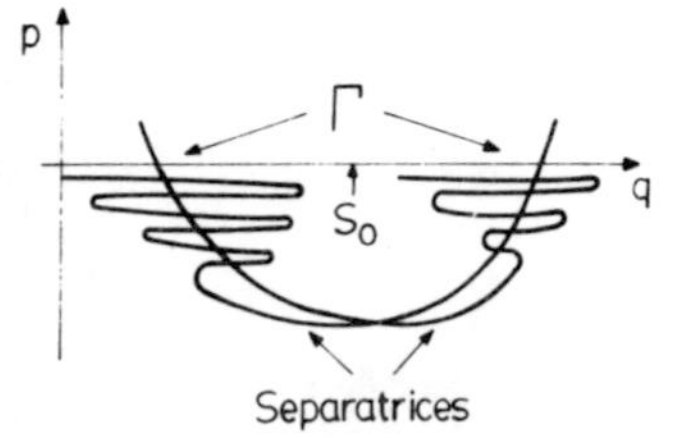

Figure 4. Phase space structure of the Hamiltonian (1.7) for the case where a smooth non-equilibrium potential does not exist (Schematically, after [2]).

2.2 A simple example

Let us consider, as an example [10,11],

$$\frac{1}{\varepsilon}\dot{x} = (x - x^3)(1 + a\cos y) + \xi(t)$$
$$\dot{y} = 1 \qquad (2.1)$$

with real x,y and real Gaussian white noise $\xi(t)$ with intensity coefficient η. The Hamiltonian (1.7)

$$H = \frac{1}{2}p_x^2 + \varepsilon p_x(x - x^3)(1 + a\cos y) + p_y \qquad (2.2)$$

is integrable for a = 0. The Poincaré section H = 0, y = 0 is shown in fig. 5. The separatrix is given by $p_y = 0$, $p_x = -2\varepsilon(x - x^3)$, and the potential therefore is

$$\phi_0 = -\varepsilon(x^2 - \frac{1}{2}x^4) \qquad (2.3)$$

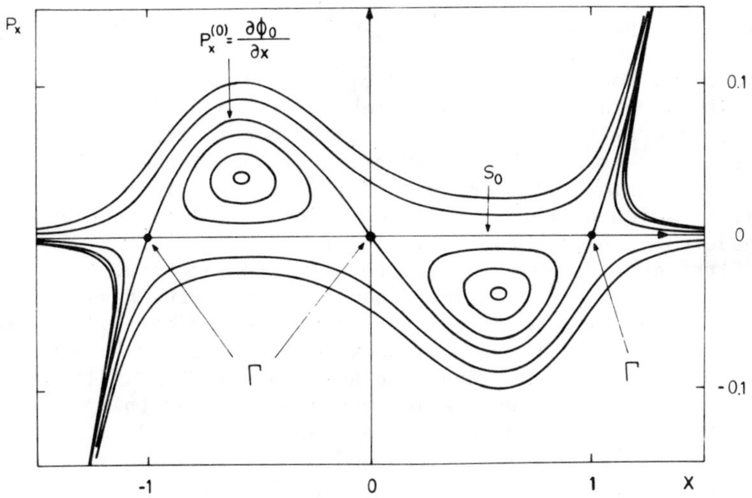

Figure 5. Poincaré cross section H = 0, y = 0 of the system (2.2) for a = 0 (from [10]). (With permission from Plenum Publ. Co.)

For $a \neq 0$ the Hamiltonian (2.2) is no longer integrable, as is evident from the Poincaré cross section of fig. 6 for a = 0.1.

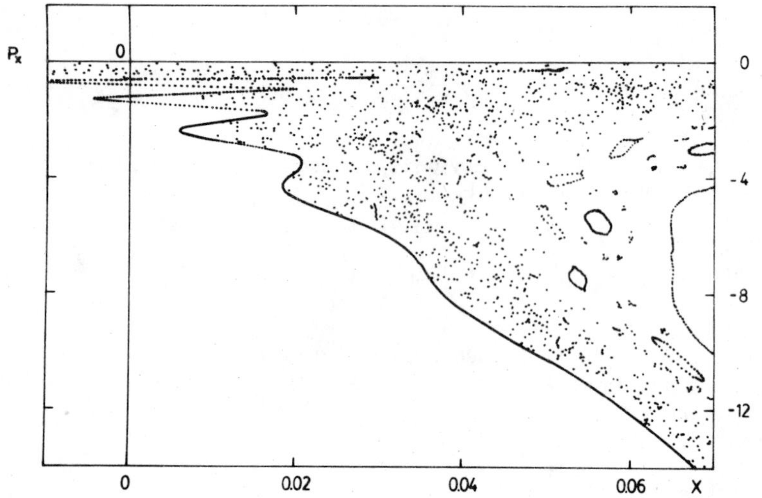

Figure 6. Poincaré cross section H = 0, y = 0 of the system (2.2) for a = 0.1 (from [10]). (With permission from Plenum Publ. Co.)

2.3 Search for special integrable cases

As we saw above, smooth non-equilibrium potentials exist in systems for which the Hamiltonian (1.7) is integrable for H = 0. Therefore it is interesting to consider methods which allow one to identify the integrable special cases within a given class of Hamiltonians (1.7). Such methods have been investigated in ref. [29] and are further applied in ref. [30], to which we refer for details. Here, we only mention the two main methods which have been applied so far.

The first method, which proved to be quite succesful in practice, consists in identifying an obvious special case of the Hamiltonian for which a smooth separatrix exists. An example is optical bistability treated in section 1.6, where obviously eq. (1.25), for $E_0 = 0$, admits the potential ϕ_0 of (1.27). Then the solution of the Hamilton Jacobi equation is sought as a perturbation expansion around this special case. In the first corrections to the special case considered oscillations of the separatrix may show up in corresponding oscillations of the potential. Adjusting the parameters of the system in such a way that the oscillations disappear one often discovers, in practice, those special cases where a smooth potential exists to all orders in the perturbation parameter. The latter fact has, of course, then to be proven by substitution of the smooth solution in the Hamilton Jacobi equation. The special case $\delta = \Delta$ of eq. (1.25) and the corresponding potential (1.28) has been found in [27] in this way.

A second method is based on the Painlevé conjecture which states that systems which have the Painlevé property are integrable. The solutions of the canonical equations for a given Hamiltonian have the Painlevé property if their only movable singularities in the complex time-plane are poles. In practice this is checked by solving Hamilton's equations by a Laurent-Taylor series arond a singular point and trying to adjust the parameters of the system in order to obtain enough free parameters in order to have a general solution. For the application of this method to Fokker Planck Hamiltonians of the form (1.7) we refer to the literature [29-31].

2.4 Functional integral and extremum principle

Let us now consider the general case, where a smooth separatrix of the Hamiltonian (1.7) and a corresponding smooth potential ϕ does not exist.

The solution of the Fokker Planck equation corresponding to eq. (2.1) can be written as a functional integral [15,32-34]

$$P(q|q_0,t) = \iint D\mu \exp[-\int_{q(-t)=q_0}^{q(0)=q} L(q,\dot{q},\eta)dt] \tag{2.4}$$

Here $P(q|q_0,t)$ is the conditional probability density, $D\mu$ a formal functional measure of integration, L is a certain Lagrangian. In the weak noise limit (2.4) can be evaluated in saddle point approximation and we find

$$P(q|q_0,t) \sim \exp[-\frac{1}{\eta}\min \int_{q(-t)=q_0}^{q(0)=q} L_0(q,\dot{q})dt] \tag{2.5}$$

where min denotes the absolute minimum and [35]

$$L_0 = \frac{1}{2}Q^{-1}_{\nu\mu}(\dot{q}^{\nu} - K^{\nu}(q))(\dot{q}^{\mu} - K^{\mu}(q)) \tag{2.6}$$

is the Lagrangian corresponding to the Hamiltonian (1.7). For $t \to \infty$ we obtain from (2.5) an expression of the form (1.9) with [11,12]

$$\phi(q) = \min \int_{q(-\infty)=q_0}^{q(0)=q} L_0(\dot{q}(t),q(t))dt \tag{2.7}$$

The min condition in (2.7) ensures that the right hand side, indeed, becomes independent of q_0: A minimizing path from q_0 to q in an infinite time-interval necessarily proceeds first from q_0 to the attractor in whose domain q_0 in chosen; along this part of the path L_0 vanishes, since

$\dot{q}^\nu = K^\nu$ is satisfied, and there is no contribution to (2.7). Then the path proceeds from the attractor to the final point q. Thus eq.(2.7) can be replaced by

$$\phi = \min_{(A_i)}[\min \int_{A_i}^{q} L_0(q,\dot{q})dt + C(A_i)] \tag{2.8}$$

where the attractors of the system are denoted by A_i. The constants $C(A_i)$ are determined from the balance of the probability flow between attractors [35,11].

Eq. (2.8) determines the non-equilibrium potential also in the case of an oscillating separatrix. Let the oscillating separatrix beginning in a given attractor A be $p_\nu = p_\nu(q)$ where $p_\nu(q)$ may be multivalued at least for some q (see fig. 7). The condition H = 0 and the relation between L_0 and H may be used to rewrite eq. (2.8) as

$$\phi(q) = \min \int_{A}^{q} p_\nu(q)dq^\nu \tag{2.9}$$

The minimum condition in eq. (2.9) now clearly leads to the appearance of a continuous but merely piecewise differentiable potential according to the construction given in fig. 7. The oscillations of the separatrix are interpolated by Maxwells' rule.

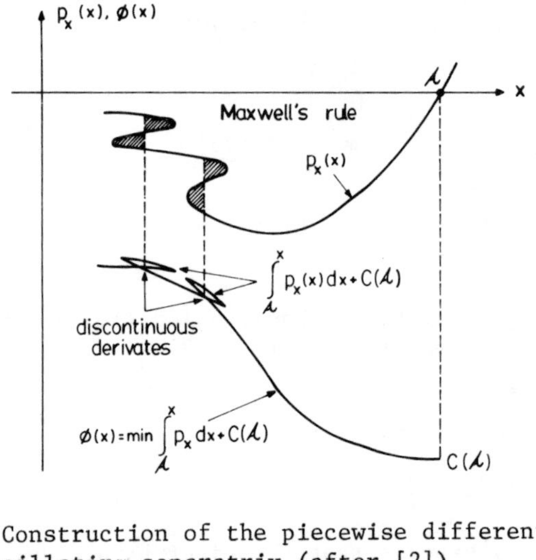

Figure 7. Construction of the piecewise differentiable potential $\phi(x)$ from the oscillating separatrix (after [2]).

All points of configuration space which can be reached from the attractor A along more than one most probable path are points where the non-equilibrium potential has discontinuous first derivatives.

Just as the oscillations of the separatrix pile up as the separatrix tries to approach a saddle so do the surfaces of non-differentiability of ϕ. However, the discontinuities of the first derivatives of ϕ also become smaller as we approach a saddle, such that in the limit the reduced separatrix approaches again a smooth curve, which is nothing but the separatrix coming out of the saddle and oscillating near the attractor.

For the case of the simple example of section 2.2 this is shown in fig. 8.

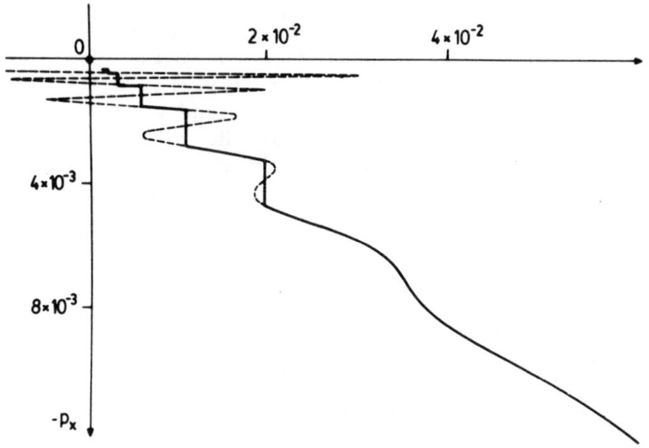

Figure 8. Reduction of the oscillating separatrix of fig. 4 according to Maxwell's rule (from [11]).

2.5 Coexisting attractors

In this final section we want to consider as another concrete example the Brownian motion defined by the equations [36,12]

$$\begin{aligned} \dot{x} &= v \\ \dot{v} &= -\gamma v - \sin x + F + \sqrt{2\eta\gamma}\,\xi(t) \end{aligned} \tag{2.10}$$

with the Gaussian white noise ξ as in (2.1). This example describes the damped stochastic motion of a particle in a periodic potential under the influence of a constant external force F. The total potential is

$$V(x) = -\cos x - Fx \tag{2.11}$$

The configuration space of this system may be taken as the surface of a cylinder

$$x_{s'} \leqq x \leqq x_s, \quad -\infty < v < +\infty \tag{2.12}$$

where $x_{s'} < x_s = x_{s'} + 2\pi$ are two consecutive saddles of the potential (2.11), and points (x,v) and $(x + 2\pi, v)$ are identified.

If $F = 0$, the system (2.10) corresponds to thermodynamic equilibrium and

$$\phi_0 \equiv v(x) + \frac{v^2}{2} \tag{2.13}$$

is the thermodynamic potential. For $F > 0$, however, ϕ_0 is not a non-equilibrium potential, at least not globally, since it is not periodic in x as required. We note that, depending on the value of F the system without noise shows different dynamical behavior. For $0 < F < F_c(\gamma)$ it has a single attracting fixed point $P_0 = (x_0,0)$ and a saddle $S' = (x_{s'},0)$, $S = (x_s,0)$ at the boundary of the x-interval. Here $F_c(\gamma)$ is some critical value, depending on γ. For $F_c(\gamma) < F < 1$ there occurs coexistence of the attracting fixed point with an attracting limit cycle $v = v_c(x)$, $x_{s'} \leqq x \leqq x_s$, while for $F > 1$ the saddles S', S and the fixed point P_0 dissapear and only the limit cycle remains.

The Hamiltonian (1.7) is of the form

$$H = \gamma p_v^2 + p_v(-\gamma v - \sin x + F) + p_x v \tag{2.14}$$

For $H = 0$ there exists the time-dependent conserved phase-space function [12,29]

$$A(x,p_x,v,p_v,t) \equiv \left(\frac{v}{p_v} - 1\right)e^{\gamma t} \tag{2.15}$$

This already ensures that an oscillating separatrix of the type considered in section 1 does not appear here. However, the existence of topologically distinct paths in the present multiply connected configuration space here introduces new complications and leads, in fact, again to the appearance of lines of non-differentiability, however, without these lines piling up towards the saddle point. The non-equilibrium potential for the present example has been constructed in [12] to which we refer. An approximating theory was given in [36]. In figs. 9 - 11 contour line plots ϕ = const. are presented for cases $0 < F < F_c(\gamma)$, $F_c(\gamma) < F < 1$, $1 < F$, respectively [12].

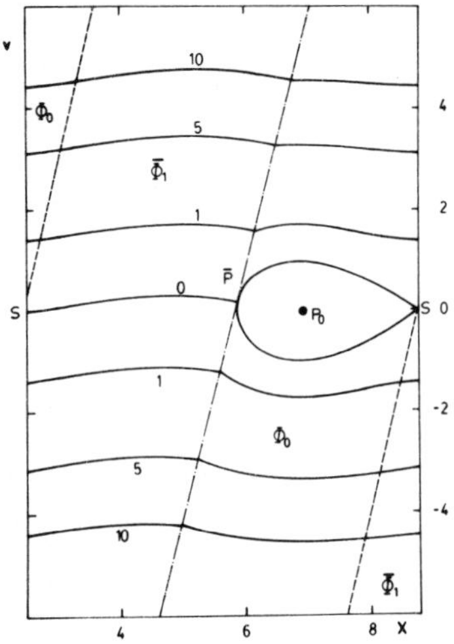

Figure 9. Equipotential lines in the case $0 < F < F_c(\gamma)$ for $\gamma = 5$, $F = 0.6$ (cf. text, from [12]).

In fig. 9 it is shown that, for $0 < F < F_c(\gamma)$, ϕ consists locally of two functions $\phi_0(x,v)$ and $\bar{\phi}_1(x,v)$ respectively. $\phi_0(x,v)$ is given by eq. (2.13) and is generated according to eq. (2.8) by trajectories of the Hamiltonian system starting in P_0 and proceeding from there to the final point (x,v). $\bar{\phi}_1(x,v)$ is a second function, periodic in x, which is generated according to eq. (2.8) by trajectories starting in P_0 but passing arbitrarily close near $S = S'$ before going to the final point (x,v) It is clear from this construction that $\bar{\phi}_1$ must be constant along the unstable manifold of S' of the deterministic system ($\xi = 0$), as the integrand L_0 of (2.8) vanishes along this line. This is the contour connecting S' and the point $\bar{P}$ in fig. 9. The dashed dotted curve through $\bar{P}$ is the line where $\bar{\phi}_1 = \phi_0$ but with discontinuous first derivative. The dashed curves through S' and S in fig. 9 are related with the stable manifold of S, S' of the deterministic system by the reflection $v \to -v$, $x \to x$. As shown in [12] along these curves $\bar{\phi}_1 = \phi_0$ holds with continuous first derivative.

In fig. 10 we consider the case $F_c(\gamma) < F < 1$, where the fixed point P_0 and the limit cycle coexist. Only equipotential lines below or at the saddle point value are shown. Around P_0 up to the saddle point value (normalized to zero) ϕ is given by ϕ_0 of eq. (2.13). Around the limit cycle up to the saddle point value ϕ is given by a periodic function, generated according to eq. (2.8) by the trajectories of the Hamiltonian system starting in the limit cycle and leading to the final point (x,v). The lines where the two local pieces of ϕ are continuously joined are located in regions where ϕ is above its saddle point value, and are not shown.

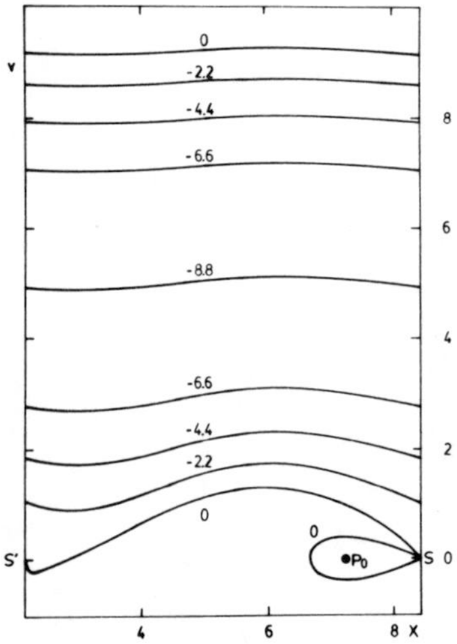

Figure 10. Equipotential lines in the case $F_c(\gamma) < F < 1$ for $\gamma = 0.13$, $F = 0.83$ (cf. text, from [12]).

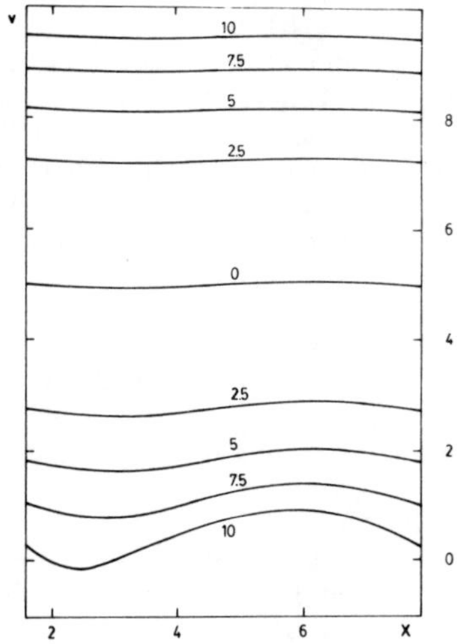

Figure 11. Equipotential lines in the case $F > 1$ for $\gamma = 0.16$, $F = 1.25$ (cf. text, from [12]).

In fig. 11, finally, the case $F > 1$ is shown where ϕ consists of a single periodic function generated via eq. (2.8) by trajectories starting in the limit cycle and leading to the final point (x,v). No lines with discontinuous derivatives occur in this case. The limit cycle corresponds, of course, to a continuously degenerate minimum of ϕ.

2.6 Concluding remarks

We have seen in these lectures how the notion of a coarse grained thermodynamic potential can be generalized for non-equilibrium systems with weak fluctuations. Extremum (minimum) principles analogous to thermodynamics determine the non-equilibrium potential from the dynamics and the noise according to eq. (2.8), and determine the steady states (i.e. the attractors) of deterministic system from the non-equilibrium potential. However, we have also seen that smooth differentiable non-equilibrium potentials are exceptional, and are structurally unstable if they occur in exceptional special cases. In general, the non-equilibrium potential is only piecewise differentiable with surfaces of discontinuous first derivatives piling up near saddles of the deterministic system. Further discontinuities of first derivatives of ϕ may appear in multiply connected configuration spaces and for coexistence of several attractors. We have only considered discrete systems in these lectures (an exception is the example considered in 1.5). An extension of the considerations of section 2 to fields would be desirable and interesting. The examples considered were still rather simple. Recently we have given an analysis of non-equilibrium potentials near codimension 2 bifurcations [37]. An extension of the analysis to even more complex dynamical systems is a matter of the future.

REFERENCES

[1] R. Graham, in Stochastic Nonlinear Systems, ed. L. Arnold, R. Lefever (Springer, New York 1981).

[2] R. Graham, T. Tél. "Macroscopic Potentials of Dissipative Dynamical Systems", to appear in Proceedings of the Bi BoS - Symposium Stochastic Process-Mathematics and Physics, Bielefeld, December 1985.

[3] M.S. Green, J. Chem. Phys. 20, 1281 (1952).

[4] R. Graham. H. Haken, Z. Physik 243, 289 (1971); 245, 141 (1971).

[5] R. Graham, in Coherence and Quantum Optics, ed. L. Mandel, E. Wolf (Plenum, New York 1973).

[6] H. Grabert, M.S. Green, Phys. Rev. A19, 1747 (1979).

[7] H. Grabert, R. Graham, M.S. Green, Phys. Rev. A21, 2136 (1980).

[8] R. Graham, A. Schenzle, Z. Physik B52, 61 (1983).

[9] R. Graham, T. Tél. Phys. Rev. Lett. 52, 9 (1984).

[10] R. Graham, T. Tél. J. Stat. Phys. 35, 729 (1984); 37, 709 (1984). (Addendum).

[11] R. Graham, T. Tél. Phys. Rev. A31, 1109 (1985).

[12] R. Graham, T. Tél. Phys. Rev. A33, 1322 (1986).

[13] H. Haken, Phys. Rev. Lett. 13, 326 (1964).

[14] H. Risken, Z. Physik 186, 85 (1965).

[15] R. Graham, Springer Tracts Mod. Phys. 66, 1 (1973).
[16] J. Guckenheimer, Ph. Holmes, Nonlinear Oscillations, Dynamical Systems and Bifurcations of Vector Fields, (Springer, New York 1983).
[17] R. Graham, H. Haken, Z. Physik 237, 31 (1970).
[18] V. De Giorgio, M.O. Scully, Phys. Rev. A2, 1170 (1970).
[19] A. Schlüter, D. Lortz, F.H. Busse, J. Fluid Mech. 23, 129 (1965).
[20] A. Newel, J. Whitehead, J. Fluid Mech. 38, 279 (1969).
[21] L.A. Segel, J. Fluid Mech. 38, 203 (1969).
[22] R. Graham, Phys. Rev. Lett. 31, 1479 (1973); Phys. Rev. A10, 1762 (1974).
[23] H. Haken, Phys. Lett. A46, 193 (1973); Rev. Mod. Phys. 47, 67 (1975).
[24] J. Swift, P.C. Hohenberg, Phys. Rev. A15, 319 (1977).
[25] M.C. Cross, Phys. Rev. A25, 1065 (1982).
[26] R. Graham, in Fluctuations, Instabilities and Phase Transitions, ed. T. Riste, (Plenum, New York 1975).
[27] R. Graham, A. Schenzle, Phys. Rev. A23, 1302 (1983).
[28] P. Talkner, P. Hänggi, Phys. Rev. A29, 768 (1984).
[29] R. Graham, D. Roekaerts, T. Tél, Phys. Rev. A31, 3364 (1985).
[30] D. Roekaerts, F. Schwarz, to be published.
[31] J. Hietarinta, J. Math. Phys. 26, 1970 (1985).
[32] R. Kubo, K. Matsuo, K. Kitahara, J. Stat. Phys. 9, 51 (1973).
[33] R. Graham, Z. Physik B26, 281 (1977).
[34] F. Langouche, D. Roekaerts, E. Tirapegui, Functional Integration and Semiclassical Expansion (Reidel, Dordrecht 1982).
[35] M.I. Freidlin, A.D. Wentzell, Random Perturbations of Dynamical Systems (Springer, New York 1984).
[36] E. Ben-Jacob, D.J. Bergmann, B.J. Matkowsky, Z. Schuss, Phys. Rev. A26, 2805 (1982).
[37] R. Graham, T. Tél. "Non-equilibrium potentials for local codimension two bifurcations of dissipative flows", preprint 1986.

FLUCTUATIONS IN "METASTABLE" SYSTEMS

R. Dandoloff *

and

Y. Pomeau **

ABSTRACT

We study time-dependent fluctuations in a simple electrical circuit made of a resistance and a capacitor. The capacitance has intrinsic dynamics, as a result -for instance- from polarization fluctuations of a dielectric medium inside the capacitor plates. In a generalized Langevin theory, the requirement of time reversal symmetry imposes strong constraints on the description of the fluctuations in this system. If a battery is inserted into it, time reversal symmetry of the fluctuations is broken during the time the battery keeps charged, despite the absence of macroscopic energy loss.

More than one century ago Poincaré and Zermelo, in their famous controversy with Boltzmann pointed out this remarkable property of equilibrium thermal fluctuations: from any measurement of these fluctuations, as a function of time, no distinction can be made between the two possible directions of time. As a consequence, any two-time correlation function at equilibrium must satisfy the equality:

$$\langle A(0)B(t)\rangle = \langle B^T(0)A^T(t)\rangle \qquad (1)$$

In this formula A(.) and B(.) are observables of the physical system, i.e. functions of position q and momenta p, and they depend on time through these variables only, so that A(t') is for A(q(t'),p(t')). Furthermore, the T-superscript in (1) is for time reversed observables: $A^T(q,p) = A(q,-p)$ (we assume for simplicity that there is no steady external magnetic field, otherwise one should also reverse this field in the T-operation). Finally the average < > in (1) is done over a Gibbs-Boltzmann ensemble of initial conditions.

* Lab. Phys. Sol., CNRS, F 92195 MEUDON Principal CEDEX, and Lab. Phys. Sol., Bat. 510, Université Paris-Sud, F 91405 ORSAY CEDEX, France
** C.E.N./Saclay, Service de Physique Théorique F 91191 GIF SUR YVETTE CEDEX, France

E. Tirapegui and D. Villarroel (eds.), Instabilities and Nonequilibrium Structures, 291–297.

The symmetry implied by eq. (1), as well as other similar conditions, puts [1] very strong conditions on any mathematical description of the time dependent fluctuations at equilibrium. In particular the so called detailed balance condition is required if one demands [2] time reversal symmetry in Markoff chains. In the Langevin approach to the theory of fluctuations, this condition of time reversal symmetry, can be expressed as follows [1,3]. Let X_i be the set of fluctuating quantities in the physical system, "i" being a discrete index so that X_i can be written as a vector. The generalized Langevin equation for this system, after convenient scalings, reads:

$$\dot{X}_i = F_i(X) + \xi_i(t) \tag{2}$$

where the dot is for time derivation; $F_i(X)$ is a vector field in the $\vec{X}$-space and $\xi(\ )$ is a Gaussian white noise term: $\langle\xi_i\rangle = 0$, $\langle\xi_i(t)\xi_j(t')\rangle = k_B T\delta_{ij}\delta(t-t')$, k_B is the Boltzmann constant, T is the absolute temperature, $\delta_{ij} = 1$ if $i = j$ and 0 otherwise, $\delta(t-t')$ is the Dirac δ-function. If $X_i^T = X_i$, $\forall i$, as we shall assume it, equation (2) describes a time reversible fluctuation of the force field $F_i(\vec{X})$ satisfies

$$\frac{\partial F_i}{\partial X_j} = \frac{\partial F_j}{\partial X_i} \tag{3}$$

which is equivalent (at least for elementary situations, to which we shall restrict ourselves) to say that an uniform potential (or "energy") function V(X) exists, such that

$$F_i = \frac{\partial V}{\partial X_i} \quad .$$

From the works of Kolmogoroff [2] and others, this is a sufficient condition too for time reversal symmetry.

In the present paper, we examine how this picture is transformed in a "metastable" situation. We do not claim to have a general description of fluctuations in all possible metastable situations, but would merely show, by the examination of a simple "gedanken" experiment that thermal fluctuations in "metastable" systems may lack time reversal symmetry. Our gedanken experiment will be measurement of thermal noise in an electric open circuit, where a charged battery maintains a potential difference across a capacitor. This system can be considered as metastable in this sense that slow chemical relaxation will discharge the battery. However thermal activation is very inefficient for relaxing the battery toward its actual equilibrium state -the discharged one. The time needed for the chemical relaxation will be assumed to be very long. However our arguments point to the lack of time reversal symmetry in thermal fluctuations of this system on much shorter time scales when the battery keeps charged. It is true that this slow "chemical" relaxation as well as eventual spontaneous recharge of the battery will be absent from our mathematical description, and so one can see our model as representing a system maintained in a steady -but out of equilibrium- state without macroscopic dissipation. In the usual terminology of statistical mechanics, this state should be dubbed "metastable". But this denomination is ambiguous, because it refers to the comparison of the energies of various possible macroscopic steady states, and our basic point will be that -for our system- a non-equilibrium steady state can exist with a dynamics that is not the one of a potential flow i.e. this dynamics does not describe the transition of the phase point representing the system with the charged battery to the real equilibrium state. Thus no energy can be found in order to compare it with the ground state energy and to decide whether a steady state is metastable or not. To recover time reversal symmetry in fluctuations, one should average over time scales, much longer than the chemical relaxation time, including spontaneous recharge and discharge of the battery.

Let us consider first a classical RC-loop, and let U be the voltage difference across R and across C. This is the same voltage because there is no other element in the circuit. Later on we shall have to make a distinction between those two voltages differences (across R and across C). In the absence of triggering by thermal fluctuations, a non zero initial value of U tends to zero as time goes on following the simple relaxation equation:

$$\dot{U} + \frac{1}{RC}U = 0 \tag{4}$$

As usual, the thermal fluctuations are added to this picture by putting on the r.h.s. of eq. (4) a Gaussian white noise $\xi_u(t)$[4], so that eq. (4) becomes:

$$\dot{U} + \frac{1}{RC}U = \xi_u(t) \tag{4'}$$

The Gaussian white noise is normalized in such a way that the mean value of the energy stored in the capacitor is given by the Ehrenfest relation:

$$<\frac{1}{2}CU^2> = k_B T$$

Suppose now that we want to supplement this by including the effect of thermal noise on the properties of the capacitor. This could arise, for instance, from the polarization fluctuations of a dielectric material inserted between the capacitor plates. For simplicity, let us assume that these fluctuations obey a Langevin equation with a single relaxation time τ:

$$\dot{C} + \frac{1}{\tau}(C - C_0) = \xi_c(t) \tag{5}$$

In this equation C_0 is the value of the capacity in the absence of noise and $\xi_c(.)$ is a Gaussian white noise. If one wants to describe thermal fluctuations in this RC-circuit by the coupled equations (4') and (5), this would violate the requirement of time reversal symmetry as expressed by condition (3). This symmetry is recovered (for instance) by adding to the l.h.s. of (4') a new term, in order to put (4') and (5) into the form

$$\dot{C} + \partial V/\partial C = \xi_c(t)$$

and

$$\dot{U} + \partial V/\partial U = \xi_u(t) \quad ,$$

V(C,U) being some well definite function of U and C. The form of (4') and (5) suggests to take:

$$V(C,U) = \frac{U^2}{2RC} + \frac{1}{2\tau}(C - C_0)^2 \tag{6}$$

which leads to the following pair of Langevin equations:

$$\dot{U} + \frac{1}{RC}U = \xi_u(t) \tag{7'}$$

and

$$\dot{C} + \frac{1}{\tau}(C - C_0) - \frac{U^2}{2RC^2} = \xi_c(t) \tag{7''}$$

Note that the inclusion of the U^2-term in (7") leads to the Taylor expansion of the mean capacitor energy, as a function of a constant U as

$$\frac{1}{2}C_0U^2 - \tau U^4/RC_0^2 + \ldots$$

Actually it is a little disturbing to have R in this expansion, as there is apparently no reason why the capacitor fluctuations should depend on the resistance in the circuit, if the voltage across the capacitor is held constant. This apparent phenomenon is due to our normalization choices and would disappear by taking as fluctuating quantity the charge of the capacitor. A more natural choice of energy would seem to be

$$V(U,C) = \frac{1}{2}CU^2 + \frac{1}{2\tau}(C - C_0)^2$$

but it leads to difficulties because of the possibility of having C negative and so an unboundedly large negative V(U,C). To avoid this, we have chosen to keep the above form (6) for V(U,C) as it leads to the equations of motion (7). This yields at least a formally consistent analysis and without a detailed microscopic description, this approach is as justified as others having the same internal consistency.

The equilibrium statistical weight for the fluctuations described by (7) is proportional to

$$\exp\{-\frac{1}{k_B T} V(U,C)\}$$

and so it vanishes as well as any derivative with respect to C near C=0 except for U = 0. Accordingly, the transition rate toward negative (and unphysical) values of C is zero and one may safely assume that C is positive. The (possible) singularity at U = 0 can be avoided by changing the potential V to:

$$V_\varepsilon(U,C) = \frac{U^2 + \varepsilon^2}{2RC} + \frac{1}{2\tau}(C - C_0)^2$$

and leaving at the end $\varepsilon \to 0$. However we shall not need the recourse to this computational trick.

Consider now a simple modification in our electric circuit: a battery is inserted in between the capacitor and the resistance to maintain a constant potential U_0 across the capacitor (in the absence of a contribution from the thermal noise to this potential). Let U be still the potential difference across the resistance, so that eq. (7') remains correct for describing the fluctuations of the potential across the resistance. But eq. (7") has to be changed, because the fluctuations of C are obviously sensitive to the potential difference across the capacitor, that is to $(U + U_0)$, and in (7") one must replace U^2 by $(U + U_0)^2$. Thus the generalized Langevin equations for this system are eq. (7') and:

$$\dot{C} + \frac{1}{\tau}(C - C_0) - \frac{(U + U_0)^2}{2RC^2} = \xi_c(t) \qquad (8)$$

that replaces eq. (7"). There does not seem to be any reason for changing the properties of the fluctuating force $\xi_c(.)$ because of the polarization of the capacitor by the battery. In particular, if we assume that U is maintained constant (and zero for instance if R = 0), the potential -as a function of C- is:

$$\frac{1}{2\tau}(C - C_0)^2 + \frac{(U + U_0)^2}{2RC}$$

so that a Gaussian white noise $\xi_c(.)$ at temperature T gives the correct equilibrium probability distribution for the fluctuations of C. The non trivial property of (7') and (8) is that they do not satisfy the condition (3) of detailed balance, so that a measurement of the kind envisaged in ref. [1] [i.e. for instance -the measurement of correlation functions given by the difference between the left and right member of eq.

(1)] could put into evidence that fluctuations described by (7') and (8) do not satisfy the time reversal symmetry of equilibrium fluctuations. Note however that this symmetry is restored if $U_0 = 0$ and if $|U_0| >> < U^2 >^{\frac{1}{2}}$, because in the latter case, fluctuations of C and U become uncoupled.

Now let us try to estimate the order of magnitude of the expected violation of the time symmetry. We think of measuring a correlation function as $S_{fg}(t) = < f[C(0)]g[C(t)] - f[C(t)]g[C(0)] >$ with f not proportional to g.

The symmetry violation will reach a maximum when

$$U_0 \sim <U^2>^{\frac{1}{2}} \sim \left(\frac{k_B\theta}{C_0}\right)^{\frac{1}{2}}$$

as it vanishes if $U_0 = 0$ and if $U_0 >> <U^2>^{\frac{1}{2}}$. It will also reach a maximum when the time scales τ and RC_0 have the same order of magnitude, as we shall assume it. Thus, $S_{fg}(t)$ reaches a maximum for $t \sim \tau$ and this maximum will be of the order of the variation of the product f(c)g(c) near $C = C_0$ as due to a change of C coming from the non potential term in (8), i.e. a change of order

$$\Delta C \sim \frac{\tau<U^2>}{RC_0^2} \sim \frac{k_B T}{C_0^2}$$

(with our units, where C and U have the same dimension). Similar considerations lead to the following estimate for the non time symmetric part of the fluctuations of U over the common time scale $\tau \sim RC_0$ and if $U_0 \sim <U^2>^{1/2}$:

$$\Delta U \sim <U^2>^{1/2} \frac{\Delta C}{C_0} \sim \frac{(k_B T)^{3/2}}{C_0^{7/2}}.$$

To get a dimensionally correct expression for ΔU, one has to consider C_0 as a function of the thermodynamic energy stored in the dielectric, say ε, so that

$$\frac{\Delta C}{C_0} \sim \frac{1}{C} \frac{\partial C_0}{\partial \varepsilon}\Big|_\theta k_B T \quad .$$

To conclude let us say that the previous remarks could perhaps apply to other physical situations. One could think to a possible (but certainly quite weak) violation of the Onsager reciprocity relations in solids under permanent load. This violation would arise from the fact that time reversal symmetry of fluctuations is basic for proving the reciprocity relations. As it is well known, these relations explain why there is no rotation of (vector) fluxes with respect to (vector) thermodynamic forces in solids without center of symmetry, even though this rotation is allowed on the basis of geometric considerations (the "Curie" principle). Finally let us also remark that the symmetry violation shown in the RC + battery-circuit should be enhanced in circuits with large nonlinearities. One could think -for instance- to insert into the capacitor a ferroelectric at its (eventual) second order phase transition point to paraelectric.

ACKNOWLEDGEMENTS

We acknowledge discussions with Dr. A.-M.S. Tremblay.

REFERENCES

[1] Y. Pomeau, J. de Physique, 43, 859 (1982).
[2] A. Kolmogoroff, Mathem. Ann., 104, 415 (1931).
[3] R. Graham and H. Haken, Z. Physik, 245, 141 (1971).
[4] N.G. van Kampen in Stochastic Processes in Physics and Chemistry, North Holland 1981, exhibits a number of arguments against generalized use of the Langevin equation consisting in inserting into the r.h.s. of the macroscopic dynamical equations a random white noise. Indeed, it may be a rather formal device to describe in such a way the thermal fluctuations of a physical system. However, in the absence of a microscopic description of those fluctuations one may require at least to have theories consistent with the time reversal symmetry of equilibrium and with the Gibbs-Boltzmann statistical distribution of the one-time fluctuations.

NORMAL FORMS WITH NOISE

C. Elphick (*)

E. Tirapegui (**)

ABSTRACT. A method to eliminate the fast variables in stochastic differential equations near an instability is given. It is based in normal forms methods.

We shall review here some recent results concerning the elimination of fast variables in stochastic differential equations. This can be done working directly with the Fokker-Planck equation for the probability density [1, 2] or adapting to stochastic differential equations the methods used in deterministic equations. We shall proceed in the last way and prove a factorization theorem for the probability density of the stochastic process following [3].

We consider a general equation of the form

$$\partial_t \underline{U} = L\underline{U} + \underline{N}(\underline{U}) + \eta(\underline{D}(t) + L^{(1)}(t)\underline{U} + \underline{M}(t;\underline{U})) \tag{1}$$

where

$$\underline{U} = \sum_{\alpha=1}^{N} U_\alpha(t)\underline{e}_\alpha, \quad \underline{e}_1 = (1,0,\ldots 0)\ldots\ldots\underline{e}_N = (0,\ldots 0,1),$$

and L is an $N \times N$ matrix already in Jordan form. We suppose we have a critical space of dimension n generated by $(\underline{e}_1,\ldots\underline{e}_n)$ and a stable space generated by $(\underline{e}_{n+1},\ldots\underline{e}_N)$. One has

---

(*) Laboratoire de Physique Théorique, Faculté des Sciences de Nice, Parc Valrose, 06034 Nice Cedex, France.

(**) Departamento de Física, Facultad de Ciencias Físicas y Matemáticas, Universidad de Chile, Casilla 487/3, Santiago, Chile.

E. Tirapegui and D. Villarroel (eds.), Instabilities and Nonequilibrium Structures, 299–310.

$$L\underline{e}_\alpha = \sum_{\beta=1}^{n} \mathcal{J}_{\beta\alpha}\underline{e}_\beta, \qquad 1 \leq \alpha \leq n \tag{2}$$

$$L\underline{e}_\alpha = \sigma_\alpha \underline{e}_\alpha, \quad n+1 \leq \alpha \leq N, \quad \sigma_\alpha < 0,$$

and $\mathcal{J}_{\beta\alpha}$ has non zero elements only in the diagonal or of the form $\mathcal{J}_{\alpha,\alpha+1} = 1$ for some values of α. The diagonal terms are either zero or pure imaginary quantities corresponding to multiple Hopf bifurcations. We also have

$$\underline{N}(\underline{U}) = \sum_{r\geq 2} \underline{N}^{(r)}(\underline{U})$$

with

$$\underline{N}^{(r)}(\underline{U}) = \sum_{\alpha,\alpha_j=1}^{N} u^{(r)}_{\alpha,\alpha_1\ldots\alpha_r} U_{\alpha_1}\ldots U_{\alpha_r}\underline{e}_\alpha. \tag{3}$$

The terms proportional to η in (1) are the noisy terms and the situation we have described up to now for the deterministic equations $\partial_t\underline{U} = L\underline{U} + \underline{N}(\underline{U})$ corresponds to be exactly at a critical point in the parameters space where one has a critical subspace of dimension n characterized by the Jordan matrix $\mathcal{J}_{\alpha\beta}$. The unfolding of this singularity can be obtained using the methods in [4] which are precisely the methods we shall generalize to the stochastic terms in (1). The latter are proportional to η which is a parameter measuring the intensity of the noise,

$$\underline{D}(t) = \sum_{\alpha=1}^{N} D_\alpha(t)\underline{e}_\alpha,$$

$L^{(1)}(t)$ is an $N\times N$ matrix

$$L^{(1)}(t)\underline{e}_\alpha = \sum_{\beta=1}^{N} L^{(1)}(t)_{\beta\alpha}\underline{e}_\beta \quad \text{and} \quad \underline{M}(t;\underline{U}) = \sum_{r\geq 2} \underline{M}^{(r)}(t;U)$$

with

$$\underline{M}^{(r)}(t;\underline{U}) = \sum_{\alpha,\alpha_j=1}^{N} v^{(r)}_{\alpha,\alpha_1\ldots\alpha_r}(t)U_{\alpha_1}\ldots U_{\alpha_r}\underline{e}_\alpha. \tag{4}$$

Here $\{D_\alpha(t), L^{(1)}(t)_{\alpha\beta}, v^{(r)}_{\alpha,\alpha_1\ldots\alpha_r}(t)\}$ are taken as δ-correlated gaussian white noises with zero mean and given correlations. For instance

$$<D_\alpha(t)D_\beta(t')> = Q_{\alpha\beta}\delta(t-t'),\ <Q_\alpha(t)L^{(1)}(t')_{\mu\nu}> = Q^{(1)}_{\alpha,\mu\nu}\delta(t-t'),$$
$$\text{etc.,} \qquad (5)$$

where $Q_{\alpha\beta}$, $Q^{(1)}_{\alpha,\mu\nu}$, are given constants. We can imagine that the stochastic terms in (1) come from fluctuating parameters of the original problem (external noise) and this equation is the most general situation one can think of at the critical point. We shall prove the following: the probability density $p_t(U_1,\ldots U_N)$ of the stochastic process defined by (1) admits asymtotically ($t \gg \sup|\sigma_\alpha|^{-1}$) a decomposition as the product of a Gaussian distribution of the (linearly) stable variables $(U_{n+1},\ldots U_N)$ centered in the central manifold and a probability density $\bar{p}_t(U_1,\ldots U_n)$ depending only on the (linearly) critical variables.

In order to prove this result we make the ansatz that $\underline{U}$ can be expressed asymptotically in terms of n variables $(A_1(t),\ldots A_n(t))$ and that these variables obey stochastic differential equations of the form

$$\partial_t A_\alpha = \sum_{r\geq 1} f^{(r)}_\alpha(A_1,\ldots A_n) + \eta \sum_{r\geq 0} g^{(r)}_\alpha(t;A_1,\ldots A_n) \qquad (6)$$

where $f^{(r)}_\alpha$ (resp. $g^{(r)}_\alpha$) are of order r in $(A_1,\ldots A_n)$. We put

$$f^{(1)}_\alpha = \sum_{\beta=1}^{n} \mathcal{J}_{\alpha\beta}A_\beta,\ f^{(r)}_\alpha = \sum_{\alpha_j=1}^{n} f^{(r)}_{\alpha;\alpha_1\ldots\alpha_r}A_{\alpha_1}\ldots A_{\alpha_r}, \qquad (7)$$

$$g^{(r)}_\alpha(t) = \sum_{\alpha_j=1}^{n} g^{(r)}_{\alpha;\alpha_1\ldots\alpha_r}(t)A_{\alpha_1}\ldots A_{\alpha_r}, \qquad (8)$$

where $f^{(r)}_{\alpha;\alpha_1\ldots\alpha_r}$ (resp. $g^{(r)}_{\alpha;\alpha_1\ldots\alpha_r}(t)$) are constants (resp. stochastic processes) to be determined. We write the expression for $\underline{U}$ in terms of $(A_1\ldots A_n)$ as follows

$$\underline{U} = \sum_{r\geq 1}\underline{U}^{(r)} + \eta\sum_{r\geq 0}\underline{V}^{(r)}(t) \qquad (9)$$

where $\underline{U}^{(r)}$ (resp. $\underline{V}^{(r)}$) are of order r in $(A_1,\ldots A_n)$. We put

$$\underline{U}^{(1)} = \sum_{\alpha=1}^{n} A_\alpha \underline{e}_\alpha, \underline{U}^{(r)} = \sum_{\substack{\alpha_j=1\ldots n \\ \alpha=1\ldots N}} U^{(r)}_{\alpha;\alpha_1\ldots\alpha_r} A_{\alpha_1}\ldots A_{\alpha_r}\underline{e}_\alpha, \tag{10}$$

$$\underline{V}^{(r)}(t) = \sum_{\substack{\alpha_j=1,\ldots n \\ \alpha=1,\ldots N}} V^{(r)}_{\alpha;\alpha_1\ldots\alpha_r}(t) A_{\alpha_1}\ldots A_{\alpha_r}\underline{e}_\alpha. \tag{11}$$

Here $U^{(r)}_{\alpha;\alpha_1\ldots\alpha_r}$ (resp. $V^{(r)}_{\alpha;\alpha_1\ldots\alpha_r}(t)$) are constants (resp. stochastic processes) to be determined. The determination of all the unknown quantities is done through a set of homological equations [5] obtained replacing (9) in the original equation (1) and then identifying there, after use of equation (6), the terms of order (j,r) where j is the order in η (0 or 1 here since we calculate only to first order in η) and r the polynomial order in $(A_1,\ldots A_n)$. The notation $(\ldots)^{(j,r)}$ stands for the terms of order (j,r) in (...). In order to obtain the equations we note from (9) that in the left hand side of (1) one has to replace $\partial_t\underline{U}$ by

$$\sum_{\alpha=1}^{n} \partial_t A_\alpha \frac{\partial \underline{U}}{\partial A_\alpha} + \hat{\partial}_t \underline{U} \text{ with } \underline{U} = \sum_{r\geq 1} \underline{U}^{(r)} + \eta \sum_{r\geq 0} \underline{V}^{(r)}(t), \tag{12}$$

where $\hat{\partial}_t$ stands for the derivative with respect to t acting only on the t dependence contained in the functions $V^{(r)}_{\alpha;\alpha_1\ldots\alpha_r}(t)$. In the right hand side of (1) one just replaces $\underline{U}$ as given by (9) and then we obtain a set of equations writing the equality of terms of order (j,r) at both sides of (1). We note that of course when using (12) one has to replace there $\partial_t A_\alpha$ using (6) to (8). This procedure gives at order (0, 1) just a check of our choice of $f^{(1)}_\alpha$ and $\underline{U}^{(1)}$ (see (7), (10)). At order (0,r) one obtains (see [4])

$$(D - L)\underline{U}^{(r)} = \underline{I}^{(r)} - \sum_{\alpha=1}^{n} f^{(r)}_\alpha \underline{e}_\alpha \equiv \underline{I}^{(r)}_1 , \tag{13}$$

$$D \equiv \sum_{\alpha,\beta=1}^{n} \mathcal{J}_{\alpha\beta} A_\beta \frac{\partial}{\partial A_\alpha}, \tag{14}$$

$$\underline{I}^{(r)} = (\underline{N}(\sum_{s\geq 1} \underline{U}^{(s)}))^{(0,r)} - \sum_{\substack{s=2,\ldots r-1 \\ \alpha=1,\ldots n}} f^{(s)}_\alpha \frac{\partial}{\partial A_\alpha} \underline{U}^{(r-s+1)} \tag{15}$$

Equations (13) will determine the usual normal form (equation (6) for $\eta = 0$) and is solved by recursion in r since $\underline{I}^{(r)}$ depends only on $\{\underline{U}^{(s)}, f_\alpha^{(s)}\}$ for $s < r$, so one considers in (13) $\underline{I}^{(r)}$ as given and solves for $\underline{U}^{(r)}$. But in order to have a solution the right hand side of (13) must be orthogonal to the adjoint Ker $(D^+ - L^+)$ with respect to some suitable scalar product (see [4]) and this condition (Fredholm alternative) determines the $f_\alpha^{(r)}$ which we choose in a minimal way (we take them zero when there is no condition on them). Once this is done we can solve for $\underline{U}^{(r)}$ and proceed to the next order.

Let us look now at the orders (1,r). One obtains from (1) at this order

$$(\hat{\partial}_t + D - L)\underline{V}^{(r)}(t) = \underline{K}^{(r)}(t) - \sum_{\alpha=1}^{n} g_\alpha^{(r)}(t)\underline{e}_\alpha , \tag{16}$$

$$\begin{aligned}\underline{K}^{(r)}(t) &= \eta^{-1}(\underline{N}(\Sigma\, \underline{U}^{(s)} + \eta\,\Sigma\, \underline{V}^{(s)}))^{(1,r)} + \\ &+ \underline{D}(t)\delta_{r,0} + L^{(1)}(t)\underline{U}^{(r)} + (\underline{M}(t; \Sigma\, \underline{U}^{(s)}))^{(0,r)} - \\ &- \sum_{\substack{\alpha=1,\dots n \\ s=0,1,\dots r-1}} g_\alpha^{(s)}(t) \frac{\partial}{\partial A_\alpha} \underline{U}^{(r-s+1)} - \\ &- \sum_{\substack{\alpha=1,\dots n \\ s=2,3,\dots r}} f_\alpha^{(s)} \frac{\partial}{\partial A_\alpha} \underline{V}^{(r-s+1)}(t).\end{aligned} \tag{17}$$

We can solve again (16) by recursion in r starting with $r = 0$ since $\underline{K}^{(r)}(t)$ depends only on $\{\underline{U}^{(s)}, s \leq r+1\}$, $\{f_\alpha^{(s)}; s \leq r\}$ which we know already from solving (13) and on $\{\underline{V}^{(s)}(t); s < r\}$, $\{g_\alpha^{(s)}(t); s < r\}$. Let us see how this works starting at order (1,0). Then (16) becomes using (2) and (11)

$$\partial_t V_\alpha^{(0)}(t) - \sum_{\beta=1}^{n} \mathcal{J}_{\alpha\beta} V_\beta^{(0)}(t) = D_\alpha(t) - g_\alpha^{(0)}(t), \quad 1 \leq \alpha \leq n, \tag{18}$$

$$(\partial_t - \sigma_\alpha)V_\alpha^{(0)}(t) = D_\alpha(t), \quad n+1 \leq \alpha \leq N. \tag{19}$$

We solve (18) putting $g_\alpha^{(0)}(t) = D_\alpha(t)$, $V_\alpha^{(0)}(t) = 0$, $1 \leq \alpha \leq n$, and from (19) we get

$$V_\alpha^{(0)}(t) = e^{\sigma_\alpha t}\int_{-\infty}^{t} dt' e^{-\sigma_\alpha t'} D_\alpha(t'), \quad n+1 \leq \alpha \leq N, \tag{20}$$

choosing initial conditions at $t = -\infty$ equal to zero. In fact asymptotically ($t \gg \sup |\sigma_\alpha|^{-1}$) $V_\alpha^{(0)}(t)$ is independent of the initial condition and we are in this region since we are using the ansatz (9) which is only valid there. At order (1,1) ($r = 1$ in (16)) we have

$$\partial_t V_{\rho;\nu}^{(1)}(t) + \sum_{\alpha=1}^{n} V_{\rho;\alpha}^{(1)}(t)\, \mathcal{J}_{\alpha\nu} - \sum_{\alpha=1}^{n} \mathcal{J}_{\rho\alpha} V_{\alpha;\nu}^{(1)}(t) =$$

$$= K_{\rho;\nu}^{(1)}(t) - g_{\rho;\nu}^{(1)}(t), \quad 1 \leq \rho \leq n, \quad 1 \leq \nu \leq n, \tag{21}$$

$$(\partial_t - \sigma_\rho)V_{\rho;\nu}^{(1)} + \sum_{\alpha=1}^{n} V_{\rho;\alpha}^{(1)}\, \mathcal{J}_{\alpha\nu} = K_{\rho;\nu}^{(1)}(t), \; n+1 \leq \rho \leq n, \; 1 \leq \nu \leq n, \tag{22}$$

$$K_{\rho;\nu}^{(1)}(t) = 2\sum_{\beta=n+1}^{N} U_{\rho;\nu\beta}^{(2)} V_\beta^{(0)}(t) + L^{(1)}(t)_{\rho\nu} - 2\sum_{\alpha=1}^{n} D_\alpha(t) U_{\rho;\alpha\nu}^{(2)} \; . \tag{23}$$

We choose to solve (21) putting $g_{\rho;\nu}^{(1)}(t) = K_{\rho;\nu}^{(1)}(t)$, $V_{\rho;\nu}^{(1)}(t) = 0$, $1 \leq \rho \leq n$, $1 \leq \nu \leq n$. We can now solve (22) since we know $K_{\rho;\nu}^{(1)}(t)$ which is expressed in terms of known quantities and of $V_\alpha^{(0)}(t)$ which we know already (see (20)). However if we decide to consider the corrections due to stochastic noise up to terms of order (1,1), i.e. we consider the orders (1,0) and (1,1), then it is more convenient to consider together equations (19) and (22) (if we go up to order (1,r) the method will be the same [3])). These equations (19) and (22) are for the quantities

$$(V_{n+1}^{(0)},\ldots V_N^{(0)}; V_{n+1;1}^{(1)},\ldots V_{n+1;n}^{(1)},\ldots V_{n+2;1}^{(1)},\ldots V_{n+2;n}^{(1)};\ldots$$

$$\ldots V_{N;1}^{(1)},\ldots V_{N;n}^{(1)}) \equiv (q_1, q_2, \ldots q_{N'}) \tag{24}$$

where $N' = (n+1)(N-n)$, and can then be written as a set of N' linear stochastic differential equations for the stochastic processes $(q_1(t),\ldots q_{N'}(t))$ of the form

$$\dot{q}_\mu - \sum_{\nu=1}^{N'} A_{\mu\nu} q_\nu = \xi_\mu(t), \quad 1 \leq \mu \leq N', \tag{25}$$

where the $\xi_\mu(t)$ are gaussian white noises with zero mean and correlations $<\xi_\mu(t)\xi_\nu(t')> = C_{\mu\nu}\delta(t-t')$ where the $C_{\mu\nu}$ are known quantities and the $N'\times N'$ matrix $A_{\mu\nu}$ has only zeros above the diagonal (this comes from the properties of $\mathbb{J}_{\mu\nu}$ and from the numbering (24) for the q_α). The diagonal elements of $A_{\nu\mu}$ are

$$A_{11} = \sigma'_1, \ldots A_{N-n,N-n} = \sigma'_N, \ldots A_{N'N'} = \sigma'_N \tag{26}$$

where $\sigma'_j = \sigma_j + i\Omega_j$, Ω_j real coming from the purely imaginary terms which one can have in the diagonal of $\mathbb{J}_{\mu\nu}$. One has then that

$$\det(A-\lambda) = \prod_{\alpha=1}^{N'} (A_{\alpha\alpha} - \lambda)$$

and consequently all the eigenvalues of A have negative real parts ($\sigma_\alpha < 0$). This implies the existence of the stationary state of (25) and in this state the probability density $p_t(q_1,\ldots q_{N'})$ is time independent and given by the gaussian

$$p_{st}(\underset{\sim}{q}) = \frac{1}{\sqrt{(2\pi)^{N'}\det\Xi}} \exp\left(-\frac{1}{2}\sum_{\mu,\nu=1}^{N'} q_\mu \Xi^{-1} q_\nu\right) \tag{27}$$

where the positive definite (generically) matrix Ξ satisfies $\Xi A^T + A^T\Xi + C = 0$ (here A^T is the transposed matrix of A and C the correlation matrix $C_{\mu\nu}$).

Let us see now the form that our results give to the asymptotic expression (9) of $\underline{U}$ in terms of $(A_1,\ldots A_n)$. One has

$$U_\alpha = A_\alpha + \tilde{U}_\alpha(A_1,\ldots A_n), \quad 1 \le \alpha \le n, \tag{28}$$

$$U_\alpha = \tilde{U}_\alpha(A_1,\ldots A_n) + \eta\left(V_\alpha^{(0)}(t) + \sum_{\nu=1}^{n} V_{\alpha;\nu}^{(0)}(t)A_\nu + \ldots\right),$$

$$n+1 \le \alpha \le N. \tag{29}$$

where $\tilde{U}_\alpha(A_1,\ldots A_n) = \sum_{r\ge 2} U^{(r)}_{\alpha;\alpha_1\ldots\alpha_r} A_{\alpha_1}\ldots A_{\alpha_r}$. The equations for the $\{A_\alpha\}$ are of the form

$$\partial_t A_\alpha = f_\alpha(A_1,\ldots A_n) + \eta g_\alpha(t;A_1,\ldots A_n) \tag{30}$$

with $\quad f_\alpha = \sum_{\beta=1}^{n} \mathbb{J}_{\alpha\beta}A_\beta + \sum_{r\ge 2} f_\alpha^{(r)}, \; g_\alpha = D_\alpha(t) + \sum_{r\ge 1} g_\alpha^{(r)}(t)$ (see (7), (8)).

The results in the absence of noise are obtained from (28) to (30) putting $\eta = 0$ in which case $\partial_t A_\alpha = f_\alpha(A_1,\dots A_n)$ is the usual normal form and the asymptotic form of $\underline{U}$ becomes

$$U_\alpha = A_\alpha + \tilde{U}_\alpha(A_1,\dots A_n), \quad 1 \le \alpha \le n, \tag{31}$$

$$U_\alpha = \tilde{U}_\alpha(A_1,\dots A_n), \quad n+1 \le \alpha \le N. \tag{32}$$

The equation of the central manifold in the variables $(U_1,\dots U_N)$ is now obtained expressing the A_α in terms of $(U_1,\dots U_n)$ from the n equations (31) and then replacing in (32). From equations (31) we can write

$$A_\alpha(U_1,\dots U_n) = U_\alpha + S_\alpha(U_1,\dots U_n), \quad 1 \le \alpha \le n, \tag{33}$$

where S_α starts with quadratic terms. Replacing (33) in (32) and calling F_α the resulting function $F_\alpha(U_1,\dots U_n) \equiv \tilde{U}_\alpha(A_1(U_1,\dots U_n),\dots A_n(U_1,\dots U_n))$ one obtains the central manifold in the form of a set of $(N-n)$ equations (F_α starts with quadratic terms)

$$U_\alpha = F_\alpha(U_1,\dots U_n), \quad n+1 \le \alpha \le N. \tag{34}$$

Replacing (33) in (29) now we obtain

$$U_\alpha = F_\alpha(U_1,\dots U_n) + \eta\Big(V_\alpha^{(0)}(t) + \sum_{\nu=1}^{n} V_{\alpha;\nu}^{(1)}(t)U_\nu + \dots\Big), \qquad n+1 \le \alpha \le N, \tag{35}$$

where the points stand for terms at least quadratic in $(U_1,\dots U_n)$.

Let us come back now to our original problem (1). The probability density $p_t(U_1,\dots U_N)$ can always be written as (with obvious notations)

$$p_t(U_1,\dots U_N) = p_t(U_{n+1},\dots U_N | U_1,\dots U_n)\bar{p}_t(U_1,\dots U_n) \tag{36}$$

where $p_t(U_{n+1},\dots U_N | U_1,\dots U_n)$ is the probability of the linearly stable variables $(U_{n+1},\dots U_N)$ conditionned by the values of the linearly critical variables $(U_1,\dots U_n)$. From (35) we can immediately obtain an expression for this conditional probability

$$p_t(U_{n+1}\dots | U_1 \dots) = \int \prod_{\alpha=n+1}^{N} dV_\alpha^{(0)} \prod_{\substack{\rho=n+1,\dots N \\ \nu=1,\dots n}} dV_{\rho;\nu}^{(1)} \,. \tag{37}$$

$$\cdot\ p_{st}(V^{(0)}_{n+1},\ldots V^{(0)}_{N},V^{(1)}_{n+1;1}\ldots V^{(N)}_{N;n})\prod_{\alpha=n+1}^{N}\cdot$$

$$\cdot\ \delta\ (U_\alpha - F_\alpha(U_1,\ldots U_n) - \eta(V^{(0)}_\alpha + \sum_{\nu=1}^{n} V^{(1)}_{\alpha;\nu}U_\nu)),$$

where $p_{st}(V^{(0)}_{n+1},\ldots V^{(1)}_{N;n})$ is the gaussian (27) and consequently the conditional probability $p_t(U_{n+1},\ldots|U_1,\ldots)$ is time independent asymptotically. The integral in (37) is over the N' variables $\{q_\alpha\}$ defined in (24) and we make now the change of variables

$$q'_\mu = q_\mu + \sum_{\nu=1}^{n} q_{N-n+(\mu-1)n+\nu}\ U_\nu,\quad 1\le\mu\le N-n, \tag{38}$$

$$q'_{N-n+\alpha} = q_{N-n+\alpha},\quad \alpha\ge 1. \tag{39}$$

The Jacobian is unity and we put $\bar{p}_{st}(\underset{\sim}{q}') = p_{st}(\underset{\sim}{q})$, then the integral in (37) becomes

$$p(U_{n+1},\ldots|U_1,\ldots) = \int\prod_{\alpha=1}^{N'} dq'_\alpha \bar{p}_{st}(\underset{\sim}{q}')\prod_{\alpha=1}^{N-n}\delta(\lambda_\alpha-\eta q'_\alpha) \tag{40}$$

with

$$\lambda_\alpha \equiv U_{n+\alpha} - F_{n+\alpha}(U_1,\ldots U_n),\quad 1\le\alpha\le N-n. \tag{41}$$

Integrating in (40) over $dq'_1\ldots dq'_{N-n}$ one obtains

$$p(U_{n+1},\ldots|U_1,\ldots) = \eta^{n-N}\int\prod_{\alpha=N-n+1}^{N'} dq'_\alpha\ \cdot$$

$$\cdot\ \bar{p}_{st}(\frac{\lambda_1}{\eta},\ldots\frac{\lambda_{N-n}}{\eta},q'_{N-n+1},\ldots q'_{N'}).$$

We note that $\bar{p}_{st}(q')$ is a gaussian since the replacement $\tilde{q}\to\tilde{q}'$ is linear. Consequently the integral in (42) gives the marginal distribution of a gaussian and is then a gaussian in the variables

$(\frac{\lambda_1}{\eta},\ \frac{\lambda_2}{\eta}\ \ldots\frac{\lambda_{N-n}}{\eta})$ which we write as

$$p(U_{n+1},\ldots|U_1,\ldots) = \frac{1}{\eta^{N-n}\sqrt{(2\pi)^{N-n}\det\Lambda(U_1,\ldots U_n)}}\cdot$$

$$\cdot\exp\Big(-\frac{1}{2\eta^2}\sum_{\alpha,\beta=1}^{N-n}\lambda_\alpha\Lambda(U_1,\ldots U_n)^{-1}_{\alpha\beta}\lambda_\beta\Big). \tag{43}$$

From (36) we see now that we have the asymptotic decomposition

$$p_t(U_1,\ldots U_N) = \frac{1}{\sqrt{(2\pi\eta^2)^{N-n}\det\Lambda(U_1,\ldots U_n)}}\exp\Big[-\frac{1}{2\eta^2}\sum_{\alpha,\beta=1}^{N-n}\cdot$$

$$\cdot(U_{n+\alpha}-F_{n+\alpha}(U_1,\ldots U_n))\Lambda(U_1,\ldots U_n)^{-1}_{\alpha\beta}(U_{n+\beta}-F_{n+\beta}(U_1,\ldots U_n))\Big]\cdot$$

$$\cdot\,\overline{p}_t(U_1,\ldots U_n). \tag{44}$$

The problem now is the calculation of $\overline{p}_t(U_1,\ldots U_n)$. This probability is completely determined by the stochastic differential equations (30) which gives when solved $P_t(A_1,\ldots A_n)$, and from this probability using the non linear change of variables $(A_1,\ldots A_n)\to(U_1,\ldots U_n)$ given by (33) one obtains $\overline{p}_t(U_1,\ldots U_n)$, as

$$\overline{p}_t(U_1,\ldots U_n) = P_t(A_1,\ldots A_n)\cdot J(\underset{\sim}{A}|\underset{\sim}{U}) \tag{45}$$

where $J(\underset{\sim}{A}|\underset{\sim}{U})$ is the Jacobian of the transformation.

The essential result we have obtained is then the factorization (44) which is a consequence of the gaussian property of the conditional probability (43). This last property was used as an ansatz in [1] to eliminate the fast variables from the Focker-Planck equation in some simple situations, however the method which was given there led to a non positive definite matrix $\Lambda_{\alpha\beta}(U_1,\ldots U_n)$ in (44) which is inconsistent as we have discussed in [6]. The method presented here is based on the very intuitive ansatz (9) which states that the original variable $\underline{U}$ can be expressed asymptotically in terms of a set of critical variables $(A_1,\ldots A_n)$ also in the stochastic case: a proof of this assertion can be found in [3,6]. On the other hand we have obtained for these variables a set of closed equations (30) which however depend on the $\underline{V}^{(s)}(t)$ up to $s=r-1$ if we stop in $g_\alpha(t;A_1,\ldots A_n)$ at terms of polynomial order r in $(A_1,\ldots A_n)$. If we go up to $r=1$ in the stochastic terms one would have

$$\partial_t A_\alpha = f_\alpha(A_1,\dots A_n) + \eta(D_\alpha(t) + \sum_{\nu=1}^{n} g^{(1)}_{\alpha;\nu}(t)A_\nu), \tag{46}$$

where $g^{(1)}_{\alpha;\nu}(t) = K^{(1)}_{\alpha;\nu}(t)$ (see (23)) depends on $V^{(0)}_\beta(t)$, $n+1 \leq \beta \leq N$, which are coloured noises satisfying equations (19). However these coloured noises can be replaced asymptotically ($t \gg \sup|\sigma_\alpha|^{-1}$) in a first approximation by white noises as is explained in [6] thus turning (46) in an ordinary stochastic differential equation on the central manifold (this replacement works in the same way if one goes up to order r). One would then call equations (46) (or (30) if we go to higher order) the normal form of the original system but for this it would be necessary to show that there is no non linear change of the variables $(A_1,\dots A_n)$ giving the stochastic terms there a simpler form (we know that this is the case for the deterministic terms $f_\alpha(A_1,\dots A_n)$ since they are the usual normal form in the absence of noise). One can show that in the cases of the ξ instability [6] (one zero eigenvalue, i.e. the matrix $\mathbb{J}$ is one by one and equal to zero) and of the Hopf instability [7] all possible terms appear in the stochastic part of (30) and this is probably the general situation.

In order to complete this discussion let us consider the unfolding of the instability which we can obtain by the methods of [4] and will change in general (30) to

$$\partial_t A_\alpha = B_\alpha + \sum_{\beta=1}^{n} B'_{\alpha\beta}A_\beta + f_\alpha(A_1,\dots A_n) + \eta g_\alpha(t;A_1,\dots A_n), \tag{47}$$

where $\{B_\alpha, B'_{\alpha\beta}\}$ are the small parameters describing the unfolding ($B'_{\alpha\beta}$ is the Jordan-Arnold unfolding of the matrix $\mathbb{J}$). One could then think that the stochastic normal form can be obtained giving the parameters of the unfolding of the deterministic normal form a stochastic character. This is true for the ξ instability but false for the Hopf bifurcation since there the normal form near the bifurcation point is of the form $\partial_t z = zF(|z|^2)$ while in the stochastic terms one finds z^3, $\bar{z}^3$, etc, (stochastic resonance) [7]. The relevance of these new terms for the stationary probability has been explicitly shown in [8]. We can then conclude that in general the stochastic normal form will contain new terms not appearing in the deterministic normal form.

REFERENCES

[1] E. Knobloch, K. A. Wiesenfeld, J. Stat. Phys. 33, 611 (1983).

[2] C. van den Broeck, M. Malek Mansour, F. Baras, J. Stat. Phys. 28 575 (1982).

[3] O. Descalzi, C. Elphick, C. Flores, E. Tirapegui: Reduction to the central manifold in stochastic differential equations (preprint Universidad de Chile).

[4] C. Elphick, E. Tirapegui, M. E. Brachet, P. Coullet, G. Iooss: A simple global characterization for normal forms of singular vector fields, preprint Université de Nice, Physique Théorique, NTH 86/6.
[5] V. Arnold: Chapitres Supplémentaires de la théorie des équations différentielles ordinaires, Editions MIR (1980).
[6] C. Elphick, M. Jeanneret, E. Tirapegui: Adiabatic elimination in the presence of noise, preprint Fac. Cs. Fís. y Mat. Univ. de Chile (1986).
[7] P. H. Coullet, C. Elphick, E. Tirapegui, Phys. Letters 111A, 277 (1985).
[8] C. Nicolis, G. Nicolis: Comment on the normal form analysis of stochastically forced dynamical systems, preprint Faculté des sciences, Univ. Libre de Bruxelles (1986).

ADIABATIC NUCLEATION, STABILITY LIMIT AND THE GLASS TRANSITION TEMPERATURE

Erich Meyer
Inst. de Física da Univ. Federal do Rio de Janeiro
Bl. A. - 4. andar, 21945 Cidade Universitária,
Rio de Janeiro, Brasil

ABSTRACT

The conditions under which the first step of a spontaneous irreversible process (e.g. nucleation) occurs are studied. It is concluded that such a process can be supposed to occur in a large adiabatic system, at constant external pressure and with increasing entropy, if the corresponding experimental conditions are satisfied. A simple adiabatic nucleation model shows a definite stability limit for the liquid-solid phase transition and is in excellent agreement with experimental data. The fact, whether the glass transition temperature of a given liquid is lower or higher than the stability limit temperature, decides, whether homogeneous nucleation is probable or not.

BASIC THERMODYNAMICS

The entropy change of a system can be defined in the following way [1],

$$\Delta s = \Delta_e s + \Delta_i s \qquad (1)$$

$$\Delta_e s = \Delta q/T \quad , \qquad \Delta_i s \geqq 0$$

where Δs, $\Delta_e s$, $\Delta_i s$, Δq and T denote the total change of entropy, the part of this change due to interaction with the surroundings, the part of this change due to phenomena taking place inside the system, the heat exchange and the absolute temperature, respectively. From relation (1) follows then, with the first law of thermodynamics, for an adiabatically enclosed system,

$$\Delta q = 0, \quad \Delta u = -p\Delta v \qquad \Delta s \geqq 0 \qquad (2)$$

where only the mechanical work $-p\Delta v$ was considered and where u, p and v are the internal energy, the external pressure and the volume of the sys-

E. Tirapegui and D. Villarroel (eds.), Instabilities and Nonequilibrium Structures, 311–315.

tem, respectively.

Two special cases can be obtained from relation (2),

$$\left.\begin{array}{l}\Delta q = 0\\ \Delta v = 0\end{array}\right\} = \begin{array}{l}\Delta u = 0\\ \Delta v = 0\end{array} \qquad \Delta s \geqq 0 \qquad (3)$$

and

$$\left.\begin{array}{l}\Delta q = 0\\ \Delta p = 0\end{array}\right\} = \begin{array}{l}\Delta h = 0\\ \Delta p = 0\end{array} \qquad \Delta s \geqq 0 \qquad (4)$$

where h is the enthalpy of the system.

Further, if the temperature of the system is maintained constant, one obtains [1],

$$\begin{array}{l}\Delta T = 0\\ \Delta v = 0\end{array} \qquad \Delta f \leqq 0 \qquad (5)$$

and

$$\begin{array}{l}\Delta T = 0\\ \Delta p = 0\end{array} \qquad \Delta g \leqq 0 \qquad (6)$$

where f and g are the free energy and Gibbs' function, respectively.

For equilibrium, relations (3-6) are all rigorously correct, with $\Delta s = 0$, $\Delta f = 0$ and $\Delta g = 0$. For a dynamic, irreversible process however, one has to ask, whether the conditions $\Delta q = 0$, $\Delta v = 0$, $\Delta p = 0$ and $\Delta T = 0$ really can be (experimentally) imposed to the system? It helps little, if one works with relations, which are mathematically correct, but which physically cannot be realized.

At first sight it seems to be easier to realize $\Delta T = 0$, than $\Delta q = 0$. But, as Δq is the heat exchange with the surroundings, it is enough to choose a large system, for studying the first step of an irreversible process (e.g. nucleation). It is not necessary to enclose the system with adiabatic walls, it is enough that the system is large, compared to the observed process. On the other hand, it is not sufficient to put the system in contact with a heat bath, for obtaining $\Delta T = 0$, as many authors believe [1]. This is because T is not an "external" temperature, but the "local" temperature everywhere in the system and there is no experimental possibility, to fix the temperature in a spontaneous irreversible process. Temperature variations may many times not be easily observable, if the system is in contact with a heat bath, because the system is then at the beginning and at the end at the same temperature, but not during the process. Because of this, it is erroneous to consider Δg and Δf as the "driving force" of the irreversible process and one sees that the relations (5) and (6) are not very useful for studying a spontaneous irreversible process.

Without doubt it is experimentally easier to impose to the system, $\Delta p = 0$ instead of $\Delta V = 0$. The latter condition can only exceptionally be realized, in the case of gases. Most processes in nature proceed with $\Delta p = 0$, where one has to consider that $\Delta p = 0$ does not mean that the pressure has to be constant everywhere in the system. "p" is the external pressure of the system and for many processes or steps of processes it is sufficient to choose a large system for being able to realize $\Delta p = 0$. So one can conclude that relation (4) is the best indicated, for studying an irreversible process in an adiabatically enclosed system or for studying the first step on a spontaneous irreversible process in an adiabatically or isothermically enclosed large system.

These conclusions are suprising, if one considers that relation (4) is lacking in most treatises of thermodynamics (for exceptions see refs. [1,2]) and that, to the knowledge of the author, this relation never has been applied to a concrete problem. On the other hand, it is known since a long time that the "best" 3-dimensional diagram, for reporting equilibrium properties (specially of gases and liquids), is Mollier's diagram, which uses exactly the same variables as relation (4); p, h and s. From such a diagram, all other useful properties of the system can easily be deduced.

ADIABATIC NUCLEATION AND THE STABILITY LIMIT OF THE LIQUID-SOLID PHASE TRANSITION

A typical concrete case, for the considerations above, is nucleation, which generally and a priori has been considered to be an isothermal phenomenon (see e.g. ref. [3]). Calculating in contrast to this with relation (4) has the advantage, that the temperature is not artificially maintained constant, as is done, working with Gibb's function. Treating the enthalpy as a function of the entropy at constant pressure, for the solid and liquid phase, the stability limit for adiabatic nucleation is found, being the intersection of the enthalpy of the solid phase, corrected by the capillary pressure of the nucleus, and the enthalpy of the liquid phase. Passing this limit (by temperature fluctuations), the nucleus appears adiabatically and with increasing entropy, in full agreement with the second law of thermodynamics [4]. The extensive calculations can be found in ref. [4] and are valid for low viscosity and homogeneous materials and the simplest approximation of the result is

$$T^{-}/T_M = x(1{,}67 - 0{,}26/\sqrt{Q})^{-1}(e^x - 1)^{-1} \qquad (7)$$

$$x = \Delta h_L / c_p T_M$$

where T^{-}, T_M, c_p, Δh_L and Q are the absolute maximum super-cooling temerature, the absolute melting temperature, the mean specific heat of the solid and liquid phase at the melting point, the latent heat of melting and the number of atoms per molecule of the solid phase, respectively.

Comparing eq. (7) with experimental data one finds in general

$$T^{-}_{exp} \geqq T^{-} \qquad (8)$$

where T^{-}_{exp} is the experimental maximum supercooling temperature of low viscosity liquids. For P, Sn, Ga, Hg, Pb and Bi a good agreement with the experimental values is found [4],

$$T^{-}_{exp} \cong T^{-} \qquad (9)$$

Eq. (7) can be approximated for large molecules (large Q) to,

$$T^{-}/T_{M} = 0,6\,x(e^{x} - 1)^{-1} \qquad (10)$$

and comparing this equation with revised values of the glass transition temperatures T_g of polymers [5], one finds,

$$T_g/T_M > T^{-}/T_M \qquad (11)$$

in all 27 cases where the necessary thermodynamic data are available, except polyoxymethylene,where T_g is still disputed ($0.41 \leqq T_g/T_M \leqq 0.58$) and where $T^{-}/T_M = 0,51$. This suggests that pure amorhpous polymers do not cyrstallize (by homogeneous nucleation), because T_g is attained before T^{-} during the freezing process, so that the amorphous structure freezes in before there is a chance for homogeneous nucleation.

DISCUSSION AND CONCLUSIONS

Similarly to the results above, Zanotto could demonstrate that ceramic glasses show homogeneous nucleation or not, depending on $T_g < T^{-}$ or $T_g > T^{-}$, respectively [6]. Further, in analogy to this, one can speculate that metallic glasses will freeze in or not, during quenching, depending on $T_g > T^{-}$ or $T_g < T^{-}$, respectively. As T_g is an (increasing) function of the quenching speed, both possibilities can exist for the same material. However, the experimental thermodynamic data for proving this last relation are not yet available.

The good general agreement of the present adiabatic nucleation model with most experimental data, suggests that nucleation is not an isothermal, but an adiabatic phenomenon and also that nucleation may not be preceded by heterophase fluctuations, but by common statistical temperature fluctuations.

ACKNOWLEDGEMENTS

The author thanks to Professor Edgar D. Zanotto for the communication, prior to publication, of the conclusions of the test of the present nucleation model on ceramic glasses and the institutions CNPq, FINEP and CEPG for financial support.

REFERENCES

[1] E.A. Guggenheim, Thermodynamics (North-Holland, Amsterdam, 1959).
[2] F. Mandel, Statistical Physics (Wiley, New York, 1977).
[3] J.W. Christian, The Theory of Transformations in Metals and Alloys (Pergamon, Oxford, 1965).
[4] E. Meyer, J. Crystal Grwoth, 74, 425 (1986). (Note the missing fraction bar in eq. (49), corresponding to eq. (7) in this issue).
[5] E. Meyer, J. Cystal Growth (submitted).
[6] E.D. Zanotto, J. Non-Crystalline Solids (submitted).

MAGNETIC ISLANDS CREATED BY RESONANT HELICAL WINDINGS

A.S. Fernandes*, M.V.A.P. Heller, I.L. Caldas**
Instituto de Física, Universidade de São Paulo,
C.P. 20516, 01498 São Paulo, SP, Brazil

ABSTRACT

The triggering of disruptive instabilities by resonant helical windings in large aspect-ratio tokamaks is associated to destruction of magnetic surfaces. The Chirikov condition is applied to estimate analytically the helical winding current thresholds for ergodization of the magnetic field lines.

1. INTRODUCTION

Disruptive instabilities limit the tokamak operation and their causes are still unclear. There are experimental evidences that in several cases these instabilities are preceded by some plasma perturbations modes [1,2]. To investigate it, dirruptive instabilities can be triggered by resonant magnetic fields created by helical windings [2]. In this paper, we suppose that magnetic surface break-up, caused by these resonances, triggers the tokamak disruptions. We estimate analytically helical winding current thresholds for these disruptions in a large aspect-ratio tokamak with circular cross-section. As an example, we calculate these thresholds for the brazilian TBR tokamak [3].

2. MAGNETIC SURFACES

Plasma equilibrium in tokamaks is described by MHD equations. The differential equations for the magnetic field lines are

$$\frac{dr}{dz} = \frac{B_r}{B_z} \quad , \quad \frac{d\theta}{dz} = \frac{B_\theta}{rB_z} \quad , \tag{1}$$

---

* On leave from Setor de Ciencias Exatas, UFP, Curitiba, PR, Brazil.
** Partially supported by Conselho Nacional de Desenvolvimento Científico e Tecnológico (CNPq).

E. Tirapegui and D. Villarroel (eds.), Instabilities and Nonequilibrium Structures, 317–324.

where r, θ and z are cylindrical coordinates. For helical symmetry equilibria these lines lie on magnetic surfaces given by the equation

$$\vec{B}\cdot\nabla\psi = 0 \quad , \tag{2}$$

where $\psi = \psi(r,u)$, $u = \theta - \alpha z$ and $\alpha = \frac{n}{mR}$ is a constant. In this work n and m are rational numbers.

We consider a large aspect-ratio tokamak with circular cross-section, represented by a periodical cylinder with length $2\pi R$, and assume the tokamak scaling

$$\frac{B_z}{B_\theta} \cong \frac{R}{a} >> 1 \quad , \tag{3}$$

where R and a are respectively the major and minor plasma raddi. The unperturbed equilibrium is determined by B_z and the plasma current density $\vec{j}$:

$$\nabla B_z = 0 \quad , \quad \vec{j} = j_0(1 - r^2/a^2)^\gamma \hat{e}_z \quad , \tag{4}$$

where j_0 and γ are constants. In this case magnetic line helicities depend on r and are characterized by the safety factor q given by

$$q = \frac{rB_z}{RB_\theta} \quad . \tag{5}$$

At $r = a$, $q(a) \sim B_z/I_p$, where I_p is the plasma current. At rational surfaces with $q(r_{m,n}) = m/n$, the magnetic field lines close on themselves after m trips along the cylinder and n trips in the poloidal direction.

The unperturbed function ψ_0 is [4]

$$\psi_0 = \frac{nr^2B_z}{2mR} - \frac{\mu_0 I_p}{2\pi}\int_0^r \frac{dr'}{r'}\left[1 - (1 - r'^2/a^2)^{\gamma+1}\right] \tag{6}$$

The magnetic field created by electrical currents I flowing in m pairs of helical windings, equally spaced, with radius b wounded on a circular cylinder (corresponding to a large aspect-ratio tokamak) exhibits helical symmetry. This field depends on the coordinates r and u. The constant α characterizes the helicity of the windings. The stream function ψ_1 for I flowing in opposite directions in adjacent windings, is [4,5]

$$\psi_1(r,u) \cong \frac{\mu_0 I}{\pi}\left(\frac{r}{b}\right)^m \cos mu \quad . \tag{7}$$

We consider $|\psi_1/\psi_0| << 1$ and the linear superposition of an unperturbed equilibrium described by $\psi_0(r)$ with a resonant helical perturbation described by $\psi_1(r,u)$. Within this approximation, the stream function

$$\psi(r,u) = \psi_0(r) + \psi_1(r,u) \tag{8}$$

satisfy the eq. (2). This perturbation in resonance with the period of the equilibrium magnetic lines creates m islands around the rational magnetic surface with $q(r_{m,n}) = m/n$. This approximation is not valid for marginally stable states when the plasma response should not be neglected.

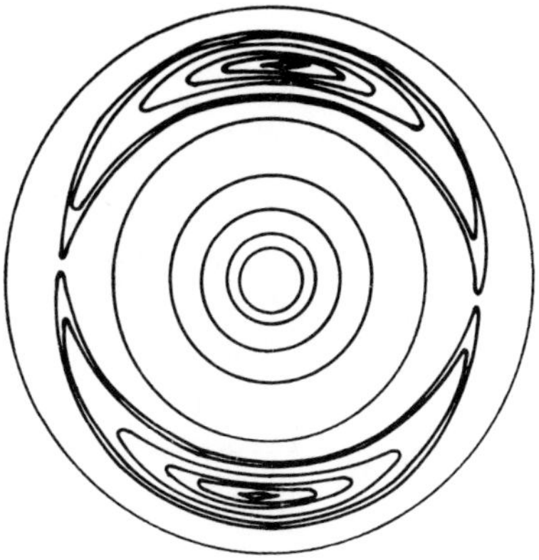

Figure 1. Island structures produced by two pairs of helical windings ($m = 2$, $n = 1$, $q(a) = 3$, $\gamma = 2$, $I_p = 10$kA and $I = 100$A).

Fig. 1 ilustrates the structures of the magnetic islands created by $m = 2$ pairs of helical windings ($n = 1$). The diagram shows the intersections of the magnetic surfaces with a plane $z = \text{const}$. All magnitudes were adjusted to fit TBR data [3]. Thus $a = 0{,}08$m, $b = 0{,}11$m, $R = 0{,}3$m.

Expanding ψ near $r = r_{m,n}$, the radius of resonant magnetic surfaces, we obtain a formula for the small island half-width $\Delta_{m,n}$

$$\Delta_{m,n} \cong \left[\frac{4\mu_0 I}{\pi\psi_0''(r_{m,n})}\left(\frac{r_{m,n}}{b}\right)^m\right]^{\frac{1}{2}} . \tag{9}$$

To obtain the island width, taking into account the radial dependence of the perturbation ψ_1 and the change in the safety factor q over the island width, we use the equation for the island separatrix

$$\psi_0(r) + \psi_1(r,u) = \psi_0(r_{m,n}) + \psi_1(r_{m,n},u_0) . \tag{10}$$

The accuracy of the formula (9) is illustrated by the comparisons, in Figs. (2), (3) and (6), of helical current threshold calculated using eqs. (9) and (10). The agreement is found to be good, specially for smaller islands.

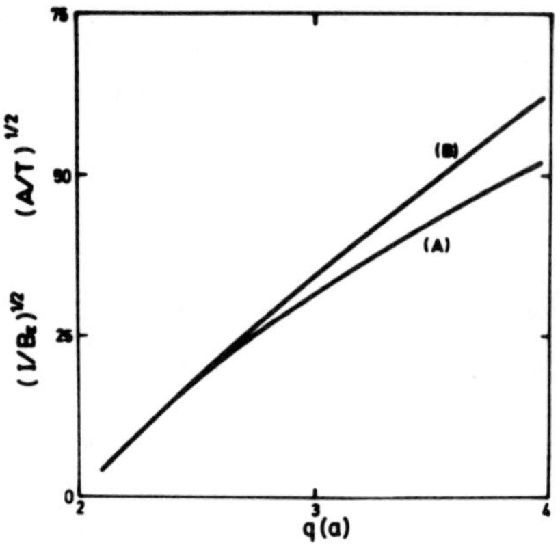

Figure 2. Helical winding current for the m = 2 islands touch the limiter. The curves A and B were obtained using, respectively, the Eqs. (9) and (10) ($q(0) = 1$, $I_p = 10$kA).

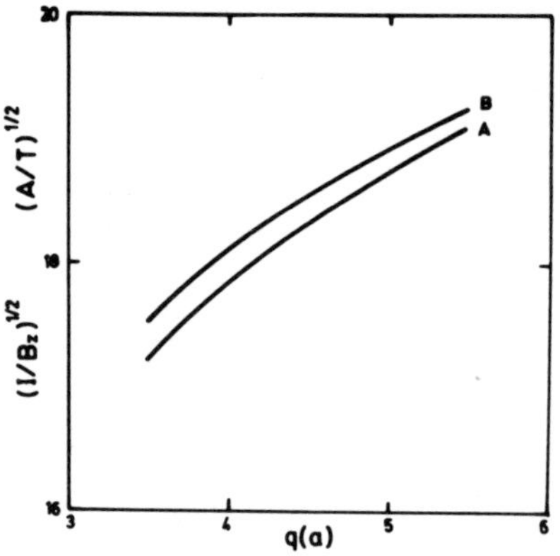

Figure 3. Helical current in two sets of helical windings for overlap of $m_1 = 2$ and $m_2 = 3$ resonances. The curves A and B were obtained using, respectively, the Eqs. (9) and (10). ($q(0) = 1$, $I_p = 10$kA).

3. THE BREAK-UP OF THE MAGNETIC SURFACES

There are experimental evidences that hard disruptive instabilities observed in tokamaks can be provocated by the interaction of the $m = 2$ magnetic islands with the plasma edge (the limiter is at $r = a$) [2]. We calculated the current I in $m = 2$ pairs of helical windings that causes the contact between the magnetic islands and the limiter. The dependence of I with q(a) is shown in the Fig. 2. The perturbation due to this current would trigger hard disruptions in the TBR tokamak.

We consider also the hypothesis that soft disruptive instabilities observed in tokamaks are caused by the ergodic wandering of magnetic field lines. Magnetic surfaces break-up occurs due to the destruction of the system symmetry. As a symmetry breaking perturbation due to magnetic islands created by different resonant helical fields grows, the magnetic surfaces are destroyed and a disruption may occur. The degree of ergodic behavior depend upon the strength of this perturbation. The helical winding current threshold for ergodization of the magnetic field lines was estimated applying the Chirikov condition [6]. Two of magnetic islands with mode numbers (m,n) and (m',n') are ergodized when

$$\Delta_{m,n} + \Delta_{m',n'} > |r_{m,n} - r_{m',n'}| \quad . \tag{11}$$

This happens for I greater than the values plotted in the Fig. 2 as a function of q(a). In this example $m = 2$ and $m' = 3$ ($n = n' = 1$). The helical current threshold increases with q(a) although the distance between the rational magnetic surfaces with $q = 2$ and $q = 3$ remain almost the same, as one can see in Fig. 4. This happens because the width islands decrease when the islands move inward (see Fig. 5), where the shear of the magnetic field lines, represented by $\psi_0''(r_{m,n})$ in eq. (9), is smaller.

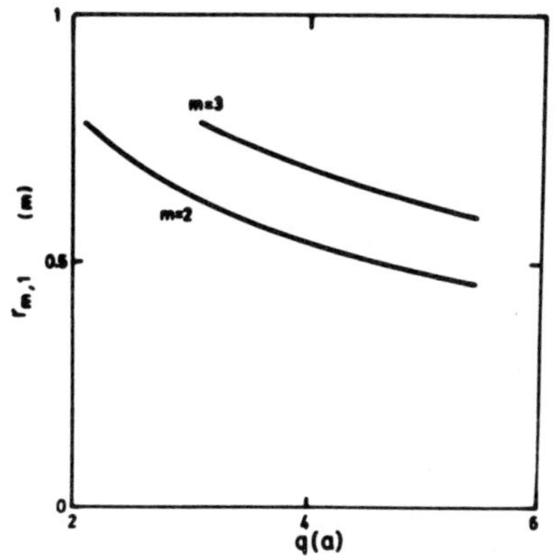

Figure 4. Rays of the $q = 2$ and $q = 3$ magnetic surfaces as a function of q(a) for $I = 97A$, $I_p = 10kA$ and $q(0) = 1$.

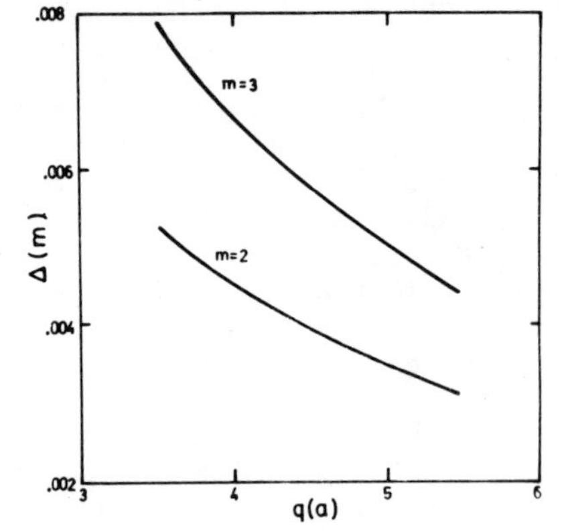

Figure 5. Islands widths of the $m_1 = 2$ and $m_2 = 3$ resonances ($n = 1$) as a function of q(a) for the same numerical values of Fig. 4.

When a single helical perturbation is superimposed upon an equilibrium with toroidal symmetry both symmetries are broken, and therefore the magnetic surfaces should be expected to disappear. However for small helical perturbation, i.e., $I/I_p \ll 1$, the major effect, on a large-aspect ratio tokamak, is the appearance of secondary magnetic islands on the rational magnetic surfaces with $q = m \pm 1$.

We calculate the helical current I on a single winding set required for ergodization of the magnetic field lines. The toroidal effect was considered multiplying the unperturbed B_z in eq. (1) by the factor $(1 + \frac{r}{R}\cos\theta)^{-1}$. Expanding the resulting expressions, neglecting terms of the order $(a/R)^2$ or higher and selecting the resonant terms at the magnetic surfaces with $q = m' = m \pm 1$ we obtained

$$\vec{B} \cdot \nabla\chi = 0 \tag{12}$$

where

$$\chi = \int^{r} dr' \left(\frac{r}{a}\right)^{m'-m} \left| nB_z - \frac{m'RB_\theta}{r} \right| + \frac{m\mu_0 I r^{m'}}{2\pi a^{m'-m} b^{m}} \cos u' \quad , \tag{13}$$

$u' = \theta - \frac{nz}{m'R}$. This function was used, in the same manner as the stream function ψ, to calculate the width of m' magnetic islands at the $q = m'$ surfaces. The ratio between the widths of the primary and secondary islands is proportional to $(a/R)^{1/2}$. An example of winding currents required for ergodization is plotted in the Fig. 6 as a function of q(a).

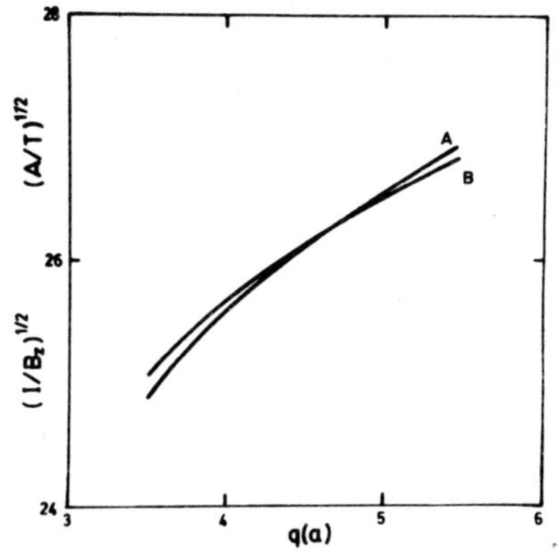

Figure 6. Helical current for overlap of m = 2 (primary) and m' = 3 (secondary) resonances. The curves A and B were obtained using, respectively, the Eqs. (9) and (10). q(0) = 1, I_p = 10kA.

4. CONCLUSIONS

It was assumed that magnetic surface destruction caused by resonant helical windings trigger the disruptive instabilities in tokamaks. Within this hypothesis the helical current that causes the contact between the m = 2 magnetic island and the limiter was calculated. The helical current threshold for magnetic surface break-up caused by overlapping of magnetic islands with two helicities was estimated analytically. Secondary magnetic islands on neighboring rational surfaces, due to the action of toroidicity, were also considered. The calculations can be applied to tokamak experiments with resonant helical windings.

ACKNOWLEDGEMENTS

We would like to thank Mr. A.N. Fagundes for reading the manuscript.

REFERENCES

[1] A.D. Cheethman et al., Plasma Physics and Controlled Nuclear Fusion Research 1984 (Tenth Conf. Proc. Brussels, 1980), Vol. 1, 337, IAEA, Vienna (1985).
[2] F. Karger et al., Proc. of the IAEA Symposium on Current Disruption in Toroidal Devices (Garching, 1979), Report IPP 3/51, A4 (1979).
[3] I.H. Tan et al., to appear in IEEE Trans. on Plasma Science, June (1986). O.W. Bender et al., to be published.
[4] A.S. Fernandes, I.L. Caldas, Report IFUSP/P-539 (1985).

[5] A.I. Morozov, L.S. Solov'ev- Reviews of Plasma Physics, Vol. 2, 1, Consultants Bureau Press, N. York (1966).
[6] G.M. Zaslavski, B.V. Chirikov, Sov. Phys. Usp 14, 549 (1972).

A NON-MARKOVIAN LORENTZ GAS MODEL

Manuel O. Cáceres
Centro Atómico Bariloche and Instituto Balseiro
Comisión Nacional de Energía Atómica and Univ. Nac. de Cuyo
8400 S.C. de Bariloche, Argentina

ABSTRACT

The model of coupled random walk is used for the description of correlated Brownian motion with anisotropic scattering. We have analyzed a non-Markovian behavior of a correlated Lorentz-gas model in the framework of the coupled CTRW theory by means of its associated coupled generalized master equation, using a mode-dependent exponential waiting-time.

1. INTRODUCTION

It has been emphasized that a Fokker-Planck equation can frequently serve as a basis for the discussion of a stochastic process involving continuous variables [1]. The discrete analog of that equation is the master equation, whose properties have been reviewed in [2] where one can also find reprints of many of the basic papers in the field. Some years ago a set of equivalent generalized master equations have been derived independently by several authors [3-6] starting from basic quantum mechanical equations under the assumption that at time $t = 0$ the density matrix is diagonal. The form of the generalized master equation (GME) is

$$\frac{\partial}{\partial t}P(j,t) = \int_0^t \Omega(t-\tau)\{\sum_{j'} W(j,j')P(j',\tau) - P(j,\tau)W(j',j)\}d\tau \tag{1}$$

In another context if initially a random walker is at the origin so that $P_{n=0}(j) = \delta_{j,0}$, it can be easily shown using the Laplace transformation, that the solution of Eq. (1) is connected with the standard lattice Green's function of the random walk description, provided that the waiting-time density of the walker is connected with the memory of the GME in an univocal form [7].

The specific form of this relation shows us that there exist only one waiting-time density which gives a δ--Dirac memory kernel. In other

E. Tirapegui and D. Villarroel (eds.), Instabilities and Nonequilibrium Structures, 325–332.

words only the exponential waiting-time density yields a markovian master equation. This shows us that the continuous time random walk (CTRW) theory describes a walker which is markovian only for the exponential waiting-time density. In [8] it is also shown that when the moments of the waiting-time density are all finite the GME is appropriate for a description of a CTRW at times which are large compared with a characteristic time. This means that when the waiting-time density has no finite moments one cannot construct a master equation.

Recently the problem of including internal degrees of freedom into the random walk (RW) has been studied in [9] [10] and the close connection between the CTRW with internal degrees of freedom and GME has been pointed out [9]. However this problem has only been studied for translationally invariant RW schemes, and a self-averaged (in each mode) waiting-time has been used. What we need is an explicit calculation using a mode-dependent waiting-time, for example, in order to study the Lorentz-gas model where the forward and backward scattering have a different waiting-time distribution (due for example to some internal excitation).

In this note, we show the connection between the coupled CTRW and the coupled GME associated with it for the general non translational invariant case, when the mode dependence in the waiting-time is taken into account [11]. This is an alternative framework to study the evolution of the coupled CTRW. Also we have analyzed the non-Markovian behavior of a correlated Lorentz-gas model. Furthermore, a variety of physical problems for nonequilibrium processes, other than scattering anisotropy, where a coupled recurrence relation of the CTRW form between different modes is relevant, can be treated by a description via a coupled GME as given in the following sections.

2. THE COUPLED GENERALIZED MASTER EQUATION

We start with a general set of coupled CTRW recurrence relations for $\{\vec{P}_n(j,t)\}_\ell$, the probability of the particle to be at site j with mode ℓ, at time t in the n-th step:

$$\vec{P}_{n+1}(j,t) = \int_0^t \sum_{j'} \overleftrightarrow{\psi}(j,j',t-\tau) \cdot \vec{P}_n(j',\tau)d\tau \tag{2}$$

which can be seen as a set of coupled generalized Chapman-Kolmogorov equations (for semi-Markovian situations), for each of the components of the vector $\vec{P}_n(j,t)$ (characterizing different "degrees of freedom or modes"). The transition matrix $\overleftrightarrow{\psi}(j,j',t)$ completely chatacterizes the set of coupled recurrence relations. Also $\{\overleftrightarrow{\psi}(j,j',t)\}_{\ell\ell'}dt$ is the probability that arriving at position j' and mode ℓ', the walker makes a transitions to a position j and mode ℓ in the time interval $(t,t+dt)$. Thus, the following normalization condition is fulfilled

$$\sum_j \sum_\ell \int_0^\infty \{\psi(j,j',\tau)\}_{\ell\ell'}d\tau = 1 \tag{3}$$

If we define the vector generating function by

$$\vec{R}(j,t,z) \equiv \sum_{n=0}^{\infty} z^n \vec{P}_n(j,t) \tag{4}$$

and assume that the initial condition for the vector probability distribution is

$$\vec{P}_0(j,t) = \delta_{j,0}\delta_{t,0^+}(\alpha_1,\alpha_2,\ldots) \tag{5}$$

where α_i are the initial normalizations for each mode, we can write a matrix equation for the Green's functions $\vec{R}(j,t,z=1)$, in the Laplace representations $(u \leftrightarrow t)$.

$$\vec{R}(j,u,z=1) = \sum_{j'} \overleftrightarrow{\psi}(j,j',u) \cdot \vec{R}(j,u,z=1) + \vec{P}_0(j) \tag{6}$$

This equation connects the matrix $\overleftrightarrow{\psi}(j,j',t)$ with the vector $\vec{R}(j,t,z=1)$, the probability per unit time to reach j in time t independently of the number of steps to arrive at j. This is the starting point to obtain the relation between the coupled CTRW Eq. (2) and the coupled GME. As an extension of the Montroll [12] relation between the probability distribution $P(j,t)$ and the generating Green's function $R(j,t)$ we assume, as in [13], the following relation between the vectors $\vec{P}(j,t)$ and $\vec{R}(j,t,z=1) \equiv \vec{R}(j,t)$:

$$P_\ell(j,t) = \sum_{\ell'} \int_0^t \{\phi(t-\tau,j)\}_{\ell\ell'} P_{\ell'}(j,\tau)d\tau \tag{7}$$

where $\phi(t,j)$ is a diagonal matrix, which is a natural generalization of the function $\phi(t)$ of Montroll (i.e. if a walker arrives at site j mode ℓ at time τ and remain there for a time $t-\tau$, $P(j,t)$ being the average over all arrival times with $0 < \tau < t$) given by:

$$\{\overleftrightarrow{\phi}(u,j)\}_{\ell\ell'} = \frac{\delta_{\ell\ell'}}{u}\{1 - \sum_{j'}\sum_{\ell''}\{\psi(j',j,u)\}_{\ell''\ell} \tag{8}$$

After Laplace transforming Eq. (7), and using the inverse of the matrix $\overleftrightarrow{\phi}(u,j)$ we arrive at the following set of coupled GME for the l-th component:

$$\begin{aligned}\frac{\partial}{\partial t}P_\ell(j,t) = \sum_{j''}\sum_{\ell''} \int_0^t \{\overleftrightarrow{\Lambda}(j,j'',t-\tau)\}_{\ell\ell''} P_{\ell''}(j'',\tau)d\tau \\ - \int_0^t P_\ell(j,\tau)\sum_{j''}\sum_{\ell''}\{\overleftrightarrow{\Lambda}(j'',j,t-\tau)\}_{\ell''\ell}d\tau\end{aligned} \tag{9}$$

where the memory kernel is:

$$\{\overleftrightarrow{\Lambda}(j,j',u)\}_{\ell\ell'} = \frac{u\{\psi(j,j',u)\}_{\ell\ell'}}{1-\sum_{j''}\sum_{\ell''}\{\psi(j'',j',u)\}_{\ell''\ell'}} \tag{10}$$

This coupled GME is a generalization of the treatment in [9] for the general situation indicated above. At this level Eq. (9) offers a formidable task if we want to solve it. In fact the memory kernel is a fourth-rank rensor and if the space domain is reduced to one-dimension $[-N,N]$, the dimension of each matrix $\{\overleftrightarrow{\Lambda}(j,j',u)\}_{\ell\ell'}$ will be $2N\times 2N$.

Nevertheless if a decoupled space-temporal model is used, a simplification is obtained [11]. Then as a special case of coupled GME, Eq. (9), we can use a decoupled space-temporal matrix $\overleftrightarrow{\psi}(j,j',\tau)$, that is:

$$\{\overleftrightarrow{\psi}(j,j',\tau)\}_{\ell\ell'} = W_{\ell\ell'}(j,j')\psi_{\ell\ell'}(\tau) \tag{11}$$

where $\psi_{\ell\ell'}(\tau)$ must satisfy the normalization condition

$$\int_0^\infty \psi_{\ell\ell'}(\tau)d\tau = 1$$

Thus, using the normalization condition Eq. (3), for Eq. (11) we obtain

$$\sum_\ell\sum_j W_{\ell\ell'}(j,j') = \sum_\ell Q_{\ell\ell'}(j') = 1 \tag{12}$$

The coupled GME then takes the form

$$\begin{aligned}\frac{\partial}{\partial t}P_\ell(j,t) &= \int_0^t \sum_{j''}\sum_{\ell''}\Omega_{\ell''\ell}(t-\tau,j'')W_{\ell\ell''}(j,j'')P_{\ell''}(j'',\tau)d\tau \\ &\quad - \int_0^t P_\ell(j,\tau)\sum_{\ell''}\Omega_{\ell''\ell}(t-\tau,j)Q_{\ell''\ell}(j)d\tau\end{aligned} \tag{13.a}$$

where the temporal memory function $\Omega_{\ell\ell'}(u,j')$ is given by

$$\Omega_{\ell\ell'}(u,j) = \frac{u\psi_{\ell\ell'}(u)}{1-\sum_{\ell''}\psi_{\ell''\ell}(u)Q_{\ell''\ell'}(j')} \tag{13.b}$$

Also for this case the temporal memory functions $\Omega_{\ell\ell'}(t,j')$ is space dependent (through the index j') and thus the simple contraction of the $2N\times 2N$ matrix $W_{\ell\ell'}$ and the $2N$ dimensional vector $P_\ell(t)$ cannot be decoupled from the temporal memory function (in the one dimensional case).

A big simplification arises if we study a translationally invariant case [11], as the one used in the next section, in order to analyze a non-Markovian Lorentz-gas model.

3. EXAMPLE: A NON-MARKOVIAN LORENTZ-GAS MODEL

In order to give an example of this approach we can show the two-mode case for the description of the motion of a one dimensional Brownian particle with anisotropic random force. The Lorentz-gas model has been used in order to study several problems and long-time tail phenomena. This has been carried out by incorporating in the model effects of disorder at the scatterer or fractal dimensional effects [15]. In this example we want to study a slight generalization of the simple persistent RW [16,17,13,18], then we must use a coupled CTRW theory where the waiting-time will be mode dependent.

Following the same notation as used in Ref. [13]

$$\vec{P}(j,t) = (P^R(j,t), P^L(j,t))$$

is the vector probability previously defined, where $P^R(j,t)$ or $P^L(j,t)$ are the conditional probabilities of finding a particle at site j and time t moving to the right or left direction, respectively, subject to the initial condition:

$$\vec{P}_0(j,t) = \delta_{j,0}\delta_{t,0^+}(\alpha_1,\alpha_2) \quad .$$

We assume in Eq. (11) for the waiting-time: $\{\psi_{RL}(t) = \psi_{LR}(t) = \psi_1(t);$ $\psi_{RR}(t) = \psi_{LL}(t) = \psi_2(t)\}$, i.e. a process where time elapsed in each step is different depending on the type of scattering, backward or forward, that will occur. Using the exponential model, we write

$$\begin{aligned} \psi_{LL}(t) &= \psi_{RR}(t) = \nu_2 e^{-\nu_2 t} \\ \psi_{RL}(t) &= \psi_{LR}(t) = \nu_1 e^{-\nu_1 t} \end{aligned} \tag{14}$$

which implies a Markovian behavior in each mode if there is no coupling among them (i.e. $W_{\ell'\ell}(j-j') = 0$, $\ell \neq \ell'$).

To complete our description we must make an assumption for the lattice structure. Analyzing the case of nearest-neighbor model we must use a translationally invariant single-step transition-probability tensor of fourth rank,

$$W(j, j') = \begin{pmatrix} W_{RR}(j-j') & W_{RL}(j-j') \\ W_{LR}(j-j') & W_{LL}(j-j') \end{pmatrix} \tag{15}$$

where we have used the index "R" and "L" to characterize each mode. The matrices $W_{\ell\ell'}(j-j')$ are given in the following way:

$$W_{RR} = W_{LL}^T = \begin{pmatrix} \cdot & \cdot & \cdot & \cdot \\ p & 0 & 0 & 0 \\ \cdot & p & 0 & 0 \\ \cdot & \cdot & p & 0 \end{pmatrix} ;$$

$$W_{LR} = W_{RL}^T = \begin{pmatrix} 0 & q & \cdot & \cdot \\ 0 & 0 & q & \cdot \\ \cdot & 0 & 0 & q \\ \cdot & \cdot & \cdot & \cdot \end{pmatrix} \tag{16}$$

The notation T indicates the transposed matrix, and p (or q) is the probability of forward (or backward) scattering relative to the flight direction before scattering $(p + q = 1)$. Using Eq. (16) we can calculate the coefficients $Q_{\ell\ell'}$ given by Eqs. (12). Then the temporal memory matrix will be

$$\Omega_{RR}(u) = \Omega_{LL}(u) = \frac{u\psi_2(u)}{1 - [\psi_2(u)p + \psi_1(u)q]} ;$$

$$\Omega_{RL}(u) = \Omega_{LR}(u) = \frac{u\psi_1(u)}{1 - [\psi_2(u)p + \psi_1(u)q]} . \tag{17}$$

Note that each element has the same Laplace structure, which can be easily antitransformed. As it was expected, it gives a δ-Dirac dependence together with a transient non-Markovian description:

$$\Omega_{LL}(t) = \nu_2[\delta(t) + q(\nu_1 - \nu_2)e^{-(\nu_1 p + \nu_2 q)t}]$$

$$\Omega_{RL}(t) = \nu_1[\delta(t) + p(\nu_2 - \nu_2)e^{-(\nu_1 p + \nu_2 q)t}] \tag{18}$$

It can be easily seen from Eq. (13.a) and using Eq. (18) that for the case $\nu_1 = \nu_2$ we recover the Markovian coupled ME description, i.e. the ME for the simplest one-dimensional stochastic Lorentz model [16,14].

This is so because when we put $\nu_1 = \nu_2$ we are averaging the waiting-time function $\psi_{\ell\ell'}(t)$ over all the modes; then we get the Montroll case for the temporal memory function independent of the mode.

Since we have used an exponential waiting-time model we get a δ-Dirac memory function, as is wellknown. Then the coupled GME will be Markovian [18]. If we presenrve $\nu_1 \neq \nu_2$ non-Markovian effects arise: this is so because the modes are coupled together. Physically our model of mode-dependent exponential waiting-time is closely connected with a non-Markovian description between each step. In other words there are different correlations between the changes of velocity directions of our Brownian particles.

We can see from Eq. (18) that the non-Markovian effect is a causal convolution of $P_\ell(j,t)$ with $\exp[-(t-\tau)/\tau_c]$, where $\tau_c = (\nu_1 p + \nu_2 q)^{-1}$ is a characteristic time, which can be used in order to define a transient non-Markovian regime. If we study this system for time $t \gg \tau_c$ we can approximate the coupled GME by one of Markovian type. We can see from this analysis that the non-Markovian effect has been quenched off for sufficiently long times and only effective factors appear in each gain-loss term of the coupled Markovian ME.

Finally we can remark that in this simple description for $t \gg \tau_c$ the effective loss-term factor will be smaller than for the Markovian approach. This is so because the mode-dependent waiting-time introduces a delay in the loss contribution if we want to write a ME. On the other hand, for the gain term the sign provided by the non-Markovian contribution will depend on the model: $\nu_1 > \nu_2$ or $\nu_1 < \nu_2$.

REFERENCES

[1] C.W. Gardiner, in Handbook of Stochastic Methods (Springer Verlag 1983)
[2] I. Oppenheim, K. Shuler and G. Weiss, The Master Equation (M.I.T. University Press, 1977).
[3] I. Prigogine and P. Resibois, Physica 27, 629 (1961).
[4] E.W. Montroll, Fundamental problems in statistical mechanics, ed. E.D.G. Cohen, (North-Holland, Amsterdam, 1962).
[5] S. Nakajima, Prog. Theor. Phys. 20, 948 (1958).
[6] R. Zwanzig, Physica 30, 1109 (1964).
[7] V.M. Kenkre, E.W. Montroll and M.F. Shlesinger, J. Stat. Phys. 9, 45 (1973).
[8] D. Bedeaux and K. Lakatos-Lindenber and K.E. Shuler, J. Math. Phys. 2, 2116 (1971).
[9] V. Landam, E.W. Montroll and M.F. Shlesinger, Proc. Natl. Acad. Sci. USA 74, 430 (1977).
[10] M.O. Caceres and H.S. Wio, Z. Phys. B54, 175 (1984).
[11] M.O. Caceres, Phys. Rev. A 00,00 (1986).
[12] E.W. Montroll and B.J. West, Fluctuation Phenomena, ed. E.W. Montroll and J.L. Lebowitz (North-Holland, Amsterdam, 1979).
[13] M.O. Caceres and H.S. Wio, Z. Phys. B58, 329 (1985).
[14] P. Grassberger, Physica 103A, 558 (1980).
[15] H. Takayasu and K. Hiramatsy, Phys. Rev. Lett. 53, 633 (1984).

[16] H.S. Wio and M.O. Caceres, Phys. Lett. 100A, 279 (1984).
[17] S. Goldstein, J. Mech. Appl. Math. 4, 129 (1950).
[18] G.H. Weiss and R.J. Rubin, Adv. Chem. Phys. 52, 363 (1983).

SUBJECT INDEX

A

B

C

D

E

L

M

N

O

P